$\mathfrak{N}$	Avogadro constant	6.0225×10^{23} mole^{-1}
k	Boltzmann constant	1.35805×10^{-16} erg/deg
R	Gas constant	8.3143×10^{7} ergs/deg mole
		1.9872 cal/deg mole
		82.054 cc atm/deg mole
c	Speed of light in vacuum	2.99792×10^{10} cm/sec
h	Planck constant	6.6256×10^{-27} erg-sec
e	Electronic charge	4.8030×10^{-10} esu
m_e	Electronic rest mass	0.9109×10^{-27} g
$\mathfrak{F}$	Faraday constant	96,487 coul/equiv

Absolute temperature at the ice point, 0°C	273.15 K
Standard atmospheric pressure (1 atm)	1.0132×10^{6} dynes/cm^2
Standard gravity (standard acceleration of free fall)	980.7 cm/sec^2

Mass	1 atomic mass unit (amu) =	1 dalton
	=	1.660×10^{-24} g
	1 lb =	453.6 g
Energy	RT, at 25°C =	592.5 cal/mole deg
	kT, at 25°C =	4.116×10^{-14} erg/molecule deg
Length	1 cm =	10^{8} Å
Speed	1 mole/hr =	44.7 cm/sec
Electrical	1 coulomb =	2.998×10^{9} esu
	1 volt =	300 electrostatic (cgs) volts

cal/mole	joules/mole	cc atm/mole	cm^{-1}		ergs/molecule
1	4.1840	41.292	0.34974	4.3365×10^{-5}	6.9465×10^{-17}
0.23901	1	9.8692	0.08359	1.0363×10^{-5}	1.6602×10^{-17}
0.024218	0.10133	1	8.470×10^{-3}	1.0501×10^{-6}	1.6823×10^{-18}
2.8593	11.963	118.07	1	1.2398×10^{-4}	1.9862×10^{-16}
23,060	96,487	9.523×10^{5}	8065.7	1	1.6021×10^{-12}
1.439×10^{16}	6.023×10^{16}	5.944×10^{17}	5.0348×10^{15}	6.2418×10^{11}	1

Barrow, Gordon M
Physical chemistry for the life sciences

Physical Chemistry for the Life Sciences

Physical Chemistry for the Life Sciences

GORDON M. BARROW

McGraw-Hill Book Company
New York St. Louis San Francisco Düsseldorf Johannesburg Kuala Lumpur London Mexico Montreal New Delhi Panama Paris São Paulo Singapore Sydney Tokyo Toronto

Physical Chemistry for the Life Sciences

1 2 3 4 5 6 7 8 9 0 MAMM 7 9 8 7 6 5 4

This book was set in Trump Medieval by York Graphic Services, Inc.
The editors were Thomas Adams and Andrea Stryker-Rodda;
the designer was Barbara Ellwood;
the production supervisor was Sam Ratkewitch.
The drawings were done by Vantage Art, Inc.
The Maple Press Company was printer and binder.

Library of Congress Cataloging in Publication Data

Barrow, Gordon M.
Physical chemistry for the life sciences.

1. Chemistry, Physical and theoretical. I. Title.
QD453.2.B373 541'.3'024574 73-18289
ISBN 0-07-003855-4

Contents

Preface

Many students who study physical chemistry have already established an interest in one of the areas, or in the collected areas, that are labeled "the life sciences." Often such interests introduce several life-science courses into the undergraduate program and, as a result, only a semester can be devoted to physical chemistry. It is for students with such interests and such time restrictions that this text is intended.

An interest in the life sciences, rather than a background in any aspect of these sciences, is assumed. I have attempted to introduce physical-chemical topics so that a student with such an interest sees the study of these topics as being reasonable and worthwhile. The dogma that physical chemistry is worthy of study because it is inherently interesting (and leads to more, and more interesting physical chemical topics) has been set aside. It is recognized that many students find physical chemistry of interest only when its relation to life science subjects is apparent.

The special nature of the physical chemistry course for which this book is intended led me also to consolidate many of the diverse topics of physical chemistry so that the student can keep more clearly in mind what he is doing and where he is going. Thus the three major areas of physical chemistry—structure, energy, and kinetics—are emphasized. The selection of topics for a text that is shorter than that for the customary year's course is a difficult matter that is not resolved by this consolidation. I have tried to select topics that, first, can be developed satisfactorily at this level and, second, lead to the opening up of significant avenues into the life sciences.

Some physical-chemical topics have been slighted because they do not have general or immediate enough application to biological science. Examples are chemical bonding theory, statistical mechanics, and phase equilibria. Others, such as light scattering, are perfectly relevant but, in view of the average student's background and the time available, cannot be developed in a very satisfying way. Still other topics that might have been considered for inclusion have been avoided because their study, it seems to me, belongs elsewhere. In this category I place many of the practical guides to the basis and operation of instruments that find extensive use in biological studies.

Hopefully this selection of topics will provide a useful base on which physical chemistry for the life sciences courses can be built. I hope to be able to respond in subsequent editions to suggestions for additions and deletions so that this text becomes a more generally useful instrument for introducing students of the life sciences to physical chemistry.

Each topic treated here is developed with a minimum of mathematics and is presented so that a sound, satisfying, and useful grasp of the topic results. The survey approach, the "scientists have shown" approach and the "formula out of a hat" approach have been minimized. The elements of physical chemistry presented should thus be of value even if the interests that led the student into the course should change.

Finally, let me emphasize, since the title "Physical Chemistry for the Life Sciences" seems to indicate otherwise, that I have not adopted the attitude that physical chemistry is a mere tool. The approaches that are characterized as physical chemical do indeed help to open up the fascinating areas of the life sciences. But these areas and the techniques that must be developed to explore them make their impact. Physical chemistry responds to the problems and demands of studies of all aspects of the physical world. Here it is the interaction between the various areas of the life sciences and physical chemistry that is central. I hope the studies of physical chemistry that this book guides are made more intriguing by this interaction and that subsequent studies in the life sciences are more satisfying as a result the physical chemical background it provides.

I am happy to acknowledge, once again, my great indebtedness to James W. Richardson of Purdue University, who reviewed the manuscript of this book as he has those of the second and third editions of PHYSICAL CHEMISTRY. I should also like to express my thanks to A. James Diefenderfer of Lehigh University, Scott C. Mohr of Boston University, Peter Smith of Duke University, and Ignacio Tinoco of the University of California, Berkeley, for their suggestions for improving the manuscript and to Professor Smith for contributing several of the problems that appear in Chapter 12. I am also grateful to the Hopkins Marine Station of Stanford University and the Naval Postgraduate School, both in Monterey, for the use of their library facilities during the preparation of this book. Finally, much of the special treatment I have given to thermodynamics in order to make it applicable to biologically important systems has been inspired by the work of Irving M. Klotz. I am grateful to him also for first making me aware that physical chemistry can be developed as a life science. Only in recent years, and with the development of this book, have I fully appreciated the added dimension that physical chemistry takes on when viewed from this standpoint.

Gordon M. Barrow

Carmel Valley
California

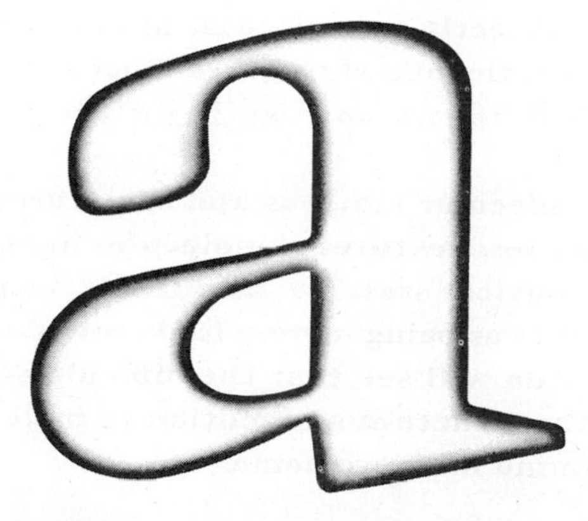

Molecular Structure

The many and varied topics of physical chemistry can be placed in the three categories:

a. Structure
b. Energy and equilibria
c. Kinetics

These constitute the subjects of Parts a, b, and c of this life-science-oriented introduction to physical chemistry.

For these three areas to encompass the topics of physical chemistry, their titles must be broadly interpreted. Thus, *structure* implies studies that lead to information on the dimensions and geometry of simple and complex molecules as found in gases, liquids, and solids.

Molecular structures cannot be thought of as rigid, imperturbable arrangements of atoms. As molecules of gases fly through the air, they twirl and wiggle. Molecules of liquids are a little more restricted, but even their structures are dynamic and respond to the influences of the surrounding molecules. Only for solids are major variations in molecular structures ruled out, leaving only vibrations of the atoms about their equilibrium positions.

Therefore, in studying molecular structures we must study the structure *and* motion of molecules. Further, thanks to the different motions that can occur in gases, liquids, and solids, this structure-and-motion study can be organized in terms of these three states of matter. The study of gases leads us most directly to molecular motions. Solids are most suitable for the study of molecular structures. Since the focus of each of these two chapters is different, you can begin with either chapter.

These introductions to molecular motions and structures culminate in Chap. 3 with studies of these features of molecules in liquids, the state of most interest in biological systems. As a result, you may take the gas- and solid-state studies as being merely background to the studies of molecules of liquids. You will see that the difficulties that stand in the way of studies of the structure and motion of molecules in liquids make all such background most welcome.

1 Molecules of Gases: Molecular Motions

Many chemical and physical properties and processes are affected by the frantic and chaotic turmoil that characterizes the molecular world. Only for the gas phase can this molecular activity be analyzed with any completeness. Although an interest in biological systems will turn most of our attention to the liquid state, an initial study of gases provides a convenient introduction to the motions and some of the properties of simple molecules.

1-1 THE *PVT* BEHAVIOR OF IDEAL GASES

In many practical situations, gases are subjected to various pressures and temperatures. If a particular sample of gas is considered, it is found that these variables affect the volume of the gas sample. For this reason an expression for the dependence of gas volume on pressure and temperature is useful in treating practical problems. Such an expression also provides a summary of gas behavior that must be accounted for when a molecular model of gases is developed. To begin with, the two variables, pressure and temperature, will be considered separately.

The volume of a sample of a gas at a constant temperature is found to be inversely proportional to the pressure on the gas—as long as the pressure is not too high or the temperature too low. This inverse proportionality, discovered for air by Robert Boyle, is known as *Boyle's law.* It expresses the volume-pressure relation of gases as

$$V \propto \frac{1}{P} \tag{1}$$

Behavior in accordance with this simple and generally applicable relation is called *ideal-gas behavior.* When a gas can be characterized (among other things) by this behavior, it is called an *ideal gas.*

A similar simple generally followed relation between gas volume and temperature can be expressed if an *absolute temperature* scale is used. The Kelvin temperature scale is such a scale. It is related to the Celsius scale by

$$T(\text{K}) = t(^\circ\text{C}) + 273.15^\circ \tag{2}$$

where K is the abbreviation for the temperature unit kelvin. When the Kelvin scale is used, we can write

$$V \propto T \tag{3}$$

This expression, known as *Gay-Lussac's law,* summarizes the observations that if the pressure is held constant, the volume of a sample of gas varies directly as the absolute temperature.

The gas laws of Boyle and Gay-Lussac separately describe the dependence of the volume of a gas sample on the pressure and temperature. The dependence on both variables can be illustrated by the surface of Fig. 1-1. Any cross section perpendicular to the T axis, i.e., a section for constant temperature, would show a Boyle's law hyperbolic curve. Any section perpendicular to the P axis, i.e., a section for constant pressure, would show the linear temperature-volume relation of Gay-Lussac's law.

For its mathematical description the surface of Fig. 1-1 requires an expression of the form

$$V = f(P,T) \qquad \textbf{4}$$

where $f(P,T)$ implies some function of the variables P and T.

The proportionality to which Eq. **4** corresponds must give the proportionalities of Boyle and Gay-Lussac as special cases, and on this basis we can write

$$V \propto \frac{T}{P} \qquad \textbf{5}$$

For a gas sample whose volume is specified at some temperature and pressure, the equality

$$V = (\text{const})\,\frac{T}{P} \qquad \textbf{6}$$

can be written to describe the surface, for that sample, on a display like Fig. 1-1.

To calculate changes in gas volume from Eq. **6** or to make estimates from a figure like Fig. 1-1, the gas sample can be specified by a statement of its volume at some indicated temperature and pressure. It is, however, often more convenient to treat the gas sample in terms of *the number of moles n* of the gas. When the amount of gas in a gas sample is measured in this way, a gas-law expression applicable to all gases that behave ideally can be set up.

The recognition of the molecular nature of matter, which is implied by the use of the mole, takes us outside purely empirical studies. We then can use Avogadro's hypothesis that equal volumes of different gases at the same temperature and pressure contain equal numbers of molecules. Introduction of the mole as a measure of an amount of gas allows us to use the generalizing statement that 1 mol of any ideal gas occupies the same volume at the same temperature and pressure.

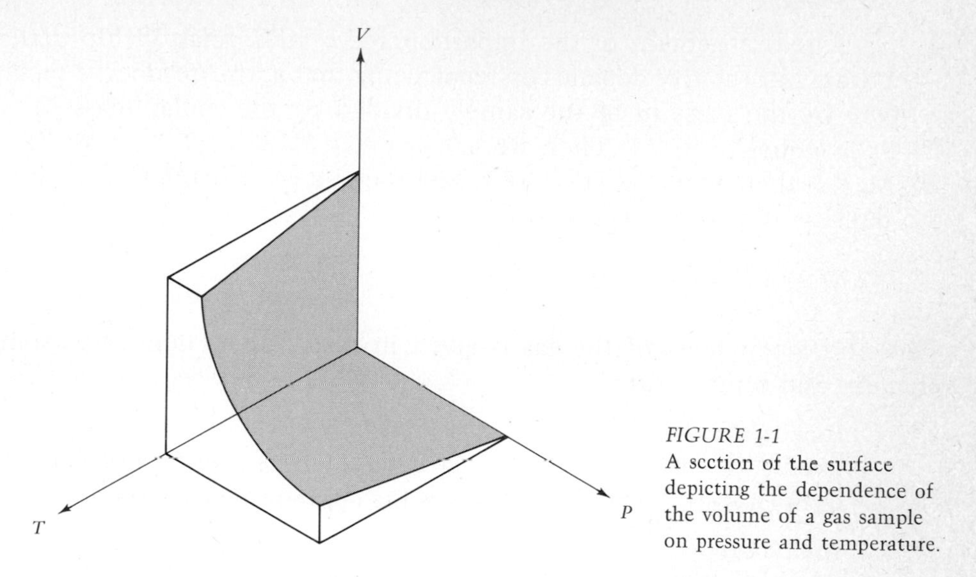

FIGURE 1-1
A section of the surface depicting the dependence of the volume of a gas sample on pressure and temperature.

Now the proportionality of Eq. **5** can be written to include the volume dependence on the amount of gas, as measured by the number of moles n, as

$$V \propto n\frac{T}{P} \qquad \textbf{7}$$

Then, with the generalization stated above and using R for the proportionality constant, we can write

$$V = Rn\frac{T}{P}$$

or $$PV = nRT \qquad \textbf{8}$$

Furthermore, R will have a single value, applicable to *all* gases that behave ideally.

This result is useful in making gas-volume calculations; moreover, it is a summary of the empirical laws of Boyle and Gay-Lussac and the hypothesis of Avogadro. The principal goal of the theory of the nature of gases, to be developed in Sec. 1-3, will be the theoretical derivation of this important result.

A numerical value for the gas constant can be obtained from the result that *at 1 atm and 0°C, 1 mol of any gas that behaves ideally occupies 22.414 liters.* Substitution of these data in the gas-law expression, Eq. **8,** gives

$$R = \frac{(1\text{ atm})(22.414\text{ liters})}{(1\text{ mol})(273.15\text{ deg})} = 0.082056\text{ liter atm/deg mol}$$

A rearrangement of the important $PV = nRT$ relationship that is sometimes informative depends on recognizing that n, the number of moles, is given by the mass m of the sample divided by the molar mass M, so that n is equal to m/M. Then we can express $PV = nRT$, first as $PV = (m/M)RT$, then as $PM = (m/V)RT$, and then as $M = (m/V)RT/P$. Since the density, d, can be expressed as m/V

$$M = d\frac{RT}{P} \quad \mathbf{9}$$

Thus the molar mass of the gas is given in terms of its density at some pressure and temperature.

Example 1-1

The density of air at a pressure of 1 atm and a temperature of 17.0°C is reported to be 1.220 g/liter. What is the apparent molar mass of air?

Solution

Substitution of the given quantities in Eq. **9** yields

$$M = d\frac{RT}{P}$$

$$= (1.220\ \text{g/liter})\frac{(0.08206\ \text{liter atm/deg mol})(290.2\ \text{deg})}{1\ \text{atm}}$$

$$= 29.05\ \text{g/mol}$$

The example of the application of $PV = nRT$ or the equivalent relation $M = dRT/P$ to air shows that we can deal either with pure gases or with gas mixtures. When gases are compressed or expanded by changes in temperature or pressure, it does not matter whether they are pure gases or gas mixtures. Often, however, we are interested in the makeup of the gas we are dealing with; measures of this makeup are treated in the following section.

1-2 SOME PROPERTIES OF GAS MIXTURES

The summary equation, $PV = nRT$, of the preceding section shows that the volume of a sample of gas at a given pressure and temperature depends on the number of moles n in the sample. There is no concern in this gas law with the properties of the molecules; even their mass is of no consequence. Thus $PV = nRT$ can be applied to a gas composed of like molecules or to one composed of a variety of molecules, i.e., a mixture of gases.

If the numbers of moles of the various components of a gas mixture are indicated by $n_1, n_2, n_3, \ldots$, the total number of moles n is

$$n = n_1 + n_2 + n_3 = \sum_i n_i$$

When this recognition of various types of molecules is inserted in $PV = nRT$ or $V = nRT/P$, we have

$$V = \left(\sum_i n_i\right)\frac{RT}{P}$$

$$= n_1 \frac{RT}{P} + n_2 \frac{RT}{P} + \cdots \qquad \mathbf{10}$$

In this way we are led to picture the total volume V of the gas as being the result of volume contributions of the components. We might write

$$V_1 = n_1 \frac{RT}{P_1}, \quad V_2 = n_2 \frac{RT}{P_2}, \quad \cdots$$

Then

$$V = V_1 + V_2 + \cdots \qquad \mathbf{11}$$

More often used are the pressure contributions, or *partial pressures*, of the individual components. We are led to these by returning to $PV = nRT$ and rewriting it as

$$P = n \frac{RT}{V}$$

$$= \left(\sum_i n_i\right)\frac{RT}{V}$$

$$= n_1 \frac{RT}{V} + n_2 \frac{RT}{V} + \cdots \qquad \mathbf{12}$$

The component terms can now be indicated as

$$P_1 = n_1 \frac{RT}{V}, \quad P_2 = n_2 \frac{RT}{V}, \quad \cdots$$

Then

$$P = P_1 + P_2 + \cdots \qquad \mathbf{13}$$

In this way, the gas mixture is viewed in terms of components, each occupying the entire container volume and each making its contribution P_i to the total gas pressure.

Equation **13** says that the pressure of a sample of gas in a container is equal to the sum of the pressures each of its components would exert if placed alone in the container. It is, in fact, an empirical deduction known as *Dalton's law of partial pressures.* It provides one of the most direct suggestions that gases do indeed consist of many small particles that are moving around independently of one another.

Notice that both the volume fraction and the pressure fraction equal the *mole fraction,* denoted by x, of a component of a gas mixture. Thus we have, for component i

$$\frac{V_i}{V} = \frac{n_i RT/P}{nRT/P} = \frac{n_i}{n} = \mathrm{x}_i$$

or

$$\frac{P_i}{P} = \frac{n_i RT/V}{nRT/P} = \frac{n_i}{n} = \mathrm{x}_i$$

Thus from information on the partial pressures or the volume fractions of the components of a gas mixture, the relative numbers of moles of the components in the sample can be deduced.

Example 1-2

If dry sea-level air is treated in terms of its three major components, the fraction of the total pressure contributed by each is N_2, 78.1 percent; O_2, 20.9 percent; and Ar, 1.0 percent. For a sample of gas at 1 atm pressure and 25°C, what are the partial pressures in millimeters of mercury of these components? What are the mole fractions of the components, and what is the effective molecular mass of air?

Solution

The partial pressures are

$$N_2\text{:}\ \frac{78.1}{100} \times 760 = 593.6 \text{ mm}$$

$$O_2\text{:}\ \frac{20.9}{100} \times 760 = 158.8$$

$$\text{Ar:}\ \frac{1.0}{100} \times 760 = 7.6$$

$$\overline{760 \text{ mm}}$$

The mole fractions $\mathrm{x}_i = n_i/n$ are equal to the pressure fractions. These are given by the original pressure percentages and lead to

$$N_2\text{: } x_{N_2} = 0.781$$
$$O_2\text{: } x_{O_2} = 0.209$$
$$\text{Ar: } x_{Ar} = 0.010$$

Finally the mass of 1 mol of the gas mixture can be calculated from the masses of the x_i mol of the components as

$$\begin{aligned} M &= x_{N_2}(28.01) + x_{O_2}(32.00) + x_{Ar}(39.95) \\ &= 0.781(28.01) + 0.209(32.00) + 0.010(39.95) \\ &= 28.9 \text{ g} \end{aligned}$$

Now let us see if the idea that gases are composed of rapidly moving small particles can lead us to the relation of Eq. **8,** which summarizes the *PVT* data for ideal gases, pure or mixed. Our interest in so doing is the refinement of our ideas about these small particles, or molecules, and their motions.

1-3 THE MOLECULAR BASIS OF $PV = nRT$

Only the molecules of gases are able to exhibit *free flight,* a term that implies translational motion only occasionally interrupted by collisions with the container walls or with one another. Here, as a result, one can study the consequences of the translational motion of molecules, and this study provides a valuable introduction to the dynamics of molecules in gases and in other physical states.

According to the molecular model, molecules move about independently and collide with each other and with the walls of the container. Let us see how this idea can lead us to the empirical expression $PV = nRT$, which sums up the experimental studies on the *PVT* behavior of gases.

Consider 1 mol of a gas in a cubic container at some temperature T. The derivation will apply to any container (as one could expect, because the pressure depends on the volume of a container and not its shape), but the analysis is easier if a cubic box is assumed. Can we deduce the pressure that must be applied to balance the effect of the collisions of the molecules with the walls of the container? How does the expression we obtain compare with $PV = nRT$?

We begin by analyzing the effect of one molecule that is assumed to be moving in the x direction and as a result bounces back and forth between wall A of Fig. 1-2 and the opposite wall. (In fact, each molecule

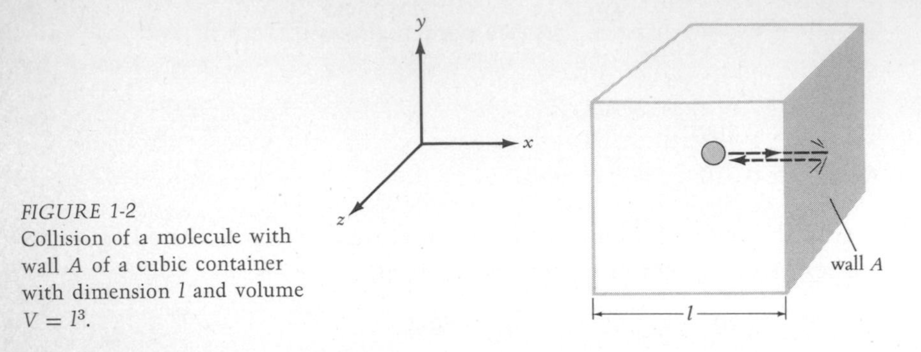

FIGURE 1-2
Collision of a molecule with wall A of a cubic container with dimension l and volume $V = l^3$.

has a complicated path as a result of collisions with other molecules but the simplified analysis gives the same answer as that obtained by assuming a more realistic trajectory.) The molecule travels with a speed u_x; that is, the molecule travels a distance of u_x cm in the x direction each second. The length of the box is l, and so the molecule must travel a distance of $2l$ between collisions with side A. Thus, the number of impacts this molecule makes per second with side A is $u_x/2l$.

The force the walls of a gas container must exert to confine the gas can be related to the bombardment of these walls by the gas molecules. Consider a molecule that strikes wall A and rebounds from it. The force the wall exerts on the molecule can be calculated from Newton's $f = ma$ relation, where f is the force that gives to a particle of mass m an acceleration a. Acceleration is the rate with which the speed changes, i.e., the change of speed per unit of time. A convenient form of Newton's relation is obtained by using this description of a to give

$$\begin{aligned} f_A &= ma \\ &= m \times \text{rate of change of } u_x \end{aligned}$$

Then, because m is constant, this can be rewritten as

$$f_A = \text{rate of change of } mu_x$$

The product mu_x is known as the *momentum* in the x direction. With this term, Newton's law can be written in the often more useful form

$$\text{Force} = \text{rate of change of momentum} \qquad \textbf{14}$$

Think of the collision of a molecule with the container wall as occurring by two successive steps. First, the particle is brought to a stop by the wall. The momentum of the particle drops from mu_x to 0. Then, the particle seems propelled by the wall and is accelerated from zero momentum to momentum $-mu_x$. The minus sign indicates a momentum in the opposite direction. The net result, corresponding to changing the direction of the particle's motion, is a total momentum change of $2mu_x$. Combining this momentum change per impact with the fact that there are $u_x/2l$ impacts per second, we obtain for the rate of change of momentum, or force, at side A

$$f_A = \frac{u_x}{2l} 2mu_x = \frac{mu_x^2}{l}$$

Thus, the pressure, which is f_A/l^2, is

$$P = \frac{f_A}{l^2} = \frac{mu_x^2}{l^3} = \frac{mu_x^2}{\text{V}} \qquad \textbf{15}$$

where $\text{V} = l^3$ is the volume of the container. A small capital V is used to show that we are dealing with the volume of 1 mol of gas.

So far the effect of only one molecule moving in the x direction has been considered. Let $\mathfrak{N}$, Avogadro's number, represent the number of molecules per mole. For 1 mol of gas there are $\mathfrak{N}$ molecules moving in all directions and colliding with all the walls at various angles. The net effect on side A is obtained by assuming that the pressure due to the one molecule considered above is average and multiplying this average effect by $\mathfrak{N}$. The pressure on wall A, which is no different from the pressure on the other walls, is thus

$$P = \mathfrak{N}\left(\frac{mu_x^2}{\text{V}}\right)_{\text{av}} \qquad \textbf{16}$$

The only quantity to be averaged, if the mass of all the molecules in the sample is the same, is u_x^2; this average is denoted by writing $\overline{u_x^2}$. Equation **16** can now be written

$$P\text{V} = \mathfrak{N}m\overline{u_x^2} \qquad \textbf{17}$$

The average speed in the x direction, however, is related to the average speed in space by the pythagorean expression (which is more familiar in two dimensions) illustrated in Fig. 1-3. This leads to

$$\overline{u^2} = \overline{u_x^2} + \overline{u_y^2} + \overline{u_z^2}$$

and since $\overline{u_x^2} = \overline{u_y^2} = \overline{u_z^2}$, to

$$\overline{u^2} = 3\overline{u_x^2} \qquad \text{or} \qquad \overline{u_x^2} = \tfrac{1}{3}\overline{u^2} \qquad \textbf{18}$$

With this relation, Eq. **17** can be written

$$P\text{V} = \tfrac{1}{3}\mathfrak{N}m\overline{u^2} \qquad \textbf{19}$$

In a way, this step ends the analysis of the effect of the molecules bouncing against container walls. But one more step is necessary to show the relation to the empirical gas law $PV = nRT$.

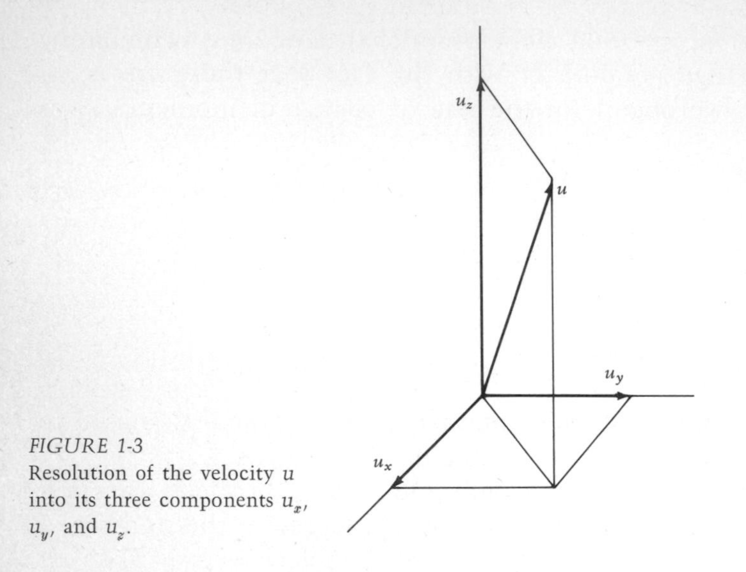

FIGURE 1-3
Resolution of the velocity u into its three components u_x, u_y, and u_z.

Putting a 2 in both the numerator and the denominator of the derived equation gives it the suggestive form

$$PV = \tfrac{2}{3}\mathfrak{N}\tfrac{1}{2}m\overline{u^2}$$

The portion $\frac{1}{2}m\overline{u^2}$ is the average kinetic energy of one molecule of the gas. Let us introduce the symbol $\overline{ke}$ for this. From the kinetic-molecular-theory derivation we then have

$$PV = \tfrac{2}{3}\mathfrak{N}\overline{ke} \qquad \textbf{20}$$

Since $\mathfrak{N}$ times the average kinetic energy of a molecule is the kinetic energy for 1 mol of gas molecules, we write this as KE and obtain

$$PV = \tfrac{2}{3}\text{KE} \qquad \textbf{21}$$

Finally, we are in a position to compare this result from theory with the experimental result, which for 1 mol of gas is

$$PV = RT \qquad \textbf{22}$$

To make the theory fit the experimental result we need only add the idea that *the kinetic energy of the molecules is related to the temperature* according to

$$\tfrac{2}{3}\text{KE} = RT$$
$$\text{KE} = \tfrac{3}{2}RT \qquad \textbf{23}$$

1-4 THE TRANSLATIONAL ENERGIES OF GAS MOLECULES

The kinetic energy of a mole of gas molecules due to their random movement is equal to $\frac{3}{2}RT$. Thus the gas constant not only enters into *PVT* calculations but also into the calculation of the energies of molecular systems. This aspect of the gas constant is less surprising when one sees that R involves the dimensions of work, or energy.

If pressure is written as force per unit area and the volume as area times length, dimensionally

$$\text{Pressure} \times \text{volume} = \frac{\text{force}}{\text{area}} \times \text{area} \times \text{length} = \text{force} \times \text{length}$$

The dimensions of force times length are those of energy, which you recall can be based on the force times the distance through which the force acts. It follows that R has the dimensions of energy per degree per mole.

Numerical values for R involving the most common energy units of ergs, joules (J), and calories (cal) can now be obtained. With the conversion factor

$$1 \text{ atm} = 1.0133 \times 10^6 \text{ dyn/cm}^2$$

we can obtain R in the units of dyne-centimeters, or ergs, as

$$\begin{aligned} R &= 0.08206 \text{ liter atm/deg mol} = 82.06 \text{ cm}^3 \text{ atm/deg mol} \\ &= 82.06 \frac{\text{cm}^3 \text{ atm}}{\text{deg mol}} \times \frac{1.0133 \times 10^6 \text{ dyn/cm}^2}{1 \text{ atm}} \\ &= 8.314 \times 10^7 \text{ ergs/deg mol} \end{aligned}$$

Further, with

$$1\,J = 10^7 \text{ ergs}$$
$$R = 8.314\,J/\text{deg mol}$$

and with

$$1 \text{ cal} = 4.184\,J$$

we have

$$R = 1.987 \text{ cal/deg mol} \tag{24}$$

The result of the kinetic-molecular-theory derivation, i.e., that the translational energy of a mole of gas molecules is equal to $\frac{3}{2}RT$, can now be used to evaluate this energy. We have, for example, at 25°C, or 298 K, the result

$$\begin{aligned} \text{KE} &= \tfrac{3}{2}(8.314)(298) = 3720\,J \\ &= \tfrac{3}{2}(1.987)(298) = 888 \text{ cal} \end{aligned} \tag{25}$$

Such numerical values are of little interest at this stage because you cannot make comparisons with other related energies. As we proceed, and as you become more familiar with the energy of molar amounts of materials, such values will be seen to be of some consequence.

Even more remote, at first, are energy quantities for individual molecules. One quantity we can now obtain is the average translational energy of a gas-phase molecule. We need only divide the energy per mole, that is, $\frac{3}{2}RT$, by the number of molecules per mole. This number, known as Avogadro's number and represented by $\mathfrak{N}$, is 6.023×10^{23}. Thus in the common molecular units of ergs, we have for 25°C the result

$$\begin{aligned}\text{Average translational energy of molecule} &= \frac{3}{2}\frac{RT}{\mathfrak{N}} \\ &= \frac{3}{2}\frac{(8.314 \times 10^7)(298)}{6.02 \times 10^{23}} \\ &= 6.17 \times 10^{-14} \text{ erg} \qquad \mathbf{26}\end{aligned}$$

Another indication of the average translational energy of gas molecules is the amount of this total that can be ascribed to motion in the three perpendicular directions, x, y, and z, used in studies of motion in space. Because, as Fig. 1-3 shows, the velocity in space is expressed in terms of component velocities in these directions by

$$u^2 = u_x^2 + u_y^2 + u_z^2 \qquad \mathbf{27}$$

the kinetic energy can be similarly resolved. Thus if Eq. **27** is multiplied by $\frac{1}{2}m$, where m is the mass of the molecule, we obtain

$$\tfrac{1}{2}mu^2 = \tfrac{1}{2}mu_x^2 + \tfrac{1}{2}mu_y^2 + \tfrac{1}{2}mu_z^2$$

If ϵ represents an energy of an individual molecule, we can express this as

$$\epsilon = \epsilon_x + \epsilon_y + \epsilon_z \qquad \mathbf{28}$$

Thus the translational energy of a molecule can be interpreted as the sum of the components along the three perpendicular directions. These coordinates are examples of *degrees of freedom*. Since, on the average, $\epsilon_x = \epsilon_y = \epsilon_z$, we conclude that the average molecular energy along each of these coordinates is $\frac{1}{3}\epsilon$, and we can write

$$\begin{aligned}&\text{Average translational energy of a molecule} \\ &\text{per degree of freedom} = \frac{1}{3}\left(\frac{3}{2}\right)\frac{RT}{\mathfrak{N}} = \frac{1}{2}\frac{RT}{\mathfrak{N}} \qquad \mathbf{29}\end{aligned}$$

This result and other related ones are utilized in so many considerations of the energies of molecules that it is convenient to introduce a single symbol for $R/\mathfrak{N}$. Thus we introduce k, known as the *Boltzmann constant,* defined as

$$k = \frac{R}{\mathfrak{N}} \tag{30}$$

and evaluated as

$$k = \frac{8.314 \times 10^{7}}{6.023 \times 10^{23}} = 1.380 \times 10^{-16} \text{ erg} \tag{31}$$

Note that k is the gas constant *per molecule;* thus it is the molecular counterpart of the molar quantity R.

In terms of k we can now write

$$\text{Average translational energy of gas molecule} = \tfrac{3}{2}kT \tag{32}$$

and

$$\text{Average translational energy per degree of freedom} = \tfrac{1}{2}kT \tag{33}$$

Our comparison of the results of kinetic-molecular theory with the observed behavior of gases has thus led us to an important first insight into the energies involved in molecular motions.

The following numerical values will be important in many later considerations of molecular energies:

Average translational energy of gas molecule
at 25°C = 6×10^{-14} erg

Average energy per translational degree
of freedom = 2×10^{-14} erg

Example 1-3

From the average translational energy of $\frac{3}{2}kT$ for gas molecules, deduce a representative value for the speeds of hydrogen molecules and oxygen molecules at 25°C.

Solution

For *any* gas at 25°C the average translational kinetic energy is

$$\begin{aligned}\tfrac{3}{2}kT &= \tfrac{3}{2}(1.38 \times 10^{-16})(298)\\ &= 6.17 \times 10^{-14} \text{ erg}\end{aligned}$$

This energy is the average of the $\frac{1}{2}mu^2$ values for the molecules of the gas. Thus for H_2

$$\tfrac{1}{2}m_{H_2}(\overline{u^2})_{H_2} = 6.17 \times 10^{-14}$$

$$\text{and} \quad (\overline{u^2})_{H_2} = \frac{6.17 \times 10^{-14}}{\frac{1}{2}\left(\frac{2 \text{ g/mol}}{6.02 \times 10^{23} \text{ molecules/mol}}\right)}$$
$$= 3.71 \times 10^{10} \text{ (cm/sec)}^2$$

The square root of this result will give the root-mean-square velocity, a quantity close to the average velocity. Thus

$$\bar{u}_{H_2} \cong \sqrt{3.71 \times 10^{10}}$$
$$= 1.93 \times 10^5 \text{ cm/sec}$$
$$= 4320 \text{ mi/h}$$

Similarly, for O_2

$$(\overline{u^2})_{O_2} = \frac{6.17 \times 10^{-14}}{\frac{1}{2}\left(\frac{32}{6.02 \times 10^{23}}\right)}$$
$$= 0.232 \times 10^{10} \text{ (cm/sec)}^2$$

$$\text{and} \quad \bar{u}_{O_2} = 0.482 \times 10^5 \text{ cm/sec}$$
$$= 1080 \text{ mi/h}$$

(Use has been made of the conversion factor 1 mi/h = 44.70 cm/sec.)

1-5 THE NONIDEAL BEHAVIOR OF GASES: VAN DER WAALS' EQUATION

All we can learn about the molecules of gases by developing a model that accounts for the ideal *PVT* gas laws (beyond the basic idea of very small noninteracting molecules) is their average energy and related quantities. The important additional idea that size can be ascribed to molecules comes from considering nonideal gas behavior.

The *PVT* behavior of gases has so far been presumed to follow Boyle's and Gay-Lussac's laws and, with the mole concept, to lead to the ideal-gas result $PV = nRT$. When measurements are extended to higher pressures, or even when very accurate measurements are made at ordinary pressures, deviations from these laws are found.

An actual gas exhibits, to some extent, deviations from the ideal-gas law, and when these deviations are recognized, the gas is said to behave as a *real, nonideal,* or *imperfect gas.* Examples of *PVT* curves that show such behavior are shown in Fig. 1-4.

The simple model of the kinetic-molecular theory has been shown

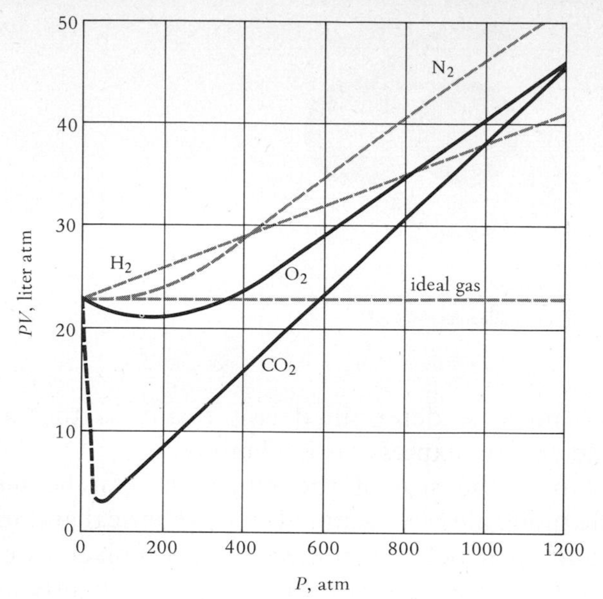

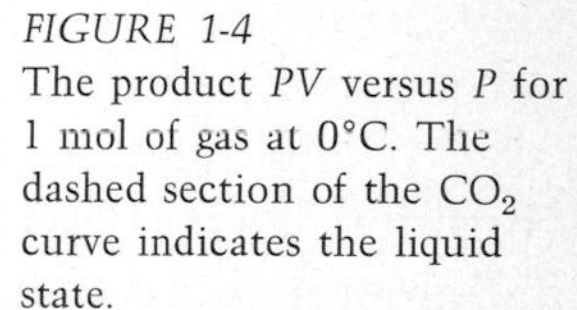
FIGURE 1-4
The product PV versus P for 1 mol of gas at 0°C. The dashed section of the CO_2 curve indicates the liquid state.

to be satisfactory in that it leads to the derivation of the ideal-gas laws. The question naturally arises whether deviations from the ideal-gas laws can be understood by the use of a more elaborate model for a gas. This can be done, and in our quest for molecular information it is of interest to investigate what refinements of the previous treatment are necessary.

In 1873, the Dutch chemist van der Waals showed that two modifications of the simple molecular model account for much of the deviation of real gases from ideal behavior. He attributed the failure of the derived $PV = nRT$ relation to describe the behavior of real gases to the neglect of (1) the volume occupied by the gas molecules and (2) the attractive forces between the molecules. The corrections introduced by these two factors will be treated one at a time.

When n mol of a gas is placed in a container of volume V, the volume in which the molecules are free to move is equal to V only if the volume occupied by the molecules themselves is negligible. The presence of molecules of nonvanishing size means that a certain volume, called the *excluded volume,* is not available for the molecules to move in. If the volume excluded by 1 mol of a gas is represented by b, then instead of writing $PV = nRT$, a more appropriate equation would be

$$P(V - nb) = nRT \qquad \textbf{34}$$

The excluded volume b is usually treated as a constant which is charac-

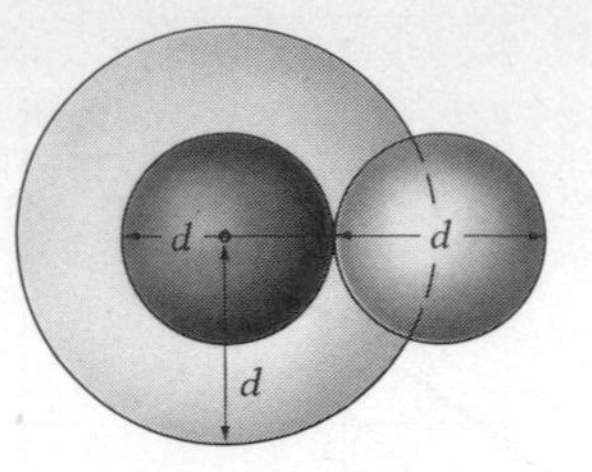

FIGURE 1-5
The excluded volume per pair of molecules according to van der Waals' treatment.

excluded volume (per pair of molecules) = $\frac{4}{3}\pi d^3$

teristic of each gas and must be determined empirically so that a good correction to the simple gas-law expression is obtained.

The relation of b to the size of the molecules can be seen by considering Fig. 1-5. The molecules are assumed to be spherical and to have a diameter d. The volume in which the centers of two molecules cannot move because of each other's presence is indicated by the lightly shaded circle in Fig. 1-5. The radius of this sphere is equal to the molecular diameter. The volume excluded per pair of molecules is $\frac{4}{3}\pi d^3$. The volume excluded per molecule is $\frac{1}{2}(\frac{4}{3}\pi d^3)$. We can compare

$$\text{Actual volume of molecule} = \tfrac{4}{3}\pi\left(\frac{d}{2}\right)^3$$

$$\text{Excluded volume per molecule} = \tfrac{1}{2}(\tfrac{4}{3}\pi d^3) = 4\left[\tfrac{4}{3}\pi\left(\frac{d}{2}\right)^3\right]$$

Therefore the excluded volume is 4 times the actual volume of the molecules. Since b is the excluded volume *per mole,* we have

$$b = 4\mathfrak{N}\left[\tfrac{4}{3}\pi\left(\frac{d}{2}\right)^3\right] \qquad \textbf{35}$$

where $\mathfrak{N}$ is Avogadro's number.

A value of b for a particular gas can be obtained by adjusting it so that the derived equation corresponds as well as possible to the observed *PVT* data. This procedure, required principally by the difficulties caused by the second correction term, results in van der Waals' equation being *semiempirical.* The form of the derived equation follows from a theoretical treatment, but the numerical values of the constants appearing in the equation are obtained from the experimental *PVT* data. Semiempirical equations are not uncommon in chemistry and frequently are considerably more satisfactory than completely empirical relations.

The second van der Waals' correction term concerns the attractive forces between molecules. That such forces exist is clearly demonstrated by the tendency for all gases to condense at temperatures low enough for these forces to overcome the kinetic energy of the molecules. Although it is clear that these attractions exist, exact knowledge of their source and quantitative values is much harder to come by. The semiempirical approach, however, requires only that a suitable term representing these attractions be inserted. Its value can be obtained by adjusting it, as is done with b, to give an equation that best fits the PVT data.

The attraction a molecule exerts on its neighbors tends to draw them in toward itself; i.e., the attraction acts with the confining pressure to hold the molecules together. Complete analysis of the consequence is difficult, but qualitatively the effect is that of reducing the independence of each molecule. The pressure needed to confine the gas is thus reduced, just as it would be if the number of independent molecules decreased. The effect of one molecule in helping to hold the gas together through these forces of attraction is proportional to the number of nearby molecules on which it can act. If there is n mol of gas in a volume V, this number is proportional to n/V, the number of moles per unit volume. Since each of the neighboring molecules is likewise attracting its neighbors, the total pulling together of the gas due to these interactions is proportional to $(n/V)^2$. The gas is confined, therefore, not only by the external pressure P but also by these intermolecular attractions, which contribute a term proportional to $(n/V)^2$. If the proportionality factor is denoted by a, van der Waals' complete equation becomes

$$\left(P + \frac{an^2}{V^2}\right)(V - nb) = nRT \qquad \textbf{36}$$

The success of this equation in fitting the PVT behavior of real gases is judged by choosing values of a and b, different for each gas and for each temperature, to give as good a fit to the observed data as possible. Although perfect agreement of calculated and observed volumes over a wide range of pressure is not obtained, the improvement over the ideal-gas-law expression $PV = nRT$ is considerable. Figure 1-6 and Table 1-1 indicate the amount of improvement in regions of very nonideal behavior. The success of van der Waals' equation in representing PVT behavior is much better than would be expected for any purely empirical expression with only two adjustable constants. The behavior in regions where both liquid and gas are present, i.e., in the two-phase region below the *critical point*, as shown in Fig. 1-6, however, cannot be followed by van der Waals' equation, and the maxima and minima must be disregarded.

Values of the van der Waals' constants a and b obtained by fitting

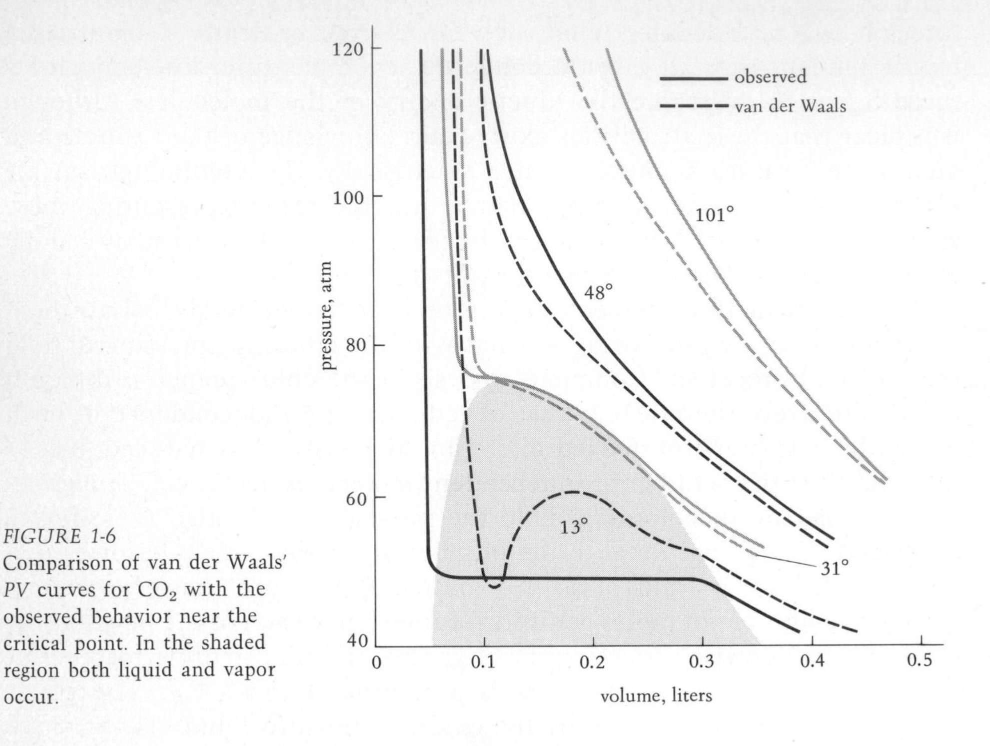

FIGURE 1-6
Comparison of van der Waals' *PV* curves for CO_2 with the observed behavior near the critical point. In the shaded region both liquid and vapor occur.

the equation to the properties of gases at the critical point are shown for some gases in Table 1-2. Included also are the molecular-diameter results deduced from the b values according to Fig. 1-5 and Eq. **35.** These data show the size that must be ascribed to these molecules if their detailed structure is ignored and they are represented by spheres. Since these spheres should be thought of as somewhat compressible, like a sponge-rubber ball, their size should depend somewhat upon the method of measurement or, in van der Waals' method, on the temperature at which the fit to the *PVT* data is made.

TABLE 1-1
Molar Volumes of CO_2 at 320 K

	V, **liters/mol**		
***P*, atm**	**observed**	**van der Waals**	**ideal**
1	26.2	26.2	26.3
10	2.52	2.53	2.63
40	0.54	0.55	0.66
100	0.098	0.10	0.26

The results, nevertheless, are valuable in providing a step into the world of molecular dimensions. The step is facilitated by the use of the unit of measure of *angstroms* (Å) where

$$1 \text{ Å} = 10^{-8} \text{ cm} \tag{37}$$

Small molecules, as you see from Table 1-2, have dimensions of a few angstroms. Angstrom units remain useful for treating very large molecules because the lengths of the bonds between adjacent atoms is typically in the 1- to 2-Å range.

Example 1-4

According to the estimate of molecular size given by van der Waals' equation, what is the fraction of the volume of oxygen gas at 1 atm and 25°C actually occupied by oxygen molecules?

Solution

According to Table 1-2, van der Waals' b parameter for oxygen leads to a value of 2.93 Å for the diameter of O_2 molecules.

Each molecule has, accordingly, a volume of

$$\tfrac{4}{3}\pi r^3 = \tfrac{4}{3}\pi\left(\frac{2.93}{2}\right)^3 = 13.2 \text{ Å}^3 = 13.2 \times 10^{-24} \text{ cm}^3$$

TABLE 1-2

Values of van der Waals' Constants a and b and the Deduced Molecular Diameter

Other values give better fits to PVT data for specific temperature ranges. Values listed here assume that an R value of 0.08206 liter atm/mol deg is used in van der Waals' equation.

gas	a, atm liters^2/mol^2	b, liters/mol	d, molecular diameter, Å
H_2	0.244	0.0266	2.76
He	0.034	0.0237	2.66
H_2O	5.46	0.0305	2.89
NH_3	4.17	0.0371	3.09
CH_4	2.25	0.0428	3.24
O_2	1.36	0.0318	2.93
N_2	1.39	0.0391	3.14
CO	1.48	0.0399	3.16
CO_2	3.59	0.0427	3.24
Ar	1.34	0.0322	2.94
C_6H_6	18.00	0.115	4.50

Now, consider 1 mol of the gas:

Volume at 1 atm and 25°C = 22,400 cm^3 $\times \frac{298}{273}$ = 24,400 cm^3

No. of molecules = $\mathfrak{N} = 6.02 \times 10^{23}$

Volume of $\mathfrak{N}$ molecules = $6.02 \times 10^{23} \times 13.2 \times 10^{-24}$ cm^3 = 7.9 cm^3

The percent of total volume occupied by O_2 molecule is

$$\frac{7.9}{24,400} \times 100 = 0.032\%$$

(Note that the volume of the oxygen molecules is not the same as the volume they exclude each other from.)

Attention has been given here to just one of the many equations, examples of *equations of state,* that seek to depict the *PVT* behavior of real gases. Van der Waals' equation has been treated because the parameters in the equation lead us to further ideas about the size and the interactions of molecules. Other *PVT* equations that have other special merits are (for 1 mol of gas):

Virial equation:

$$\frac{PV}{RT} = 1 + \frac{B(T)}{V} + \frac{C(T)}{V^2} + \cdots$$

Berthelot equation:

$$\left(P + \frac{a}{TV^2}\right)(V - b) = RT$$

Dieterici equation:

$$Pe^{a/VRT}(V - b) = RT$$

1-6 INTRODUCTION TO THE ROTATION OF MOLECULES

If molecules are treated as point particles, as they were in the kinetic-molecular theory of ideal gases, the only motion that can be associated with them is translation. But when we think of molecules as having size, as van der Waals' treatment of real gases suggests, we must also expect them to be able to undergo rotational motion. Molecules will be pictured as assemblies of atoms joined together by springlike chemical bonds. As such gas molecules fly through space, they can twirl about with various speeds of rotation. This rotational motion gets mixed with other motions in liquid and solid systems and therefore can be studied in detail only for gases. Its study here and in the following sections will provide further insight into molecular dynamics.

Molecules of a gas must be pictured as spinning about their center of gravity as suggested in Fig. 1-7. To understand this motion let us first treat the simpler problem of an ordinary-sized object, the ball on the string of Fig. 1-8, for example, twirling about a fixed point. Later, additional features will be added to enable us to deal (1) with many-particle objects that rotate about their center of gravity and (2) with such objects of molecular dimensions.

Consider the particle of Fig. 1-8. The speed of rotation can be described in terms of the particle's velocity v, or in terms of the number of revolutions per second (rev/sec), i.e., the frequency ν. Since the particle travels a distance v cm/sec and the distance for 1 rev is $2\pi r$ cm, we have the relation

$$\nu = \frac{v}{2\pi r} \tag{38}$$

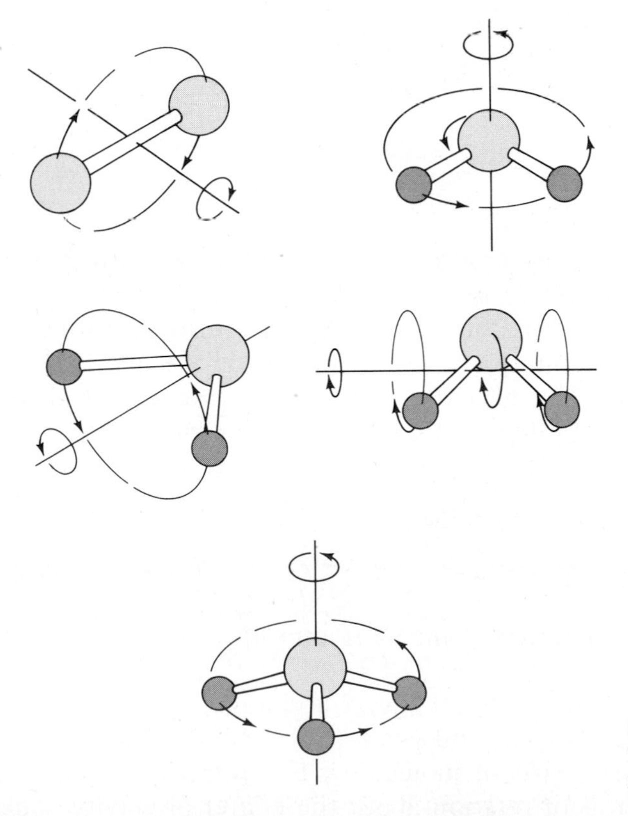

FIGURE 1-7
Examples of molecular rotations.

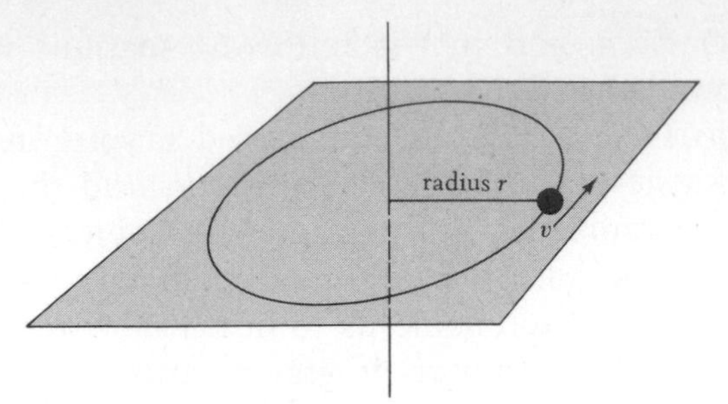

FIGURE 1-8
The rotation of a particle about a fixed point.

The speed of rotation is also described by the angular velocity ω expressed in terms of radians per second. Since there are 2π rad in a revolution, $\omega = 2\pi\nu$, and

$$\omega = \frac{v}{r} \tag{39}$$

The kinetic energy ϵ of the particle, which again is $\frac{1}{2}mv^2$, can be rearranged to a form more convenient for rotary systems by rewriting $\frac{1}{2}mv^2$ as

$$\epsilon_{\text{rot}} = \frac{1}{2}mr^2\left(\frac{v}{r}\right)^2 \tag{40}$$

Then if I, the *moment of inertia,* is introduced for the collection of terms mr^2, and if ω is inserted for v/r, we have

$$\epsilon_{\text{rot}} = \frac{1}{2}I\omega^2 \tag{41}$$

Notice that in expressions for rotary systems, compared with those for linear systems, I plays the role of m, and ω plays the role of v.

One further feature of rotating systems turns out to be important. It is the rotational counterpart of the momentum mv of a particle moving with linear motion. The idea that for rotational systems I plays the role of m and ω plays the role of v leads us to define the angular momentum as

$$\text{Angular momentum} = I\omega \tag{42}$$

If we replace I by mr^2 and ω by v/r we obtain an alternative expression

$$\text{Angular momentum} = mr^2\frac{v}{r} = mvr \tag{43}$$

Expressions for rotational energy and angular momentum, like Eqs. **41** and **42,** that make use of I and ω can readily be extended to macroscopic ball-and-stick counterparts of molecules. The rotation of such assemblies is described in terms of rotation about the center of gravity. The angular velocity ω (but not the velocities v) is the same for all the balls of the ball-and-stick assembly. The rotational-energy and angular-momentum expressions above can then be used if the moment of inertia is defined so that it sums up the contributions of each of the particles. We write

$$I = \sum_i m_i r_i^2 \qquad \textbf{44}$$

where m_i is the mass of the ith particle, r_i is the distance of that particle from the center of gravity, and the summation is over all the particles.

Now even for many-particle assemblies, if we interpret I as $\sum_i m_i r_i^2$, we can use

$$\epsilon_{rot} = \tfrac{1}{2}I\omega^2 \qquad \text{and} \qquad \text{Angular momentum} = I\omega$$

Some obstacles, as you will see, stand in the way of applying these descriptions of rotational motion to molecules, but we can get an indication of how fast molecules of gases twirl as they fly through space by assuming that the rotational energy is equal to the $\frac{1}{2}kT$ value we found for each translational degree of freedom. From masses of molecules and the sizes suggested by Table 1-2, a representative value of I for small gas-phase molecules can be estimated by assuming masses of about $10/(6.0 \times 10^{23})$ g separated by about 2 Å. Then

$$I = \frac{10}{6.0 \times 10^{23}}(2 \times 10^{-8})^2$$

$$= 7 \times 10^{-39} \text{ g cm}^2$$

Earlier we had

$$\tfrac{1}{2}kT = \tfrac{1}{2}(1.38 \times 10^{-16})(298)$$

$$= 2.06 \times 10^{-14} \text{ erg}$$

Now, from $\epsilon_{rot} = \frac{1}{2}I\omega^2$ and the assumption that $\epsilon_{rot} = \frac{1}{2}kT$ we obtain

$$\omega = \sqrt{\frac{2\epsilon_{rot}}{I}} = \sqrt{\frac{2(2.06 \times 10^{-14})}{7 \times 10^{-39}}}$$

$$= 2 \times 10^{12} \text{ rad/sec}$$

Also $\nu = \dfrac{\omega}{2\pi} \cong 0.3 \times 10^{12}$ rev/sec

Molecules lead a dizzy existence. You will find it worthwhile to note not only that molecules twirl very rapidly but also that 10^{12} is a typical molecular-rotational frequency. This number, like the 25°C value of $\frac{1}{2}kT$, will help to develop your appreciation for the molecular scale of things.

This speed of rotation and some ideas about radiation lead, as the next section will show, to a direct experimental study of molecular rotations. This study immediately reveals that the above treatment of rotations must be augmented to make it completely applicable to rotating molecules.

Example 1-5

How fast would the extremities of a representative small molecule have to travel if the molecule performs 10^{12} rev/sec? Compare the result with typical molecular translational speeds and with the speed of light.

Solution

Assume a molecule with a radius of 2 Å. If the molecule is pictured as a sphere, the distance traveled by a point on its surface in 1 rev is $2\pi r = 12$ Å $= 12 \times 10^{-8}$ cm. If 10^{12} rev are completed in 1 sec, such a point on a molecule, or an atom of a molecule, must have a speed of 12×10^{-8} cm/rev $\times 10^{12}$ rev/sec $= 12 \times 10^4$ cm/sec $= 2700$ mi/h. This speed is comparable with translational speeds and very much less than the speed of light, 3×10^{10} cm/sec.

1-7 RADIATION AND MOLECULAR ROTATIONS

One of the most powerful tools for the direct study of molecules, *spectroscopy,* depends on their interaction with light or, more generally, radiation. One illustration is the study that can be made of the rotation of molecules of a gas. Here the nature of radiation will be introduced so that the basis of such studies can be appreciated.

Two models of electromagnetic radiation must be used to account for observed phenomena. One describes radiation in terms of oscillating electric and magnetic fields; the other describes radiation as a stream of energy packets known as *photons* or *quanta*. In studies of the interaction of radiation with matter, both models are used.

The feature of the wave model of radiation that is most important here is the *electric field.* The concept of an electric field arises because charges have an effect on one another even when they are separated. It is convenient to think of an electric field as existing about a charge. A charged particle in that field is pushed or pulled by an amount that depends on the magnitude of the field to which it is exposed. In fact, the magnitude of the electric field at any point in space is given by the force that a unit positive charge at that position would experience. Since the force acts in a particular direction, the field at a point has a direction as well as a magnitude. Often these direction and magnitude features are conveyed by drawing *lines of force,* as in Fig. 1-9, and interpreting the field in terms of the concentration and direction of these lines of force.

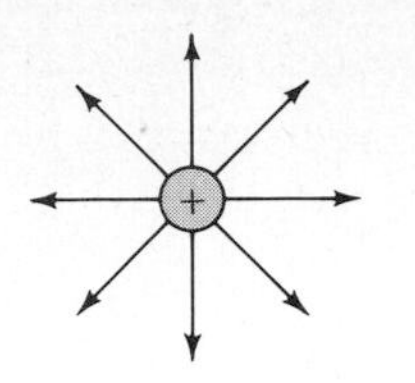
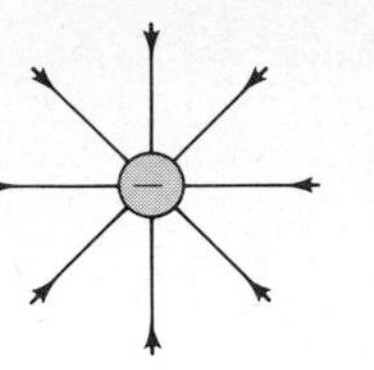
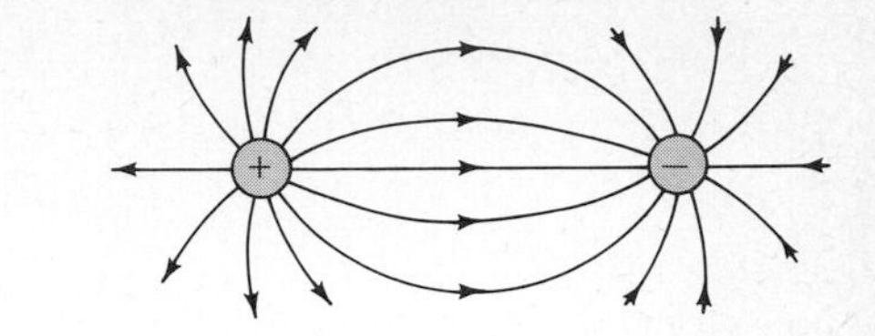

FIGURE 1-9
Example of lines of force used to depict the electric field, i.e., the force that acts on a unit positive charge.

The electric-field concept enters into studies of electromagnetic radiation because the radiation can be described in terms of oscillating electric and magnetic fields that travel through space with the speed of light. As Fig. 1-10 shows, the electric field, which is the field component of most direct interest, acts on a charged particle in the path of the radiation to produce a succession of pushes and pulls as the successive half waves of the electric field pass by.

The oscillating electric field can be described in terms of the wavelength λ of the oscillation or of the frequency ν with which the oscillation occurs as the radiation passes a fixed observation point. These two quantities are related, as shown in Fig. 1-11, through the speed of light $c = 2.9979 \times 10^{10}$ cm/sec, by

$$\nu = \frac{c}{\lambda} \qquad \mathbf{45}$$

In practicc, both ν and λ arc uscd to dcscribc radiation, as illustratcd by the radiation regions shown in Fig. 1-12. The units of ν are cycles per second, now called *hertz* (Hz).

Comparison of the frequency of rotation calculated for a representative gas-phase molecule in the preceding section and the frequency ranges of Fig. 1-12 shows that the microwave radiation has frequencies corresponding to the frequencies of molecular rotations.

A rotating molecule can be expected to withdraw energy from

FIGURE 1-10
The electric field of electromagnetic radiation (plane-polarized radiation is illustrated). The effect is to produce alternating pushes and pulls on a charged particle in the path of the radiation.

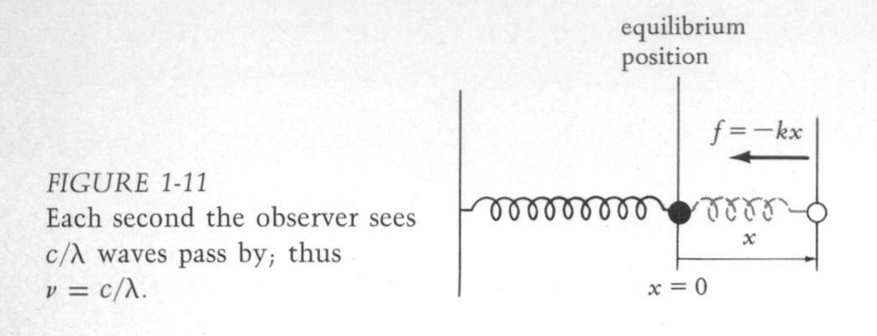

FIGURE 1-11
Each second the observer sees c/λ waves pass by; thus $\nu = c/\lambda$.

electromagnetic radiation or give up energy to the radiation if it can interact with the oscillating electric field associated with the radiation. The molecule can do this if (1) it has positive and negative ends, i.e., if it has an *electric dipole* and (2) if it rotates at such a frequency that the electric field of the radiation can give the molecule repeated pushes or pulls to speed up its rotation or slow it down. The rotating dipole provides a *coupling* with the oscillating electric field of the radiation and allows energy to be transferred from the radiation to the molecule, or vice versa. It is concluded, therefore, that a molecule must have a dipole moment in order to give rise to a rotational spectrum. One observes, in fact, that electrically symmetric molecules like H_2, N_2, and CO_2 give rise to no absorptions that can be attributed only to changes in the rotational energy of the molecules.

The energy effects of radiation are most easily treated by switching over to the second radiation model, which pictures radiation as a stream of photons or quanta. A rotating molecule can pick up, or *absorb*, a quantum of radiation energy, and thereby increase its rotational energy, or it can *emit* a quantum and thereby decrease its rotational energy. The necessary bridge between the two models of radiation is provided by *Planck's relation*. With $\Delta\epsilon$ representing the energy of a photon, it can be written as

$$\Delta\epsilon \propto \nu \qquad \textbf{46}$$

or with h representing a proportionality constant as

$$\Delta\epsilon = h\nu \qquad \textbf{47}$$

When $\Delta\epsilon$ has the energy units of ergs and ν the units of reciprocal seconds (sec^{-1}), the numerical value of h is given by

$$h = 6.626 \times 10^{-27} \text{ erg sec} \qquad \textbf{48}$$

What happens if a beam of microwaves shines on a sample of gas whose molecules might be expected to interact with the electric field of the radiation? Radiation is, in fact, absorbed by the molecules, and our idea that molecules rotate with frequencies of about 10^{12} rev/sec, corresponding

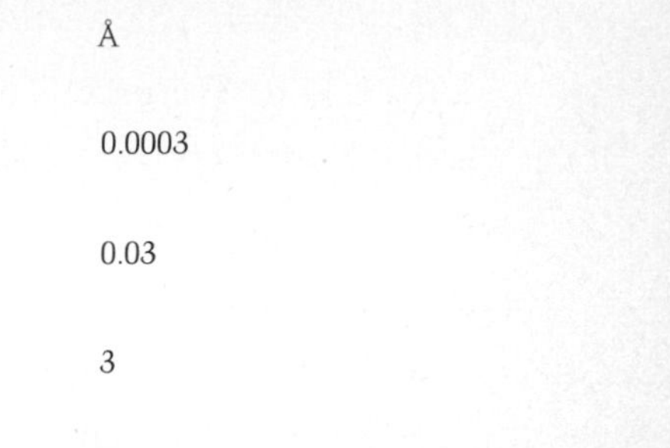

frequency ν		wavelength λ	
Hz		cm	Å
	cosmic rays		
10^{22}		3×10^{-12}	0.0003
	γ rays		
10^{20}		3×10^{-10}	0.03
	x-rays		
10^{18}		3×10^{-8}	3
10^{16}	ultraviolet	3×10^{-6}	300
	visible		
10^{14}		0.0003	30,000
	infrared		
10^{12}		0.03	
	microwave		
10^{10}		3	
10^{8}	radio	300	

FIGURE 1-12
Regions of the electromagnetic spectrum.

to the frequencies of microwave radiation, is borne out. But, unexpectedly, we find that only certain frequencies or, according to $\Delta\epsilon = h\nu$, only quanta of certain energies are absorbed. This is quite at odds with the ideas of Sec. 1-6 that led us to believe that molecules can rotate with any angular velocity and therefore with any rotational energy $\frac{1}{2}I\omega^2$. That only quanta of certain energies can be absorbed suggests that only certain rotational energies of the molecule are permitted. To understand the radiation that is absorbed, as shown by microwave spectra, we now must see how the restriction on rotational energies is expressed.

Example 1-6

Characterize by wave- and particle-model features radiation that appears blue to the eye. It has a wavelength of about 4000 Å.

Solution

Wave properties:

$$\lambda = 4000 \text{ Å} = 0.4 \times 10^{-4} \text{ cm}$$

$$\nu = \frac{c}{\lambda} = \frac{3 \times 10^{10} \text{ cm/sec}}{0.4 \times 10^{-4} \text{ cm}}$$

$$= 7.5 \times 10^{14} \text{ Hz}$$

Particle properties:

$$\begin{aligned}\Delta\epsilon &= h\nu \\ &= (6.62 \times 10^{-27} \text{ erg sec})(7.5 \times 10^{14} \text{ sec}^{-1}) \\ &= 5.0 \times 10^{-12} \text{ erg}\end{aligned}$$

Energy of a mole of photons:

$$\begin{aligned}\mathfrak{N}\,\Delta\epsilon &= (6.02 \times 10^{23})(5.0 \times 10^{-12}) \\ &= 3.01 \times 10^{12} \text{ erg/mol} \\ &= 3.01 \times 10^{5}\, J/\text{mol} \\ &= 72{,}000 \text{ cal/mol} = 72 \text{ kcal/mol}\end{aligned}$$

1-8 QUANTUM RESTRICTIONS AND ROTATIONAL SPECTRA

The behavior exhibited by particles whose motion is confined to regions as small as atoms and molecules can be understood only by imposing certain restrictions. A variety of observations, such as the microwave spectral results, require the acceptance of these restrictions. So also does our inability with the ordinary laws of motion and electrostatic interactions to explain the stability of the arrays of electrons and nuclei that constitute atoms and molecules. The restrictions we impose to deduce the allowed energies lead to satisfactory explanations of these experimental results, and it is on this basis, rather than on any derivations, that their acceptance depends.

There are two different but compatible ways of introducing these quantum restrictions into our ordinary macroscopic treatments. The way used here is applicable to systems that engage in rotational motion. The electrons of atoms, the spinning electrons themselves, and spinning nuclei provide, along with the rotation of molecules, other important areas of application. The other way of introducing quantum restrictions, which depends on the recognition of the wave nature of particles, is most directly applicable to linear motion. It will be introduced when the vibrations of molecules are dealt with in Sec. 3-3.

The basic postulates specifying quantum restrictions giving results consistent with observed behavior for systems that can undergo rotational motion can be expressed by two statements. These specify the rotational properties that the system can have. We say they describe the *allowed states* of the system.

1 The total angular momentum $I\omega$ can have only the values

$$I\omega = \sqrt{J(J+1)}\,\frac{h}{2\pi} \qquad J = 0, 1, 2, \ldots \tag{49}$$

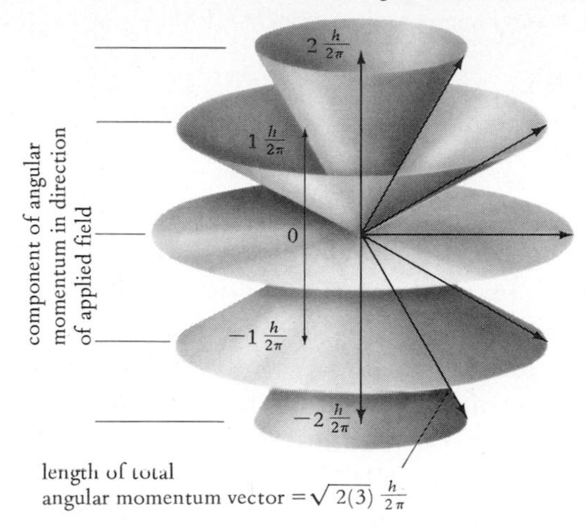

FIGURE 1-13
For a total angular momentum corresponding to $J = 2$ there are five ways the angular-momentum vector can point to give angular-momentum components in a particular direction that are integral multiples of $h/2\pi$. In general, there are $2J + 1$ values of M, and thus $2J + 1$ states for a given value of J.

The symbol J, which represents a series of integers, is an example of a *quantum number.*

2 If a direction is imposed on the rotating system, as by a neighboring particle or application of an electric or magnetic field, the angular-momentum component in that direction can have the values

$$M\frac{h}{2\pi} \qquad M = -J, -J+1, \ldots, 0, 1, \ldots J \tag{50}$$

Thus, in this approach, each allowed state of the system is characterized by its total angular momentum and the value this total would project along an imposed direction. It is customary to depict these quantities by vectors and to draw diagrams like that of Fig. 1-13 to show for each value of J the possible states that occur. Each state corresponding to a given J value is designated by means of a value of M. As Fig. 1-13 shows, there are $2J + 1$ states for each value of J.

Consider now the rotational motion of a linear molecule. The allowed rotational states have angular momenta given by

$$I\omega = \sqrt{J(J+1)}\,\frac{h}{2\pi} \qquad J = 0, 1, 2, \ldots$$

and each choice of J corresponds to $2J + 1$ states with that angular momentum. We are led directly to the rotational energies that are allowed to the molecule by writing

$$\epsilon_{\text{rot}} = \tfrac{1}{2}I\omega^2 = \frac{(I\omega)^2}{2I}$$

$$= J(J+1)\frac{h^2}{8\pi^2 I} \qquad J = 0, 1, 2, \ldots \tag{51}$$

The pattern of energies for the allowed rotational states for linear molecules is shown in Fig. 1-14, with the scale factor $h^2/8\pi^2 I$ still containing the molecular structure quantity I, which has yet to be determined. (Although this moment of inertia $I = \Sigma m_i r_i^2$ could be estimated by using bond-length data from other sources, spectral study of the rotational-energy pattern provides some of the most accurate moment-of-inertia data.)

Even when a molecule has a dipole moment and can interact with the radiation, as discussed in the preceding section, there is a restriction on the rotational transitions that can be induced. This restriction, an example of a *selection rule,* is that a molecule can increase or decrease its rotational energy only to the next higher or lower energy level when it absorbs or emits a quantum of electromagnetic radiation. This selection rule is written

$$\Delta J = \pm 1 \tag{52}$$

(The restriction can be understood in terms of the spin of radiation photons and the principle of the conservation of angular momentum, or its existence can easily be recognized from the nature of the observed rotation spectra.) Rotational spectra are almost always studied by observing the radiation absorbed by the sample. For such *absorption spectra,* the only part of the selection rule that is of interest is that of $\Delta J = +1$.

Molecules are distributed throughout many of the lower allowed rotational energy levels, such as those depicted in Fig. 1-14. The transitions which can occur,

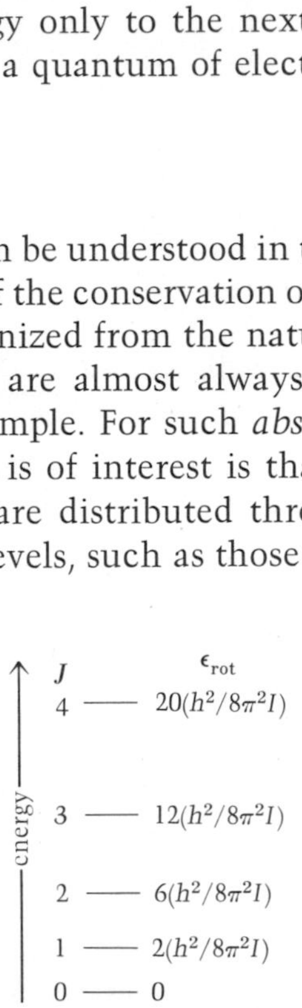

energy ↑ J	ϵ_{rot}
4	$20(h^2/8\pi^2 I)$
3	$12(h^2/8\pi^2 I)$
2	$6(h^2/8\pi^2 I)$
1	$2(h^2/8\pi^2 I)$
0	0

FIGURE 1-14
The first five allowed rotational energies for a linear molecule.

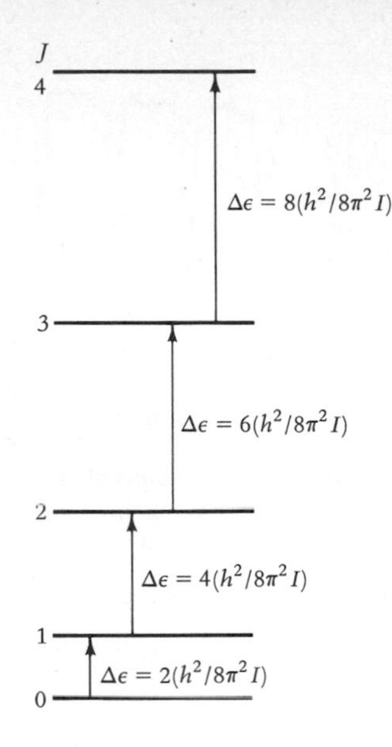

FIGURE 1-15
The energies of the quanta expected to be absorbed on the basis of the rotational-energy diagram of Fig. 1-14 and the $\Delta J = +1$ selection rule.

therefore, are between adjacent levels indicated by the $\Delta\epsilon$ terms of Fig. 1-15. These energy differences correspond, then, to the energies of quanta of radiation that can be absorbed to bring about the transition $\Delta J = +1$.

A feature of the quantum energies predicted for absorbed radiation (not the rotational energies but the transitions between them) is apparent from Fig. 1-15: the values of $\Delta\epsilon$ are spaced by a constant factor of $2(h^2/8\pi^2 I)$. Thus we expect to find in the microwave region a series of absorptions of radiation spaced by an equal energy amount, which can be identified with $2(h^2/8\pi^2 I)$, or an equal frequency amount $2(h/8\pi^2 I)$, for the molecule under investigation.

The rotational spectra of most molecules occur in the microwave spectral region. Generally, diatomic molecules have such low moments of inertia that their rotational transitions, which have energies related to $(h^2/8\pi^2 I)$, occur at the infrared end of the microwave spectrum. In fact, for rather low-moment-of-inertia molecules, like HCl, infrared spectral techniques can be used to obtain spectra like that of Fig. 1-16.

Microwave measurements generally are of much greater resolution and frequency precision. For CO, for example, such data provide a value of 0.76373×10^{-14} erg for the spectral-energy spacing term $2(h^2/8\pi^2 I)$. From this we obtain $I = 14.560 \times 10^{-40}$ g cm^2.

The moment of inertia can be related to the masses of the two atoms and the distance between them. As Fig. 1-17 shows, the moment of

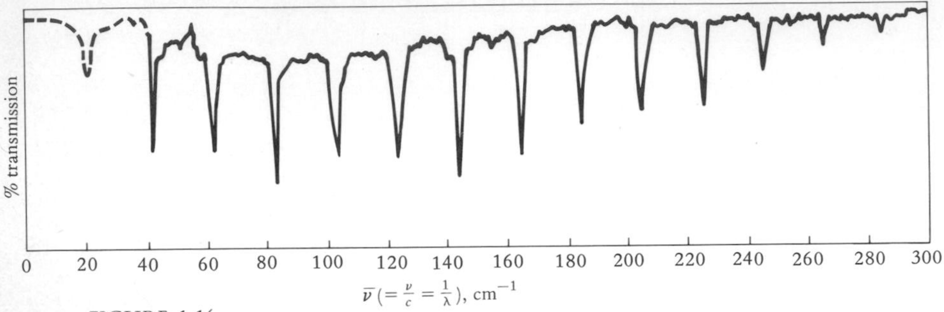

FIGURE 1-16
The observed absorption spectrum of gaseous HCl in the far-infrared spectral region. The spacing between spectral lines is approximately constant at the value of 20 cm^{-1}. (These units are known as *wave numbers.* They can be converted to frequencies by multiplication by c and then to energies by multiplication by h.)

inertia for the diatomic assembly of the CO molecule is

$$I = \frac{m_C m_O}{m_C + m_O} r^2 \qquad 53$$

Often we encounter this collection of mass terms when we deal with two-particle systems. It is customary to introduce μ, the *reduced mass,* which for CO is defined as

$$\mu_{CO} = \frac{m_C m_O}{m_C + m_O} \qquad 54$$

Insertion of atomic mass values for C and O in Eq. **54** leads to $\mu_{CO} = 1.139 \times 10^{-23}$ g. With $I = 14.560 \times 10^{-40}$ g cm^2 and $I = \mu r^2$, we obtain $r_{CO} = 1.131$ Å.

For a general polyatomic molecule, a rather complicated energy-level pattern and rotational spectrum result. The pattern and spectrum depend on three *principal moments of inertia,* and when the spectrum is interpreted in terms of the energy pattern, values can be obtained for these moments of inertia. From these results bond angles and bond distances can again be determined.

Very accurate structural data for small gas-phase molecules, as in Fig. 1-18, are the chief product of microwave studies of molecular rotations, although here the insight provided into this molecular motion is also valuable.

For example, the energy spacing between allowed states is found to be small, at or above room temperature, compared with the thermal-energy factor kT. We must expect the molecules of a gas to be spread out through many of the rotational states. The molecules of a gas sample will

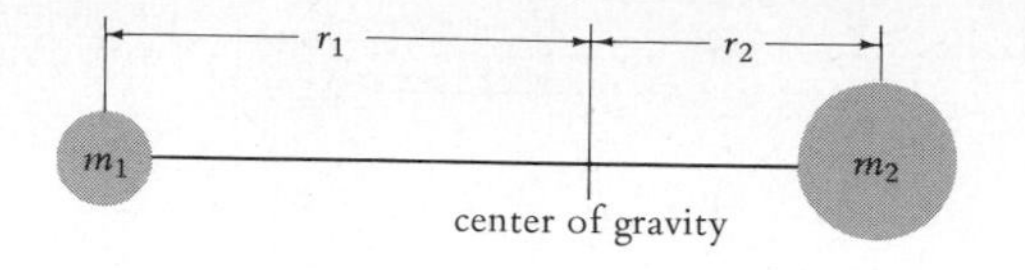

The center of gravity is located so that

$$m_1 r_1 = m_2 r_2$$

Also

$$r_1 + r_2 = r$$

Therefore

$$r_2 \frac{m_2}{m_1} + r_2 = r \qquad \text{and} \qquad r_1 + r_1 \frac{m_1}{m_2} = r$$

or $$r_2 = \frac{m_1}{m_1 + m_2} r \qquad \text{and} \qquad r_1 = \frac{m_2}{m_1 + m_2} r$$

Now

$$\begin{aligned} I &= m_1 r_1^2 + m_2 r_2^2 \\ &= \frac{m_1 m_2^2}{(m_1 + m_2)^2} r + \frac{m_1^2 m_2}{(m_1 + m_2)^2} r \\ &= \frac{(m_1 m_2)(m_1 + m_2)\, r}{(m_1 + m_2)^2} \\ &= \frac{m_1 m_2}{m_1 + m_2} r^2 \end{aligned}$$

FIGURE 1-17
The relation of the moment of inertia of a diatomic molecule to the masses of its atoms and its bond length.

therefore be twirling with a great range of rotational speeds as the molecules fly through space with an even greater range of translational speeds.

Example 1-7

What are the lowest possible and the next higher angular momenta and rotational energies for a gas-phase CO_2 molecule? The moment of inertia of the molecules is 71.1×10^{-40} g cm^2.

Solution

Possible angular momenta are given by

$$\text{Angular momentum} = \sqrt{J(J+1)}\,\frac{h}{2\pi} \qquad \text{where } J = 0, 1, 2, 3$$

Therefore, the least possible angular momentum, which occurs for $J = 0$, is zero. The next possible angular momentum, which is calculated from

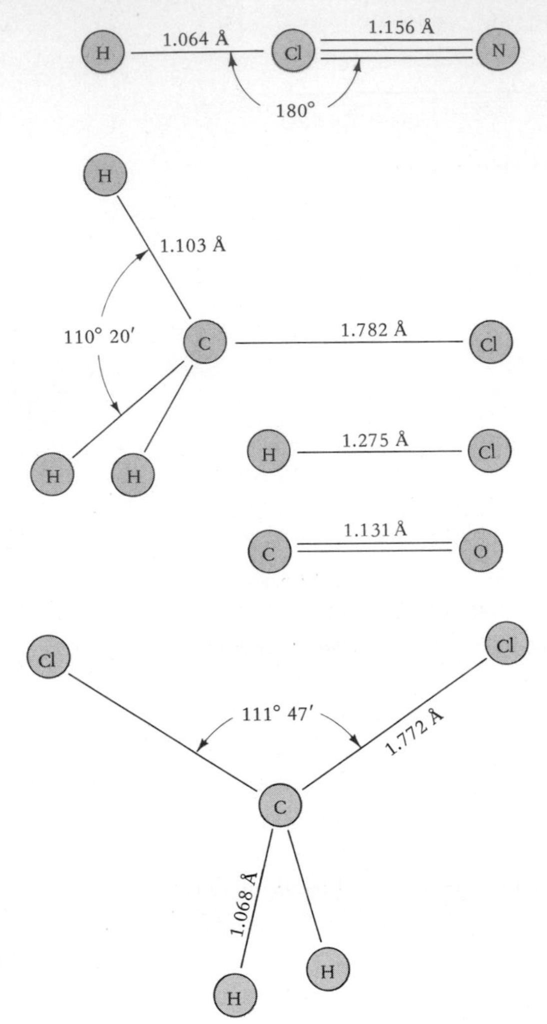

FIGURE 1-18
Some molecular-structure results from microwave spectroscopy.

$J = 1$, is

$$\sqrt{1(2)}\,\frac{h}{2\pi} = \sqrt{2}\,\frac{6.62 \times 10^{-27}}{2\pi} = 1.49 \times 10^{-27}\text{ g cm}^2/\text{sec}$$

(The units can be worked out by treating ergs as dyne centimeters and then, according to $f = ma$, dynes as gram centimeters per second per second.)

Notice that these are the two lowest angular momenta for *all* linear molecules and that the value of the moment of inertia of the CO_2 molecule has not yet been used.

The rotational energy is given by

$$J(J+1)\frac{h^2}{8\pi^2 I} \qquad \text{where } J = 0, 1, 2, \ldots$$

The lowest possible rotational energy, corresponding to $J = 0$, is zero. The next allowed rotational energy for the CO_2 molecule *does* depend on its moment of inertia and is calculated as

$$1(2)\frac{(6.62 \times 10^{-27})^2}{8\pi^2(71.1 \times 10^{-40})} = 0.0156 \times 10^{-14} \text{ erg}$$

This is considerably less than the room-temperature value of 4.1×10^{-14} erg of kT.

PROBLEMS

1 A gas sample occupies a volume of 10.0 cm^3 at a pressure of 740 mm Hg. What volume would it occupy at the same temperature at a pressure of 760 mm? (The relation between the pressure units of atmospheres and millimeters of mercury is 1 atm = 760 mm Hg.)

2 To what volume will 1 liter of gas expand if it is heated from 0°C to the physiological temperature of 37°C, the pressure remaining constant?

3 On the same graph draw curves for P versus V at 25 and 300°C for a sample of a gas that obeys Boyle's law and Gay-Lussac's law and has a volume of 100 cm^3 at 25°C and 1 atm.

4 For the gas sample of the preceding problem, draw a V-versus-T graph for the constant pressures of 1, 0.01, and 200 atm.

5 Attempt to sketch a surface that shows the PVT behavior of the gas sample of Probs. 3 and 4.

6 What is the concentration in moles per liter and molecules per cubic centimeter of an ideal gas at 25°C, 1 atm pressure?

7 A 1-liter gas bulb weighs 43.8240 g when evacuated and 44.9830 g when filled with air at 1 atm pressure and 25°C. Assuming that air behaves as an ideal gas under these conditions, calculate the effective molecular mass of air.

8 The following data have been obtained for the density of CO_2 as a function of pressure at 10°C:

P, atm	d, g/liter
0.68	1.29
2.72	5.25
8.14	16.32

By a suitable graphical extrapolation based on the expression $M = (d/P)RT$, obtain the molecular mass of CO_2.

9 Into a gas bulb of 2.83 liters are introduced 0.174 g of H_2 and 1.365 g of N_2, assumed to behave ideally. The temperature is 0°C.
a What are the mole fractions of each gas?
b What is the total gas pressure and what are the partial pressures of H_2 and N_2?

10 The air we breathe out is depleted in O_2 and augmented with CO_2 and water vapor. If the total pressure of a representative sample of exhaled air is 760 mm, the partial pressures of the five major components N_2, O_2, H_2O, CO_2, and Ar, would be 558, 115, 48, 32, and 7 mm Hg, respectively. (These values differ from normal inhaled sea level air which has the corresponding partial pressures of about 586, 157, 10, 0, and 7 mm of Hg.)
a What are the partial pressures and the mole fractions in exhaled air and in sea level air?
b What is the difference in the number of moles of N_2, O_2, H_2O, and CO_2 in 1 liter of sea-level air and 1 liter of exhaled air? Take the temperature to be approximately constant at 30°C.
c What volume of exhaled air contains the same number of moles of N_2 as 1 liter of inhaled air?

11 The density of the vapor in equilibrium with solid NH_4Cl was found [W. H. Rodebush and J. C. Michalek, *J. Am. Chem. Soc.*, **51**:748 (1929)] to be 0.1373 g/liter at a temperature of 596.9 K and a pressure of 192.3 mm Hg. From these data, deduce the nature of the vapor of ammonium chloride under these conditions.

12 Calculate and compare the average kinetic energies, speeds, and momenta of the molecules of He and Hg at 25°C. Would the comparison you have made be altered if the temperature were specified to be 1000°C?

13 The kinetic-molecular theory attributes an average kinetic energy of $\frac{3}{2}kT$ to each particle. What speed would a small but not molecular particle such as a mist droplet with a mass of 10^{-12} g have according to the kinetic-molecular theory? Assume room temperature. Compare this value with molecular speeds at this temperature.

14 For 1 mol of an ideal gas the *compressibility factor* Z defined by $Z = PV/RT$ has the value unity. For carbon monoxide this factor is found to be 0.9920 at 25°C and 75 atm and 1.7412 at 25°C and 800 atm. What values of Z would be expected on the basis of (*a*) the ideal gas law and (*b*) van der Waals' equation for carbon monoxide at these two conditions?

15 Calculate, according to van der Waals' equation, the pressure that would have to be applied to a sample of N_2 that occupies 1 liter at 25°C and 1 atm to reduce its volume to 0.01 liter at the same temperature.

16 Plot a P-versus-V curve for 1 mol of water at 100°C on the basis of van der Waals'

equation. Include the volume range up to about 40 liters. On the same graph compare the actual PV curve estimated from the facts that (*a*) the density of liquid water can be taken as 1 g/ml regardless of pressure and temperature, (*b*) the density of steam at 100°C and 1 atm pressure is about 0.0006 g/ml, and (*c*) at pressures lower than 1 atm water vapor can be assumed to behave ideally.

17 Van der Waals' b term for carbon tetrachloride vapor is given as 0.138 liter/mol. What molecular diameter is implied by this value? Express the result in centimeters and in angstroms.

18 What are the wavelength and the frequency of radiation that has quanta with energies of 3×10^{-12} erg?

19 What are the frequency and the wavelength of radiations for which a quantum has the same energy as the average translational energy of a gas molecule at 25°C? To what electromagnetic region does this correspond?

20 *Wien's displacement law* says that the wavelength at the maximum of the energy distribution emitted by a hot body is inversely proportional to the absolute temperature. If the wavelength is expressed in angstroms and the temperature is given in kelvins, the corresponding equality is

$$\lambda_{max} = \frac{2.9 \times 10^7}{T}$$

Calculate T for wavelength maxima in the visible and in the infrared regions. Discuss the results.

21 Radar used in communication systems makes use of frequencies in the microwave region, the same region in which molecules absorb radiation energy into their rotational energies. List some atmospheric gases that could not possibly interfere with radar communication. Are there any that could interfere? (In fact, radar makes use of narrow frequency bands that can avoid atmospheric absorptions.)

22 The bond length of the gas-phase NaCl molecule is 2.36 Å.

a What is the moment of inertia of the molecule?

b What are the three smallest rotational energies that the molecule can have?

c What two frequencies and wavelengths of radiation might be absorbed as a result of these energies?

d In what spectral region is such radiation found?

23 Consider an oxygen molecule in a cubic container with 1-mm dimensions. Assume a temperature of 37°C.

a How long, on the average, will it take the molecule to traverse the length, 1 mm, of the container?

b How many times per second will the molecule collide with the container walls?

c If the molecule has a rotational energy of $\frac{1}{2}kT$ and a moment of inertia of 19.3×10^{-40} g cm^2, what are the angular velocity and the frequency of rotation of the molecule? (Ignore quantum restrictions and work from $\frac{1}{2}kT = \frac{1}{2}I\omega^2$.)

d On the average, how many rotations will a molecule complete as it traverses the 1-mm distance between the container walls?

REFERENCES

Gases and the Kinetic-Molecular Theory

MOORE, W. J.: "Physical Chemistry," 4th ed., chaps. 1 and 4, Prentice-Hall, Inc., Englewood Cliffs, N.J., 1972.

DANIELS, F., and R. A. ALBERTY: "Physical Chemistry," 3d ed., chaps. 1 and 9, John Wiley & Sons, Inc., New York, 1966.

CASTELLAN, G. W.: "Physical Chemistry," 2d ed., chaps. 1 to 4, Addison-Wesley Publishing Company, Inc., Reading, Mass., 1971.

BARROW, G. M.: "Physical Chemistry," 3d ed., chaps. 1, 2, and 13. McGraw-Hill Book Company, New York, 1973.

These physical-chemistry texts often provide treatments that adopt alternative approaches or points of view. They are valuable reference works for the topics of this chapter and for most of the material of the remaining chapters. Moore and Barrow use SI, or metric, units, and the other two books use the same units that are used here.

HILDEBRAND, J. H.: "An Introduction to Kinetic Theory," Reinhold Publishing Corporation, New York, 1963. A brief, very readable discussion of the nature of ideal gases, real gases, and the kinetic-molecular theory.

BRUSH, S. G.: "Kinetic Theory," vols. 1 and 2, Pergamon Press, Oxford, 1965 and 1966. Well-chosen selections, with orienting introductory remarks, of the original contributions to the development of kinetic theory. An excellent introduction to the history of the subject.

TABOR, D.: "Gases, Liquids, and Solids," Penguin Books, Inc., Baltimore, 1969. Treatment of gases and the kinetic-molecular theory of ideal and real gases. Included also is material related to the studies of the following two chapters.

Molecular Rotations

BARROW, G. M.: "The Structure of Molecules," W. A. Benjamin, Inc., Menlo Park, 1964.

2 Molecules in Crystals: Molecular Structures

Gases give their molecules the freedom to develop fully their molecular motions. Thus gases are the most suitable state for an introductory study of these motions.

By contrast, the crystalline state is the least permissive of molecular motion. The rigidity and order imposed on the molecules of a crystal make the crystalline state most suitable for studies of molecular structures, and on this basis we now turn to the study of crystals.

Although this state is not that of most concern in studies of living systems, the structural data gathered from crystal studies are basic to our understanding of the molecular details of liquid-state biological processes.

2-1 CLASSIFICATION OF CRYSTALS AND THEIR INTERNAL STRUCTURES

All gases are to some extent the same, in that under some conditions they show *PVT* behavior in accordance with the ideal-gas-law expression $PV = nRT$. No such universal behavior is found for the crystalline forms of solids. Even so, the complex array of individuality that appears when crystals are first investigated gives way on further study to schemes that place all crystals into a handful of classes. Members of each class (which may be as diverse as simple ionic materials, minerals, or proteins) form crystals with some common characteristics.

The most obvious characteristic of the crystalline state is the smooth, flat crystal faces that are often apparent. The growth of each crystal face depends on such accidental features as obstruction by other crystals or container walls, contamination by foreign chemical substances, or depletion of the material being deposited. As a result, a given material may form crystals with various faces more or less developed; however, these crystals are found to have faces that intersect each other at fixed angles. Therefore, the angular relations between crystal faces can be used as a basis for crystal characterization and classification. These faces are shown for idealized crystals in diagrams in Fig. 2-1.

Crystals can be classified according to the symmetry exhibited by an ideal crystal or, what is equivalent, by the axes that are most convenient for describing the crystal-face directions. Although the symmetry features are more basic, the axes that are compatible with the crystal-face arrangements

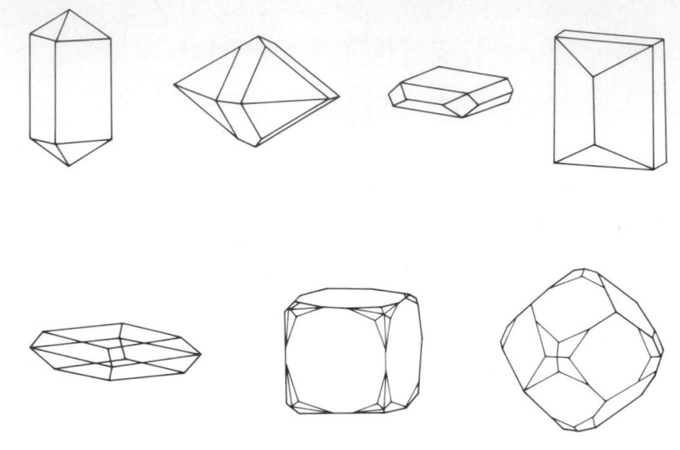

FIGURE 2-1
Idealized crystal shapes.

are adequate for our purpose. In Table 2-1 the six sets of axes used to describe the face arrangements of all the crystals that exist are shown. Any crystal with well-formed faces can be assigned to one of these classes. It is here that one finds such diverse classmates as minerals and proteins, for regardless of the internal makeup of a crystal, it must belong to one of the six crystal classes.

Some crystals are so small or poorly formed that there are not enough crystal faces to permit classification. Then the classification is made on the basis of the x-ray view of the internal structure, which is treated in the following sections.

Long before it was recognized that crystals are formed when molecules or ions pack together, the external form of crystals was interpreted in terms of the ordered arrangement of representative units of the crystal. It is still helpful to postpone consideration of the actual particles that pack together to form crystals and to consider here only the essential features of units that can be repeated in three dimensions to produce crystals with the features classified in Table 2-1.

TABLE 2-1
The Six Crystal Systems

system	axes	angles
Cubic	$a = b = c$	$\alpha = \beta = \gamma = 90°$
Tetragonal	$a = b \neq c$	$\alpha = \beta = \gamma = 90°$
Hexagonal	$a = b \neq c$	$\alpha = \beta = 90°,\ \gamma = 120°$
Orthorhombic	$a \neq b \neq c$	$\alpha = \beta = \gamma = 90°$
Monoclinic	$a \neq b \neq c$	
Triclinic	$a \neq b \neq c$	$\alpha \neq \beta \neq \gamma \neq 90°$

Consider first the essentially different arrays of points that can be laid out to form infinite, ordered two-dimensional arrays. You will find that there are only five possibilities, as shown in Fig. 2-2. Any other regular pattern would be found to be identical to one of these except for relative lengths of the point spacings in Fig. 2-2*b*, the value of the angle in Fig. 2-2*d*, and the relative spacings or angle in Fig. 2-2*e*. A convenient way of viewing the characteristics of each array is to draw representative units, as in Fig. 2-3, and to imply that the entire array follows from a repetition of this unit. Note that the representative unit need not be drawn from point to point. As long as it has the correct size and shape, it can be placed anywhere on the array and still serve as a representative unit.

Carrying this approach over to three dimensions, A. Bravais found that 14 different types of arrays of points, known as *lattices*, are enough to encompass all regular three-dimensional arrays. Representative units of each of these lattices are shown in Fig. 2-4. Although this way of drawing the lattices obscures the fact, all points of a lattice are equivalent; i.e., they have the same number of nearby lattice points and these points are arranged in the same way or with the same symmetry. This helps one understand, for example, that the three cubic lattices are indeed different, giving to each lattice point nearest-neighbor numbers of six, eight, and twelve.

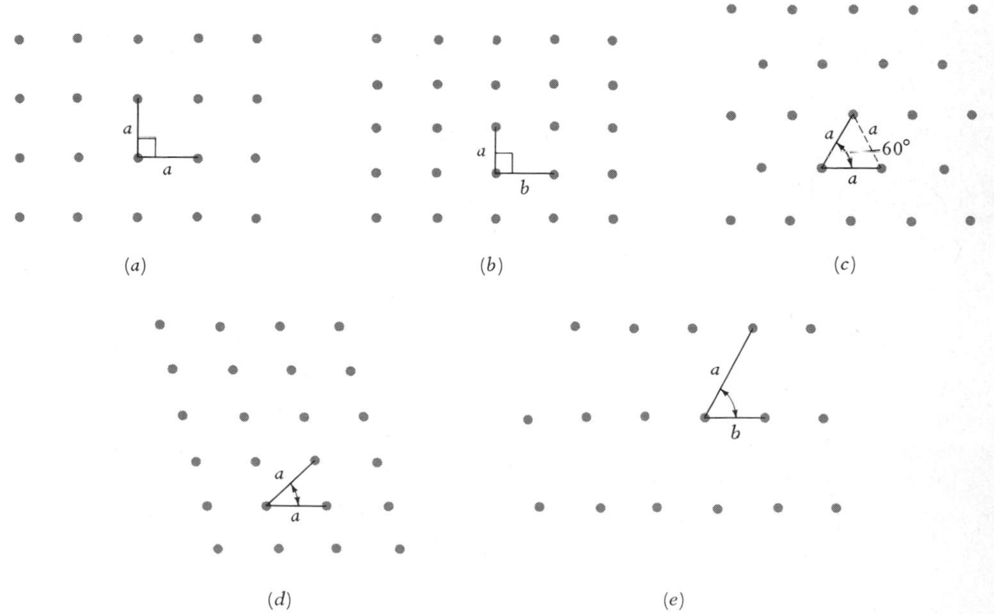

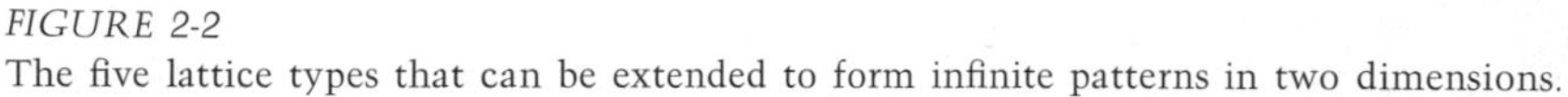
FIGURE 2-2
The five lattice types that can be extended to form infinite patterns in two dimensions.

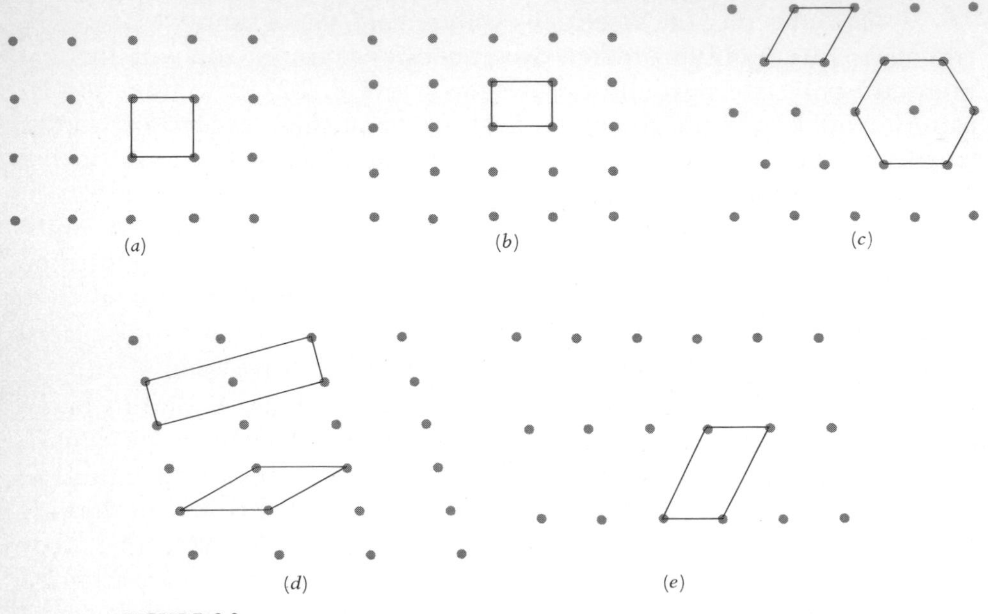

FIGURE 2-3
Representative segments, or *unit cells,* of the lattices of Fig. 2-2. In (*c*) and (*d*) it is sometimes convenient to choose a cell larger than the smallest representative unit in order to emphasize the symmetry of the lattice.

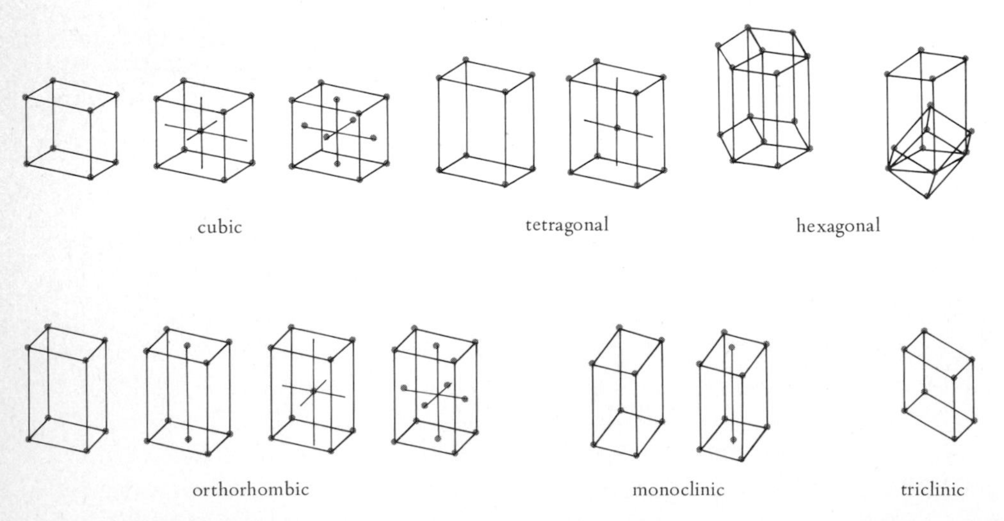

FIGURE 2-4
The 14 three-dimensional lattice types arranged according to the six systems of Table 2-1.

In Fig. 2-4, the Bravais lattices have been grouped according to the six crystal systems. An actual crystal can be assigned to a crystal system. It must, therefore, have an internal structure corresponding to one of the lattices belonging to that class. It is not necessary that molecules or ions be arranged like the points in the lattices, but they must be arranged so that equivalent points in the crystal are arranged like the points in the lattice.

The smallest representative unit of a real crystal, which corresponds to the lattice units of Fig. 2-4, is known as the *unit cell*. Its dimensions are labeled a, b, and c, and its angles are indicated by α, β, and γ. For orthogonal axes the volume of the unit cell is simply $a \times b \times c$. For non-orthogonal axes, the unit cell volume can be calculated from the formula

$$V = abc(1 - \cos^2 \alpha - \cos^2 \beta - \cos^2 \gamma + 2 \cos \alpha \cos \beta \cos \gamma)^{1/2} \qquad \mathbf{1}$$

The nature of the internal structure of a crystal and the dimensions of the unit cell of the crystal can be deduced by using x-ray-diffraction methods. Such information provides the basis for further detailed studies showing the arrangements of the molecules in the unit cell and the detailed structure of the molecules.

2-2 THE BASIS OF X-RAY DIFFRACTION AND THE DETERMINATION OF FIBER STRUCTURES

The general principle of diffraction methods depends on the phenomenon of interference, which occurs when any wave is scattered from a number of centers. This phenomenon, for example, is exhibited by visible radiation when a beam of light passes through a series of closely spaced slits, as illustrated in Fig. 2-5. If the light is monochromatic, i.e., consists of radiation of only a single wavelength, the wave motions of the light emerging from the slits will add together only in certain directions. In these directions *constructive interference* is said to occur, and in these directions a beam of diffracted light will appear. In other directions the diffracted waves will be out of phase to various extents, *destructive interference* will occur, and less light will be seen.

Studies of the internal structure of crystals depend on interference and require penetrating radiation that can enter the crystal and reveal its internal structure. Furthermore the wavelength of the radiation must be comparable with the spacing of crystal planes so that the scattering will display interference effects. X-rays meet these two requirements.

The study of structures by x-ray diffraction can be introduced by considering a set of molecular chains packed together, as in Fig. 2-6. The

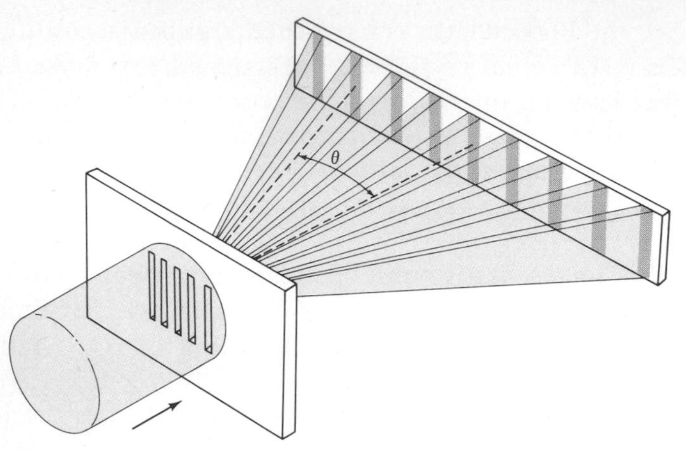

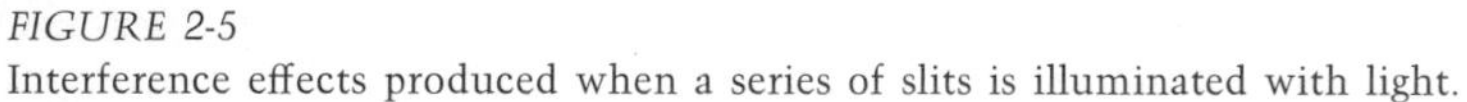

FIGURE 2-5
Interference effects produced when a series of slits is illuminated with light.

only structural characteristic is the repeat distance along each of the molecular chains of the fiber. (An actual fiber, e.g., a bundle of DNA molecules, also generally has an ordered arrangement of the molecular chains, and each chain may be characterized by more than one repeat distance. These elaborations can be ignored in an analysis of the principal features of fiber diffractions.)

Suppose that a fiber is mounted in a beam of homogeneous x-rays, i.e., x-rays of a single wavelength, and surrounded by a film, as in Fig. 2-7. Since x-rays are quite penetrating, most of the beam will pass through the fiber without change, but some of the beam will be scattered by interaction with the electrons concentrated at each atom in the molecular chains. If the atoms were not arranged periodically along the chains, the scattered portion of the x-ray beam would emerge at all angles and produce a general darkening of the photographic plate of Fig. 2-7. But, in fact, only certain angles of scattering are observed. This results (as we now show) from interference effects produced by the regular spacing along the molecular chains.

The incoming beam of x-rays can be represented as in Fig. 2-8 with all the waves in phase. The x-rays scattered from the atoms of any one chain will constructively interfere (as Fig. 2-8 illustrates) only if the waves from successive atoms of the chain are displaced by an integral number of waves. For directions in which the scattered waves are not shifted by exactly 0, 1, 2, . . . wavelengths, destructive interference will occur, and no scattering will be observed. Thus, only when the relation

$$n\lambda = d \sin \gamma \qquad \text{where } n = 0, 1, 2, \ldots \tag{2}$$

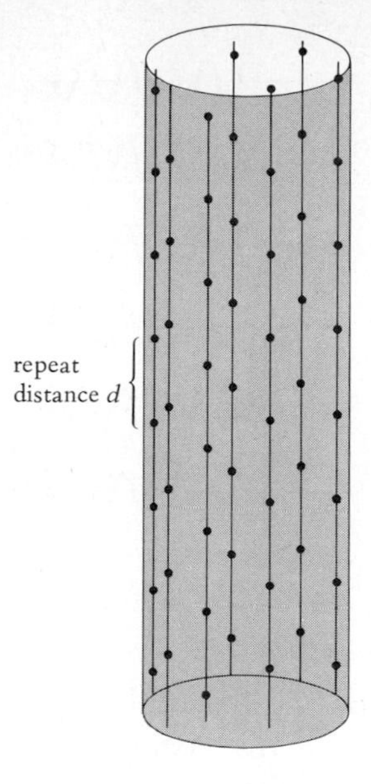

FIGURE 2-6
A hypothetical fiber whose structure is characterized only by a repeat spacing along each molecular chain.

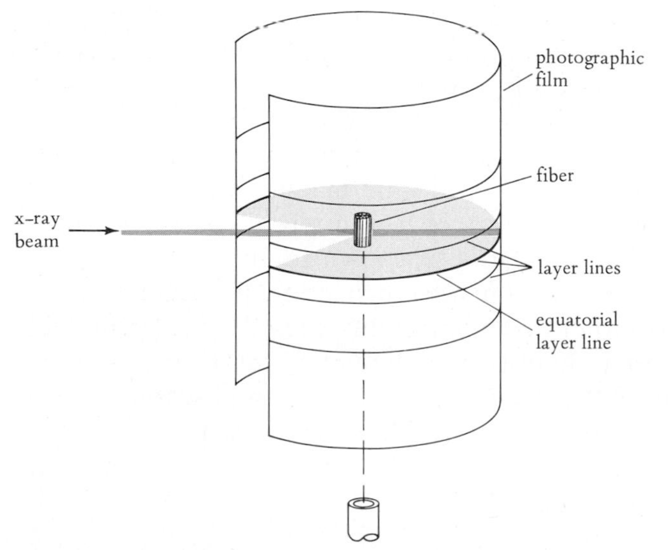

FIGURE 2-7
The arrangement for an x-ray-diffraction photograph of a fiber.

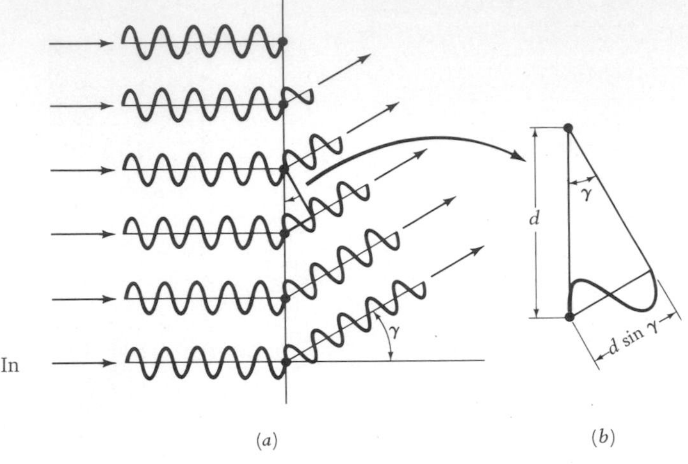

FIGURE 2-8
(a) Example of a scattering direction in which constructive interference occurs. (b) For the constructive interference shown in (a), $\lambda = d \sin \gamma$. In general for constructive interference, $n\lambda = d \sin \gamma$, where $n = 0, 1, 2, \ldots$.

is satisfied, will scattered, or diffracted, x-rays occur. These scattering angles, moreover, depend only on the spacing along each chain and not on the relative placement of the chains. Thus scattering occurs only at certain angles above and below the plane that contains the x-ray beam and is perpendicular to the fiber axis. The resulting darkenings of the photographic plate are known as *layer lines*. The middle line is the *equatorial* layer line.

The spacing d between atoms or groups of atoms along a molecular chain which x-rays show to have high scattering powers is of the order of angstroms. Thus, if we want to observe the diffracted beams for several values of n from such repeat units, we must use x-rays with a wavelength that is appropriate in view of $n\lambda = d \sin \gamma$. Since $\sin \gamma$ can at most equal 1, only values of $n\lambda/d$ up to 1 can satisfy this relation. The x-rays of wavelength 1.542 Å, known as Cu K_α x-rays and emitted when a beam of high-energy electrons strikes a copper target, are often used.

The relation $n\lambda = d \sin \gamma$ shows that if the wavelength λ of the x-rays is known, and if values of the angle γ are measured for one or more layer lines, i.e., for one or more values of n, the repeat spacing d can be deduced. An example of the calculation is shown in Fig. 2-9.

In practice the layer lines on fiber-diffraction photographs often consist of a series of spots rather than continuous lines. This is attributable to ordering of the molecular chains perpendicular to the fiber axis. Information on the structure of the fiber along the fiber cross section can be obtained from the details within each layer line, but there are more generally applicable ways of deducing three-dimensional structures. These are dealt with in subsequent sections.

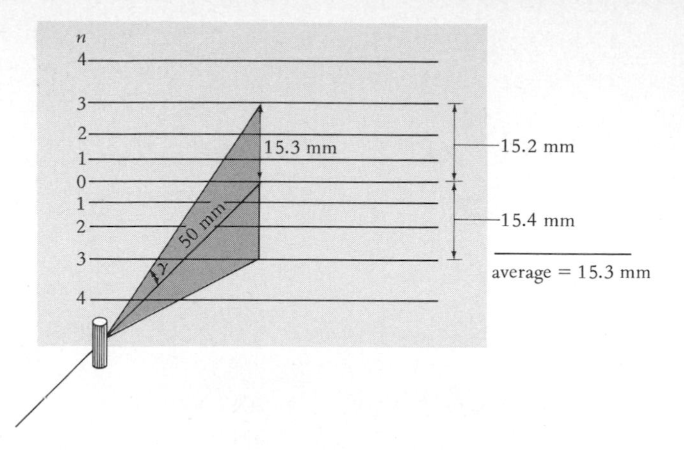

FIGURE 2-9
Sample data for determining the repeat spacing of a fiber. The calculation would be

$$\tan \gamma = \frac{15.3}{50} = 0.306$$

$$\gamma = 17°0'$$

$$\sin \gamma = 0.292$$

Given that Cu K_α x-rays are used with $\lambda = 1.542$ Å, then from $n\lambda = d \sin \gamma$

$$d = \frac{3(1.542)}{0.292} = 15.8 \text{ Å}$$

The fiber-diffraction technique is particularly important in biochemical studies because many important macromolecules are linear. The distances at which features of the molecule are repeated often provide valuable clues to the structure of the molecular chains. An example is shown in Fig. 2-10.

2-3 THE BASIS FOR THE X-RAY-DIFFRACTION DETERMINATION OF CRYSTAL STRUCTURE

When a monochromatic beam of x-rays is passed into a crystal, as in Fig. 2-11*a*, some scattering of the beam results from its interaction with the electrons of the atoms or ions of the crystal. This scattering can be analyzed by recognizing the various planes of atoms or ions that occur as a result of the repetitive character of the crystal structure. Each layer of atoms will scatter only a small part of the x-ray beam. If the atomic layers in the crystal did not have a spacing of the same order of magnitude as the wavelength of the x-rays, simple reflections and scattering of the x-rays would occur, but in fact the reflection is not simple and is greatly disturbed by the interference among the x-rays scattered from different layers.

The incoming beam of x-rays can again be represented as in Fig. 2-11*b* with all the waves in phase. The nature of the outgoing beams must be investigated. As Fig. 2-11*b* shows, the beams scattered from successive layers of crystal particles may show waves that are out of phase to some extent with the scattered waves from the other layers, in a particular direction. When this happens and the addition of the waves tends to cancel each other, destructive interference is said to occur. Only if all these

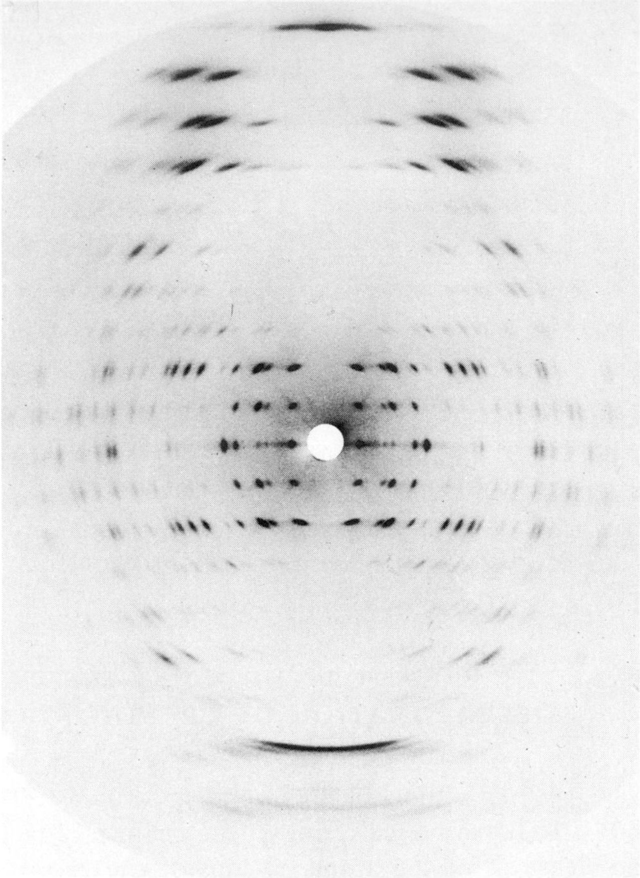

FIGURE 2-10
The fiber-diffraction photograph of a synthetic double-stranded RNA. The structure of the fiber is similar to that of the double-stranded regions of transfer RNA. The layer lines indicate a helix pitch length of 30.9 Å. [*Courtesy of Professor Struther Arnott, Purdue University.*]

scattered beams come out in phase will they add up and contribute to a net scattered beam. This constructive interference occurs whenever the phase of beams scattered from successive layers is shifted by an integral multiple of wavelengths. Reference to Fig. 2-11*c* shows that this happens when the relation

$$n\lambda = 2d \sin \theta \qquad n = 1, 2, 3, \ldots \tag{3}$$

holds. This important equation is known as *Bragg's diffraction law.*

This basic equation shows that for a given value of the x-ray wavelength λ, measurement of the angle θ or of the term $\sin \theta$ gives the information on the spacing between planes through the scattering centers that make up the crystal.

Since the relative orientation of sets of crystal or lattice planes whose spacing is revealed by the diffraction of x-rays is of great importance in

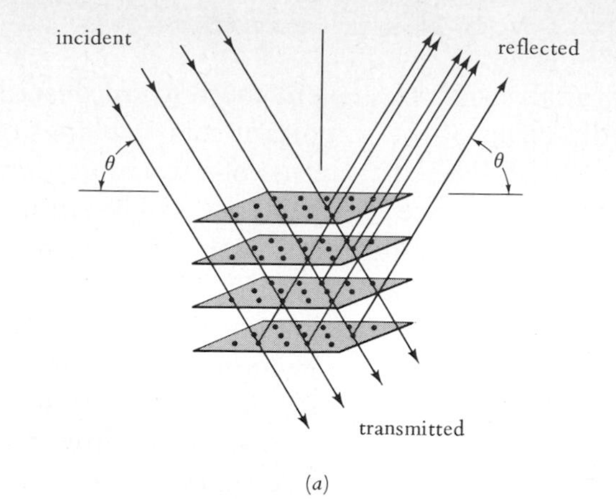

(a)

in phase
net reflection
constructive interference
out of phase
no net scattering
destructive interference

(b)

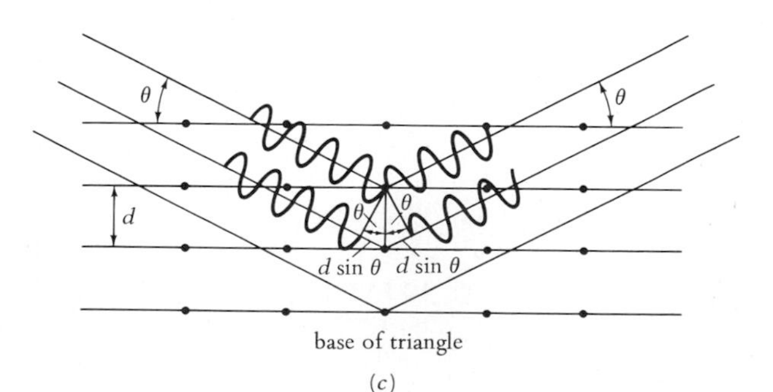

(c)

FIGURE 2-11

Reflection of x-rays from the reflecting layers of a crystal. (*a*) The mirrorlike reflection expected for all angles of incidence if interference effects are not considered. (*b*) With interference effects only certain angles of incidence will lead to reflection. (*c*) The basis for the Bragg scattering condition $n\lambda = 2d \sin \theta$.

crystal-structure analysis, a convenient method for describing these planes is needed. The direction of the important sets of planes of a lattice, as Fig. 2-12 suggests, can be described in terms of intercepts that are multiples of the unit-cell dimensions. Since crystal planes will be similarly oriented, they can be similarly described. Thus planes can be described by their relative intercepts on the x, y, and z axes as $a:b:c$, or $a:2b:\infty c$, and so forth.

Much more convenient, particularly in the analysis of diffraction data, are sets of numbers called *Miller indices*. They are obtained by taking the reciprocals of the coefficients of a, b, and c that specify, by means of intercepts, the direction of a set of planes. These three reciprocals are then cleared of fractions and reduced to the smallest set of integers. The $a:b:c$ plane becomes a 111 plane; the $a:2b:c$ plane becomes a 212 plane; the $a:b:\infty c$ plane becomes a 110 plane. The Miller indices, referred to in general by hkl, describe the relative directions of the crystal planes. Information on the coordinate system and the values of the three unit distances is given separately.

A feeling for these indices can soon be acquired if one remembers their reciprocal nature. A low first number means an intercept at a large distance on the x axis; a low second number means a large intercept on the y axis; and so forth. A plane parallel to an axis has a zero term, corresponding to the reciprocal of its intercept on that axis.

Returning to Bragg's law, we see that any set of planes described by the assignment of integers to h, k, and l in the general Miller index description hkl will have an associated interplanar spacing d and can lead

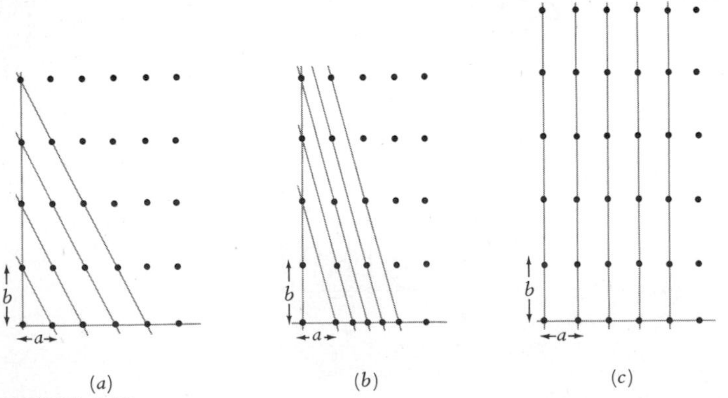

FIGURE 2-12
End view of a lattice showing how planes that pass through relatively large numbers of lattice points can be described in terms of intercepts measured in units of a and b.
(a) Planes with relative intercepts a, b, ∞; Miller indices 110.
(b) Planes with relative intercepts a, $2b$, ∞; Miller indices 210.
(c) Planes with relative intercepts a, ∞, ∞; Miller indices 100.

to diffraction according to $n\lambda = 2d \sin \theta$. As a result, if we could somehow make the identification, we could label the diffracted x-ray beam, or spot on the photographic plate, with the appropriate *hkl* to indicate its association with the set of crystal planes. This indexing procedure is, in fact, a first step in analyzing a diffraction pattern; how it is done will be shown in the following section.

First, a convenient treatment of the different orders, corresponding to different values of n, of diffraction from a given set of planes must be introduced. Writing the Bragg relation not as $n\lambda = 2d \sin \theta$ but as

$$1\lambda = 2\frac{d}{n} \sin \theta \qquad \mathbf{4}$$

shows that higher orders are diffracted as if they were first-order diffractions from planes that have interplanar spacings less than the actual spacings by a factor equal to the order of the diffraction. Thus, second-order diffraction from, say, a 110 plane can be looked on as if it were first-order diffraction from a set of planes with half this spacing. In view of the reciprocal nature of the Miller indices, we could describe this diffraction by the indices 220. This procedure is used, and instead of dealing with higher diffraction orders we deal only with the first order and think of them as arising from planes with spacing some fraction of the actual spacing. Thus the notation d_{110}, for example, implies the actual spacing between 110 planes; d_{220} will be half this spacing.

Now let us see how the diffraction of x-rays, as described by Bragg's law, Eqs. **3** and **4,** and the description of crystal planes by Miller indices are brought together to yield information about the shape and dimensions of the unit cells of actual crystals.

2-4 DIMENSIONS OF THE UNIT CELL: THE PRECESSION TECHNIQUE

One method, now widely used in crystal-structure studies, is the *precession technique.* The crystal is moved in a precession motion in the x-ray beam. The precession method spreads out all the diffraction spots, having a selected value of one of the indices h, k, or l over the film. The values of the remaining two indices can be determined. Thus the diffraction spots that result can be assigned all three indices h, k, and l. Then the shape and complete dimensions of the unit cell can be deduced.

Consider a crystal mounted in an x-ray beam as in Fig. 2-13*a* and then tilted relative to the beam so that the $l = 0$ diffractions, i.e., all those that would be indexed as *hk*0, can be produced. If the crystal

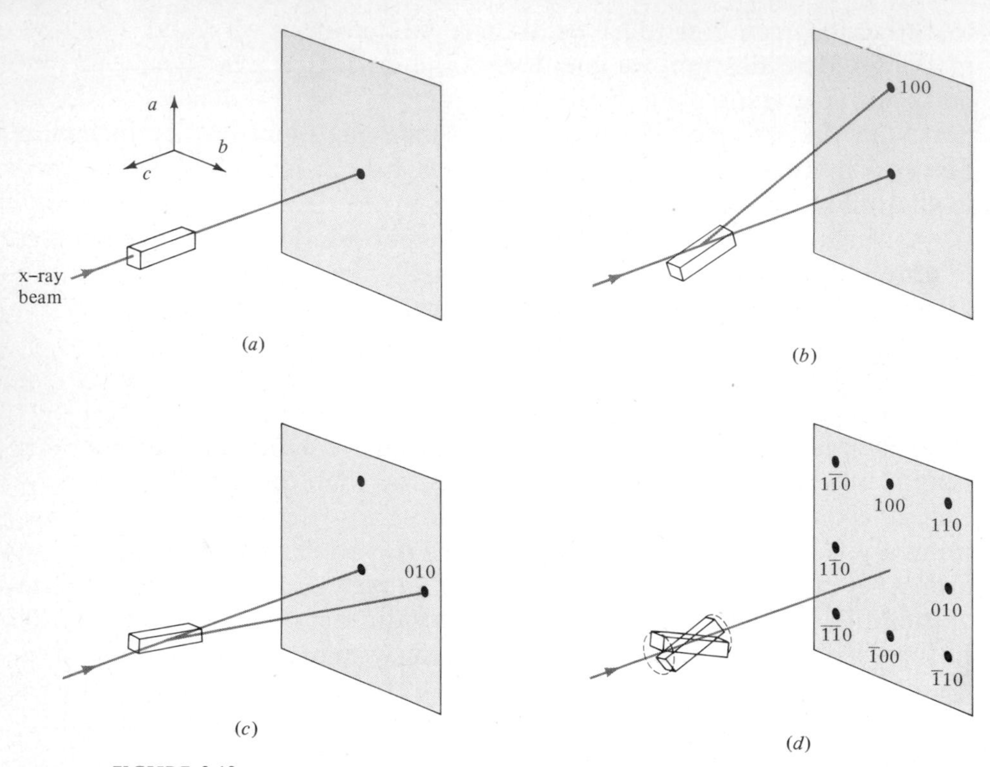

FIGURE 2-13
The relation of crystal orientation to diffraction pattern in the precession method.

has $a = b$, that is, is cubic or tetragonal, and it is positioned as in Fig. 2-13*b*, diffraction from the 100 could occur and would form the indicated spot.

Now, suppose the crystal is precessed so that the x-ray beam impinges on the crystal with the same angle as before but now strikes the side of the crystal as in Fig. 2-13*c*. A 010 diffraction will be recorded, as shown.

In a similar way, if the precession is continued, $\bar{1}00$ and $0\bar{1}0$ diffractions, the bar implying minus values, will form spots at the bottom and at the side to form the four-spot pattern produced by this set of planes. Many more diffractions would be observed with the complete range of precession implied by Fig. 2-13*d*. The additional spots can also be related to crystal planes. The 110 planes, for example, lead to spots along the diagonals of the film. Most important is the recognition that *any diffraction spot will appear on the film in a direction from the film center that is perpendicular to the set of planes that produces the diffraction.*

When rows of spots can be recognized in a precession photograph, they can be associated with sets of planes that have a common degree of tilt relative to a given direction. Thus, the $0k0$ diffractions are found to be in an equatorial layer-line array of spots. The planes that produce these spots have no tilt relative to the a axis. Other lines of spots due to planes with various amounts of tilt relative to this axis are $2k0$, $3k0$, and so forth. Similar indexing of spots due to amounts of tilt relative to the b axis give the arrays $h10$, $h20$, . . . Complete indices of the spots on the interlocking layer lines is then achieved. Thus the diffraction spots on a precession photograph like Fig. 2-14 can easily be indexed for cubic and tetragonal crystals.

The procedure is a little less obvious when, as in other crystal types, the a and b axes are not orthogonal. But the diffraction spot from a set of planes still occurs in a direction perpendicular to the plane directions. An illustration of the nonorthogonal features of the diffraction pattern that must then be dealt with is given in Fig. 2-15.

So far we have not used the information provided by the distance of the diffraction spots from the center of the film. The precession method again introduces a mechanically complicated procedure that leads to an easy interpretation of these distances. The construction of Fig. 2-16 shows that if a tilt is given the film or other detector and this tilt is synchronized with the crystal precession, the distance from the film center to an hkl diffraction spot is inversely proportional to d_{hkl}. Thus the film obtained in the precession method is a display of the orientation of the crystal planes and their spacing. The orientation depends on the fact that the *direction* of the diffraction spot from the center of the film is the direction perpendicular to the plane. The spacing results because the *distance* of the spot from the film center is proportional to the reciprocal of the plane spacing. The precession-method film is said to exhibit the *reciprocal lattice* of the crystal.

Determining the unit-cell shape and dimensions from a precession-method photograph is illustrated in Fig. 2-17. Precession is about the c axis of the crystal, and the diffractions can be attributed to $hk0$ planes. The equivalence of the photograph in the horizontal and vertical directions shows that the a and b unit-cell dimensions are identical. Further, if we take the a axis to be vertical and the b axis to be horizontal, the layer-line labels can be attached to the noticeable rows of diffraction spots as in Fig. 2-14a. The unit-cell dimensions a and b are obtained by measuring the separation between rows of spots and applying the $d_{hkl} = \lambda F/y$ relation. The unit-cell dimensions, which correspond to the d_{100} and d_{010} spacings, can best be determined by averaging the layer-line separations measured for many layer lines.

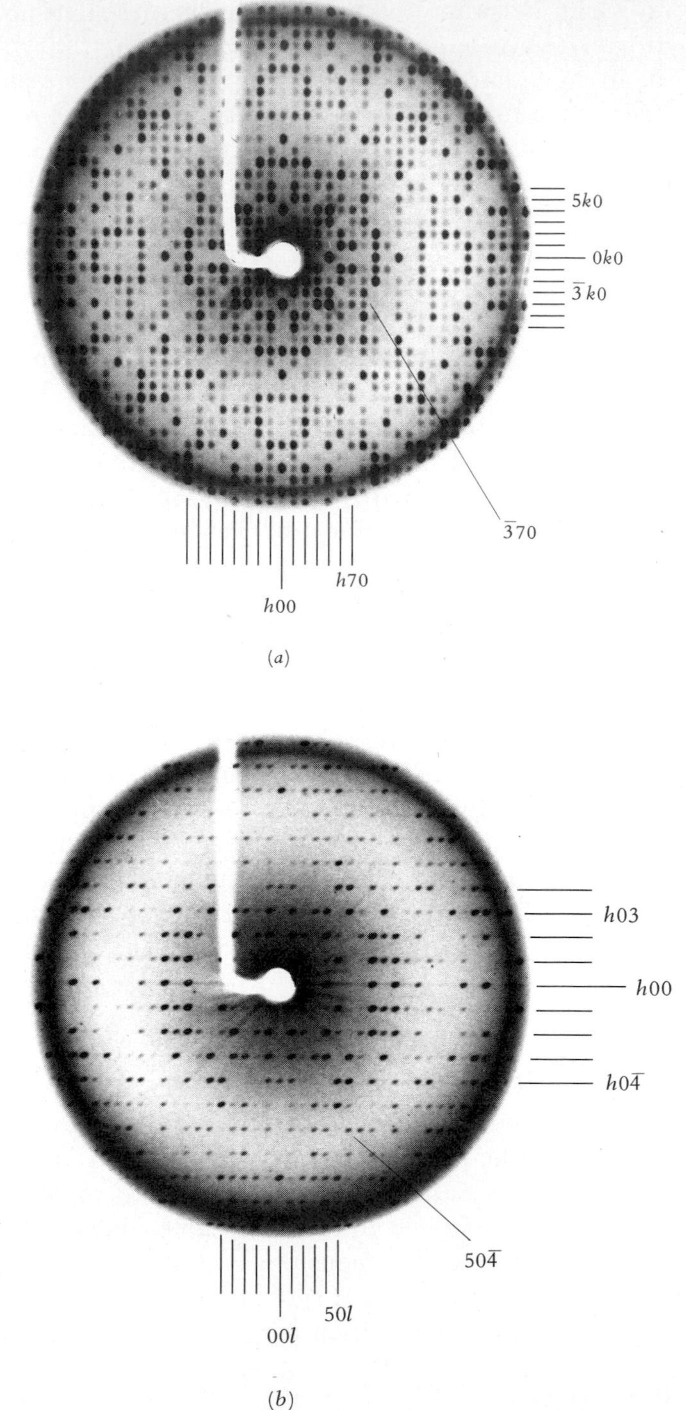

FIGURE 2-14
The precession diffraction patterns formed by the lysozyme chloride crystal and a Cu K_α x-ray beam. In (a) all spots are $hk0$; in (b) the crystal is mounted so that the spots are $h0l$. Some of the indexing is indicated. [*J. R. Knox, J. Chem. Educ.,* **49:**476 (1972), *courtesy of Professor Knox.*]

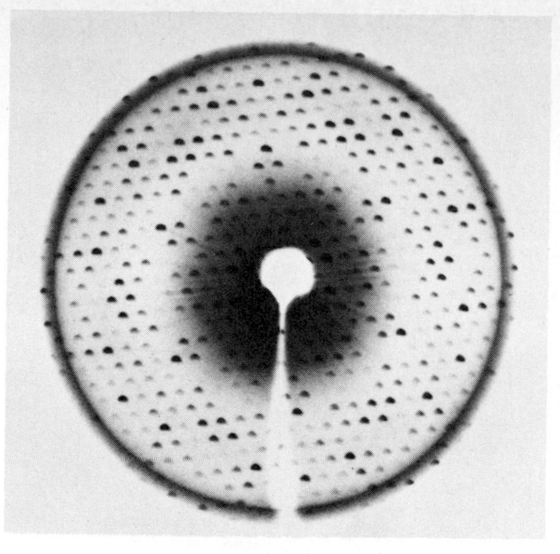

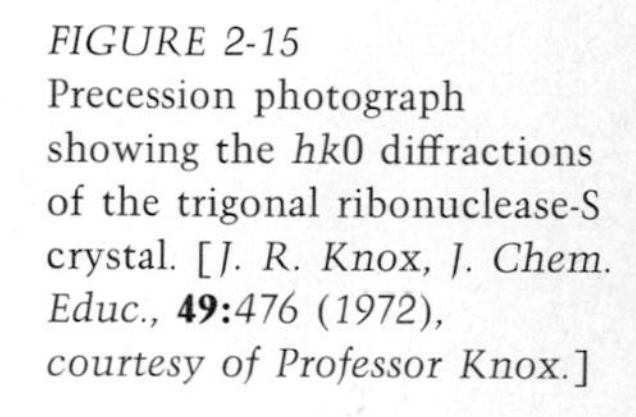

FIGURE 2-15
Precession photograph showing the *hk*0 diffractions of the trigonal ribonuclease-S crystal. [*J. R. Knox, J. Chem. Educ.*, **49:**476 (1972), *courtesy of Professor Knox.*]

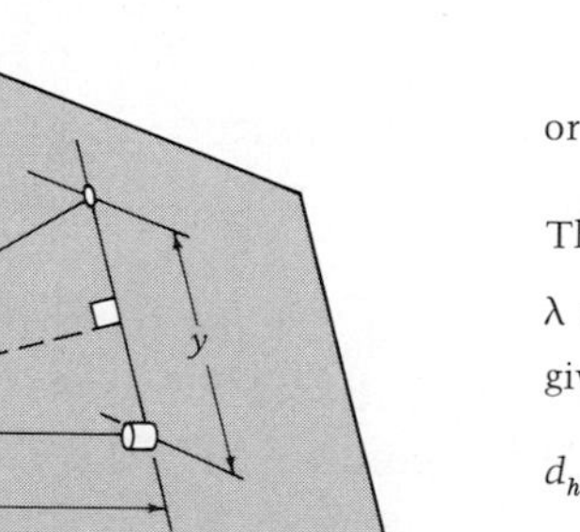

FIGURE 2-16
The relation between the position of a diffraction spot on a precession photograph and the spacing of the planes to which it corresponds. The movcmcnt of the photographic plate is synchronized with that of the crystal so that the plate remains perpendicular to the crystal axis.

$$\sin \theta = \frac{y/2}{F}$$

or $$2 \sin \theta = \frac{y}{F}$$

Then

$$\lambda = 2d_{hkl} \sin \theta$$

gives

$$d_{hkl} = \frac{\lambda}{2 \sin \theta} = \frac{\lambda F}{y}$$

Thus, given λ, the wavelength of the x-rays, and *F*, the distance from crystal to photographic plate, measurement of *y* for an *hkl* spot leads to a value of d_{hkl}. Note also that $y \propto 1/d_{hkl}$.

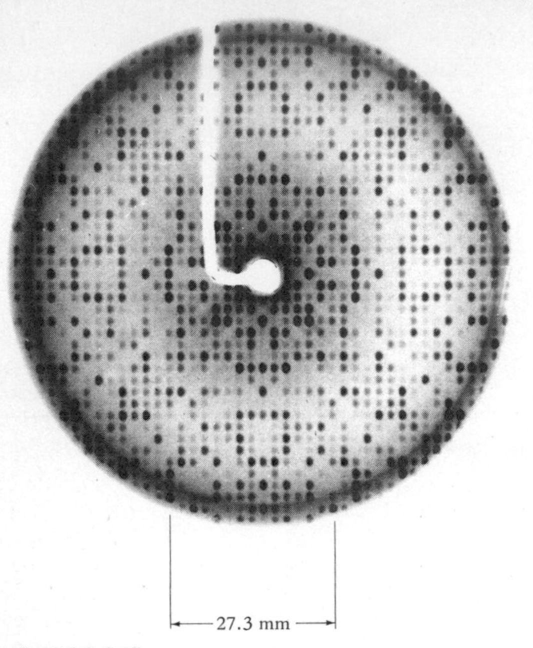

FIGURE 2-17
Calculation of the d_{100} and d_{010} of the lysozyme chloride crystal from the data of Fig. 2-14 and camera length F of 100 mm:

$$2y_7 = 27.3 \text{ mm}$$

$$y_7 = 13.65$$

$$d_{700} = d_{070} = \frac{(1.542 \text{ Å})(100 \text{ mm})}{13.65 \text{ mm}}$$

$$= 11.29 \text{ Å}$$

$$d_{100} = d_{010} = 11.29 \times 7 = 79.1 \text{ Å}$$

Figure 2-14*b* shows the pattern for a lysozyme crystal mounted so that diffractions from planes parallel to the *b* axis are detected, i.e., *h*0*l* reflections are shown. The *c* axis is vertical, and again the rows of spots can be indexed. The value of the unit-cell parameter *c* is obtained as 37.9 Å. We thus have the unit-cell dimensions 79.1, 79.1, and 37.9 Å and can calculate the unit-cell volume as 237,100 $Å^3$.

2-5 THE MOLECULAR MAKEUP OF THE UNIT CELL

Once the unit-cell dimensions of a crystal are determined, we proceed toward the goal of a detailed structural map of the contents of this representative crystal unit by determining the number of molecules that are accommodated by the unit cell. Consider specifically a protein crystal.

We introduce the following symbols:

V = volume of unit cell
ρ = density of crystal
x_p = fraction of crystal that is protein†
z = number of protein molecules in unit cell
M = gram-molecular mass of protein
$\mathfrak{N}$ = Avogadro's number

The mass of an individual protein molecule can be expressed as $M/\mathfrak{N}$ and the protein mass of the unit cell as $zM/\mathfrak{N}$. This mass is also given by the product of the protein fraction, the density of the crystal, and the volume of the unit cell, i.e., by $x_p \rho V$. Thus we can form the equation

$$\frac{zM}{\mathfrak{N}} = x_p \rho V$$

or

$$z = \frac{x_p \rho V \mathfrak{N}}{M} \qquad \textbf{5}$$

Generally an approximate value of M is available from studies of solutions of proteins. A value of x_p can be estimated by weighing a crystal before and after drying. Additional allowance for retained salt must sometimes be made. The crystal density is usually determined by observing the position adopted by a crystal in a density-gradient tube. Then if a value of V, the unit-cell volume, has been obtained from x-ray studies, the number of molecules per unit cell can be calculated.

Example 2-1

For the lysozyme example of the preceding section, a crystal density of 1.24 g/cm^3 and a protein fraction of 0.67 were obtained. The mass of 1 mol of lysozyme is estimated from osmotic-pressure measurements and from sedimentation and diffusion studies to be about 15,000 g. The mass of a single molecule is thus 15,000 atomic mass units. (In biochemical literature, atomic mass units are frequently referred to as *daltons.*) How many molecules are there per unit cell?

Solution

From the estimated molar mass of 15,000 g, the unit-cell volume of 237,100 Å^3 from the preceding section, and the relation of Eq. **5,** we obtain

† The remainder, which may constitute as much as one-third of the total, is made up of water and salt retained in the crystal.

$$z = \frac{x_p \rho V}{M}$$

$$= \frac{0.67(1.24)(237{,}100 \times 10^{-24})(6.02 \times 10^{23})}{15{,}000}$$

$$= 7.9$$

Thus, we deduce that z, the number of molecules per unit cell, has the value 8.

In fact, lysozyme from egg white has 129 amino acid residues and a molecular mass of 13,930 rather than the estimated value of 15,000.

2-6 INTENSITIES OF DIFFRACTIONS AND THE STRUCTURE WITHIN THE UNIT CELL

The crystal symmetry and the unit-cell dimensions sometimes give enough information to determine the exact structure of a simple crystal. For most crystals, however, the problem of determining the arrangement of the atoms or ions within each unit cell remains. The information necessary for deducing these atomic positions is contained in the *intensities* of the diffraction spots.

Let us now consider a crystal that could belong to any of the classes with orthogonal axes and let us indicate the unit-cell dimensions along the x, y, and z axes as a, b, and c.

To simplify, let us consider a two-dimensional view in which planes are considered to be parallel to the z axis, as in Fig. 2-18, the z axis being taken perpendicular to the plane of the paper. Such planes have indices $hk0$, and the edges of the two planes shown are for the 210 and 310 planes. Now let us find the effect of the B atoms which are displaced from the A atoms by distances x and y, measured in units of the unit-cell dimensions a and b, on the intensities of the diffraction lines.

Waves scattered by the planes of B atoms will interfere with any waves that are scattered by the planes through the A atoms. The scattering from the successive A planes in Fig. 2-18 is displaced by exactly 2π and leads to constructive interference. Now, however, the planes of B atoms interfere with the scattering from the planes of A atoms by providing scattered beams that are out of phase by an amount that depends on the positions of the B atoms. For the 210 reflections the B atoms provide waves that are out of phase by $(x/\frac{1}{2})2\pi$ along the x axis and $(y/\frac{1}{1})2\pi$ along the y axis. Consideration of the 310 reflections shows, as Fig. 2-18 suggests, that for these reflections the B atom contributions would be out of phase by $(x/\frac{1}{3})2\pi$ and $(y/\frac{1}{1})2\pi$ along the x and y axes. In general, the phase difference

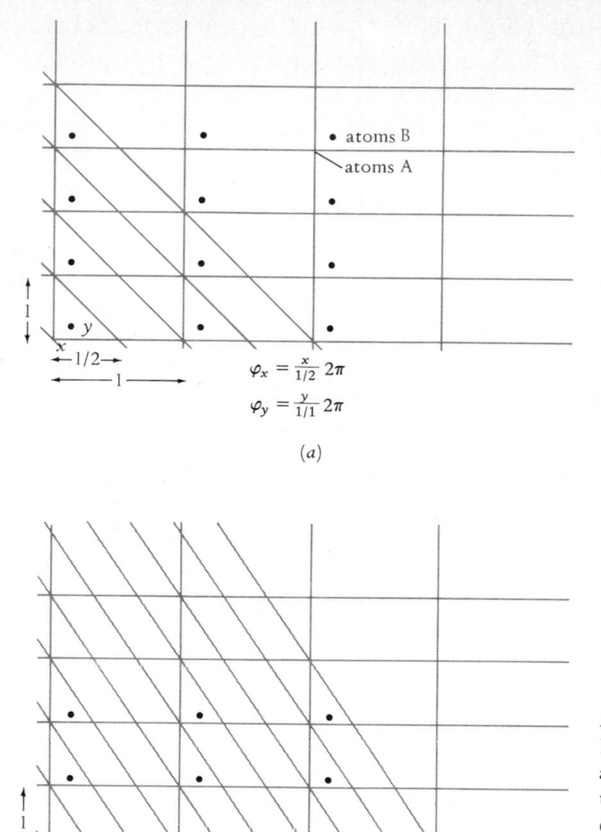

FIGURE 2-18
Phase shifts φ_x and φ_y for B atoms when the origins of the unit cells are oriented for constructive interference from (*a*) 210 and (*b*) 310 planes. Distances are reported in terms of *a* along the horizontal, x axis, and in terms of *b* along the vertical, *y* axis.

of the beam resulting from the displacement of the B atoms from the origin is

$$2\pi\left(\frac{x}{1/h} + \frac{y}{1/k}\right) = 2\pi(hx + ky)$$

For a general plane in three dimensions, this phase difference would be

$$2\pi(hx + ky + lz)$$

It should be remembered that *h*, *k*, and *l* are the Miller indices of the reflection planes and of the diffraction spot being considered and that x, *y*, and *z* are the coordinates of the atom B expressed as fractions of the unit cell of dimensions *a*, *b*, and *c*.

The net scattering from an hkl plane requires the summing up, allowing for the phase differences, of the scattering amplitudes from all atoms in the unit cell. The summation of waves, all with the same frequency but with different amplitudes and phases, can be accomplished by representing the amplitude and phase of each wave by the diagrams of Fig. 2-19. Then if two scattered waves are considered, as in Fig. 2-20, the square of the amplitude of the resultant wave is obtained as

$$(f_1 \cos \varphi_1 + f_2 \cos \varphi_2)^2 + (f_1 \sin \varphi_1 + f_2 \sin \varphi_2)^2 \qquad 6$$

A similar expression can be written when the effect of more than two atoms must be shown. Then we let j be an index for the atoms of the unit cell. According to the phase-difference expression $2\pi(hx + ky + lz)$, the jth atom contributes to the hkl diffraction a scattered wave with phase $\varphi_j = 2\pi(hx_j + ky_j + lz_j)$. The amplitude factor f_j can be taken as being approximately proportional to the number of electrons of the atom, i.e., to the atomic number. (In fact, the scattering power of an atom varies, but in a known way, with the angle of scattering.)

The intensity of an x-ray beam is proportional to the square of the

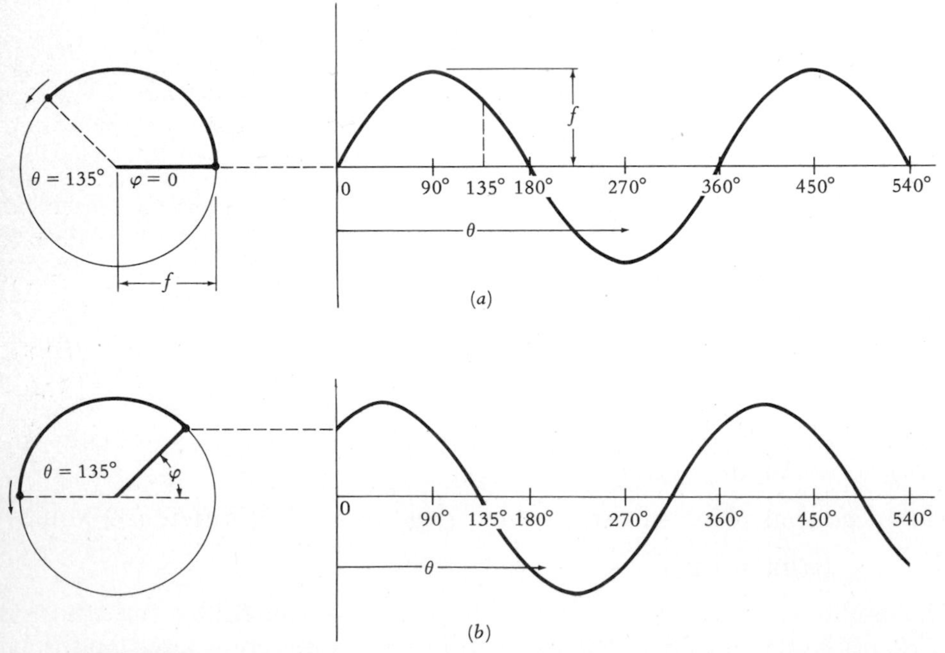

FIGURE 2-19
The generation of a wave by the rotation of a point at a radius equal to the amplitude of the wave for (a) phase angle zero and (b) phase angle φ.

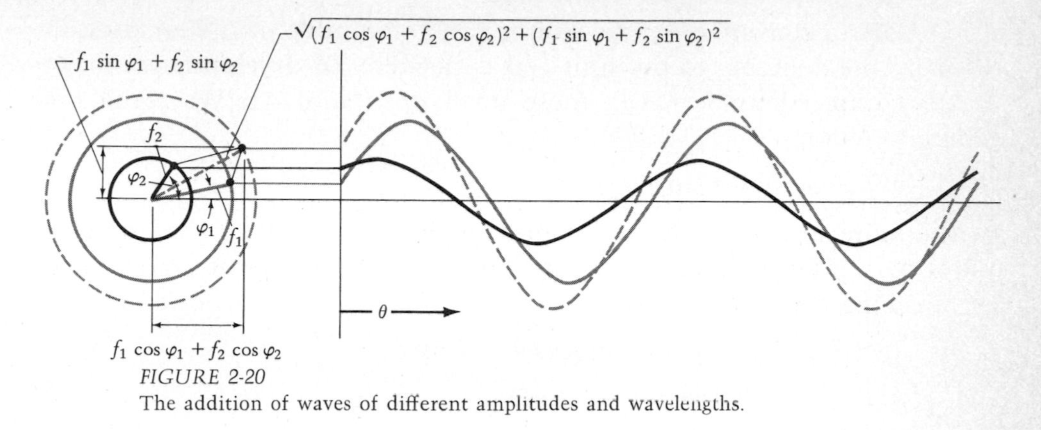

FIGURE 2-20
The addition of waves of different amplitudes and wavelengths.

amplitude of the beam and thus, by analogy with Eq. **6,** we can write

$$I(hkl) \propto \left[\sum_j f_j \cos 2\pi(hx_j + ky_j + lz_j)\right]^2 + \left[\sum_j f_j \sin 2\pi(hx_j + ky_j + lz_j)\right]^2 \qquad 7$$

This result provides a route from a detailed crystal structure, as given by the x, y, and z coordinates of each of the atoms of the unit cell, to the relative intensities of each of the diffraction spots, as specified by values of h, k, and l. One can be content with this result, which shows that with sufficient trial-and-error steps a structure for a molecule in a crystal can be obtained. There are, furthermore, many diffraction spots that can be measured, and a structure which leads to a correct intensity calculation for all the observed spots is almost unquestionably the correct structure.

Preferable (and for large molecules almost essential) is a method for deducing a molecular structure more directly from the observed diffraction pattern without having to resort to this trial-and-error procedure. An indication of the problems that arise when this is attempted and some of the methods that are adopted to overcome these difficulties can now be given.

2-7 STRUCTURES FROM FOURIER SYNTHESIS

One of the most important developments in x-ray-diffraction techniques was Sir William Bragg's realization that the approach of the preceding section could to some extent be turned around.

First a convenient expression for the amplitude of the x-rays scattered by the contents of the unit cell is needed. To develop it we make use of complex numbers, i.e., those involving $i = \sqrt{-1}$. We recall the general relation

$$e^{i\theta} = \cos\theta + i\sin\theta \tag{8}$$

Furthermore, the square of a complex quantity is equal to the complex quantity times its complex conjugate. Since this latter quantity is obtained by replacing i by $-i$, we have

$$\begin{aligned} |e^{i\theta}|^2 &= (\cos\theta + i\sin\theta)(\cos\theta - i\sin\theta) \\ &= \cos^2\theta + \sin^2\theta \end{aligned} \tag{9}$$

Comparison of the form of this equation with the squared amplitude and intensity expressions given in Eqs. **6** and **7** suggests that the desired complex-number amplitude expression, known as the *structure factor* $F(hkl)$ should be written as

$$F(hkl) = \sum_j f_j e^{2\pi i(hx_j+ky_j+lz_j)} \tag{10}$$

In view of Eq. **9** this yields

$$|F(hkl)|^2 = \left[\sum_j f_j \cos 2\pi(hx_j + ky_j + lz_j)\right]^2 + \left[\sum_j f_j \sin 2\pi(hx_j + ky_j + lz_j)\right]^2$$

The square of the structure factor correctly gives the intensity of the scattered beam obtained above as Eq. **7**. Thus the complex-number expression of Eq. **10** does represent the amplitude of the x-ray beam scattered by the contents of the unit cell.

The structure-factor summation is over all the atoms of the unit cell. If the positions of these atoms were known, a value of the structure factor could be deduced for each hkl reflection. In practice we must proceed not *from* but *to* the atomic positions.

The procedure used depends upon Bragg's suggestion that the crystal be looked upon not as a set of discrete scattering points but as a three-dimensional distribution of varying electron densities. X-rays are, in fact, affected by atoms because they are regions of high electron density, and therefore high scattering power, compared with that of the surrounding regions. Instead of attempting to deduce the coordinates of the atoms in the crystal, one can look for an electron-density function $\rho(xyz)$ that represents the electron-density distribution in the crystal.

The deduction of the relation between $\rho(xyz)$ and the intensities of the diffraction spots requires some knowledge of Fourier transforms. Students without that knowledge may omit this material and proceed directly to Eq. **17.** That expression is the desired result, i.e., the relation of $\rho(xyz)$ to the structure-factor terms $F(hkl)$, each $F(hkl)$ being proportional to the square root of the intensity of the *hkl* diffraction.

The electron-density function $\rho(xyz)$ is quite complicated if the molecules of the crystal are large. Even in such cases, however, we know that the function will be periodic, repeating itself just as the crystal as a whole repeats the unit cell. Periodic functions, no matter how complex, can be represented by sums, or series, of appropriate sine and cosine terms known as *Fourier series.* A simple illustration of a periodic function and its approximation by a series of such terms is given in Fig. 2-21. More generally we need to use both sine and cosine terms, and a general periodic function $f(x)$ can be then represented as

$$f(x) = \sum_{n=0}^{\infty} (a_n \cos 2\pi nx + b_n \sin 2\pi nx) \qquad \mathbf{11}$$

Adjustment of the many constants a_n and b_n can produce a series with the same form as the original function $f(x)$.

In view of the complex-number–trigonometric relation of Eq. **8,** this Fourier series representation can also be expressed in complex-number form. The counterpart of Eq. **11** is then found to be

$$f(x) = \sum_{n=-\infty}^{+\infty} c_n e^{2\pi inx} \qquad \mathbf{12}$$

Again, adjustment of sufficient constants c_n makes it possible to describe the periodic function $f(x)$.

Thanks to the periodic nature of any crystal, with periodicities a, b, and c, the electron-density function can be formally represented by a Fourier series according to

$$\rho(xyz) = \sum_{p,q,r=-\infty}^{+\infty} A(pqr) e^{2\pi i(px+qy+rz)} \qquad \mathbf{13}$$

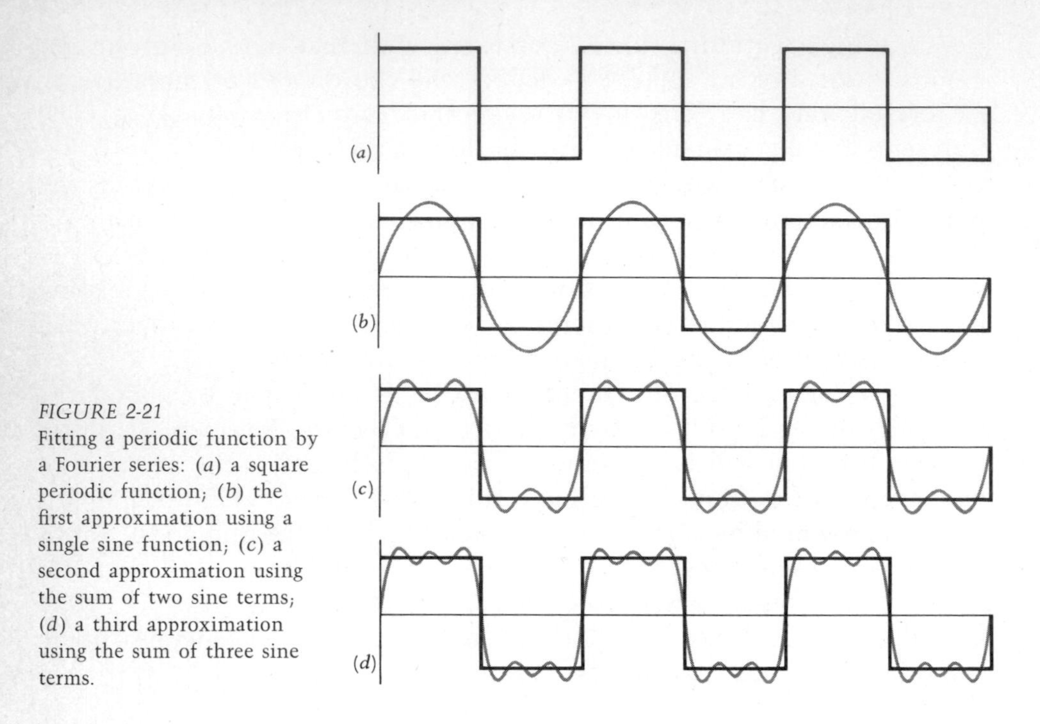

FIGURE 2-21
Fitting a periodic function by a Fourier series: (*a*) a square periodic function; (*b*) the first approximation using a single sine function; (*c*) a second approximation using the sum of two sine terms; (*d*) a third approximation using the sum of three sine terms.

where p, q, and r take on all integral values from $-\infty$ to $+\infty$ and $A(pqr)$ are the coefficients (to be determined) of the many Fourier series terms. Finding the coefficients $A(pqr)$ is the goal of the derivation. If these coefficients were known it would be possible to draw the electron density and thus to locate the atoms in the unit cell.

This interpretation of the unit-cell contents by an electron-density function rather than by atoms occupying various positions can be inserted into the structure-factor expression. Then in place of Eq. **10** one writes the integral expression

$$F(hkl) = (\text{const}) \int\int_0^1\int \rho(xyz) e^{2\pi i(hx+ky+lz)}\, dx\, dy\, dz \qquad \mathbf{14}$$

A valuable result is obtained by inserting the Fourier series interpretation of $\rho(xyz)$ given by Eq. **13** into Eq. **14,** giving

$$F(hkl) = (\text{const}) \int\int_0^1\int \sum_{p,q,r} A(pqr) e^{2\pi i(px+qy+rz)} \times e^{2\pi i(hx+ky+lz)}\, dx\, dy\, dz$$

$$= (\text{const}) \int\int_0^1\int \sum_{p,q,r} A(pqr) e^{2\pi i[(h+p)x+(k+q)y+(l+r)z]} \times dx\, dy\, dz \qquad \mathbf{15}$$

All terms of the type $\int_0^1 e^{2\pi i\theta}\,d\theta$ are zero, as can be easily seen by expressing the exponential term in trigonometric form. This fact eliminates all terms of the series except those for which $p = -h$, $q = -k$, and $r = -l$. For all these, the exponential is $e^{2\pi i(0)} = 1$, and the net result of all the summations and integrations of Eq. **15** is

$$F(hkl) = (\text{const})A(-h,-k,-l)$$

or $$A(-h, -k, -l) = \frac{F(hkl)}{\text{const}} \quad \text{or} \quad A(p, q, r) = \frac{F(hkl)}{\text{const}} \qquad \textbf{16}$$

Substitution of this result in the Fourier series expression, Eq. **13,** for the electron density gives, finally,

$$\rho(xyz) = \frac{1}{\text{const}} \sum_{h,k,l=-\infty}^{+\infty} F(hkl)e^{-2\pi i(hx+ky+lz)} \qquad \textbf{17}$$

The power of this elegant result is seen when it is recalled that $F(hkl)$ is the amplitude of the wave scattered by the hkl plane. The measured intensity of the beam scattered by the hkl plane, furthermore, is proportional to $[F(hkl)]^2$. For "centrosymmetric" crystals the $F(hkl)$ terms are expressed by real numbers and square roots of the measured intensities lead to relative values of $\pm F(hkl)$. Except for the undetermined sign on the values of $F(hkl)$, the expression of Eq. **17** gives the desired result, a method for using the intensities of the diffraction spots to deduce the crystal structures, as represented by $\rho(xyz)$. The electron density at any point (x,y,z) in the unit cell can be determined by performing the summation of Eq. **17,** for the chosen value of x, y, and z, over all values of hkl, that is, over all the observed diffraction spots. The more intense diffraction spots correspond to the numerically greater diffraction amplitudes and structure-factor terms, so that at first one need extend the summation only to the more intense diffraction spots. Such summations can be performed for various points, specified by values of x, y, and z, in the unit cell. By determination of the electron density at sufficient points, an electron-density map can be drawn showing the positions of the atoms by regions of high electron density.

The undetermined sign, which results because the intensities and not the amplitudes of the diffraction beams are obtained, turns out to be very troublesome, and a number of techniques have been suggested to make use of Eq. **17** in spite of this difficulty.

These approaches usually depend on methods for obtaining the approximate shape of the molecule and its approximate position in the unit cell. Some molecules, for instance, contain a heavy atom which has a high scattering probability and can be located, whereas the complexities of the remainder of the molecule can be ignored initially. The signs of the most important structure factors can then be guessed. These few structure-factor values can then be used in Eq. **17** to deduce a crude electron-density map. This first-approximation map can help us estimate the position of more atoms. Then the signs of more structure factors can be deduced from Eq. **10,** or **14.** In this way, one refines a structure by working in the information from more and more diffraction spots until a structure of the desired detail is obtained.

Complete deduction of the structure of a complicated molecule by the x-ray-diffraction technique is a major research problem, but no other method provides such a wealth of reliable structural information. The number of large molecules whose structure in the solid state has been completely worked out continues to grow and to provide valuable basic information for many areas of the life sciences.

2-8 STRUCTURE RESULTS AND SOME ASPECTS OF PROTEIN STRUCTURES

Crystal and molecular structures deduced from x-ray-diffraction studies show a great individuality. Each substance has molecules with a characteristic shape and the molecules pack together according to this shape. Some generalities can be recognized.

First we can note the *primary* structural features of bond lengths and bond angles. Values for a few molecules have already been mentioned in Chap. 1. Spectroscopic and diffraction studies have provided such results for many simple and complex molecules. For many purposes representative bond-length and bond-angle data can be used instead of the actual results. Some values of importance in organic and biological systems are given in Fig. 2-22. An alternative procedure, which raises some problems, assigns covalent radii to each atom so that a covalent bond length can be deduced from the sum of the radii of the bonded atoms. Covalent-radii values are given in Table 2-2.

The spatial demands that atoms in a molecule make can be deduced from the closest packing observed. If a variety of crystals are investigated, the minimum contact distances between pairs of atoms given in Table 2-3 are obtained. These, as the table shows, are related to the molecular dimensions deduced from van der Waals' b parameter. These dimensions are not as rigidly adhered to as the covalent-bond lengths and angles. Under exceptional circumstances a contact distance smaller than these values by as much as 0.1 Å apparently can occur.

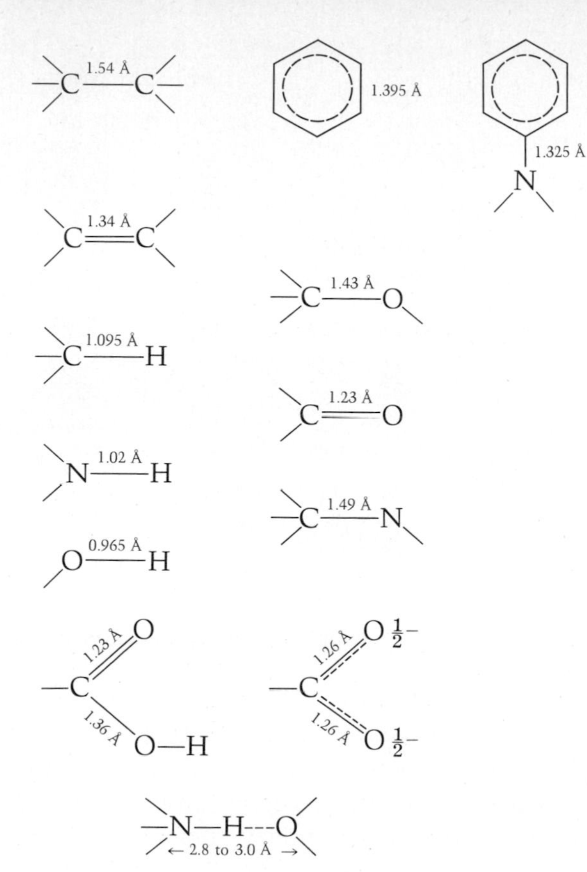

FIGURE 2-22
Some average bond lengths.

Similar primary-structure data are available from diffraction studies of ionic crystals. One characteristic is the packing arrangement adopted by the various ionic species. Although this affects the distance between neighboring ions to some extent, it is possible to deduce from the structures of many ionic crystals a fairly self-consistent set of ionic radii. One set of such values is shown in Fig. 2-23.

TABLE 2-2
Covalent Radii for Atoms Involved in Single-bonded Compounds, Å

H	C	N	O	F
0.37	0.77	0.70	0.66	0.64
	Si	P	S	Cl
	1.17	1.10	1.04	0.99
	Ge	As	Se	Br
	1.22	1.21	1.17	1.14
	Sn	Sb	Te	I
	1.40	1.41	1.37	1.33

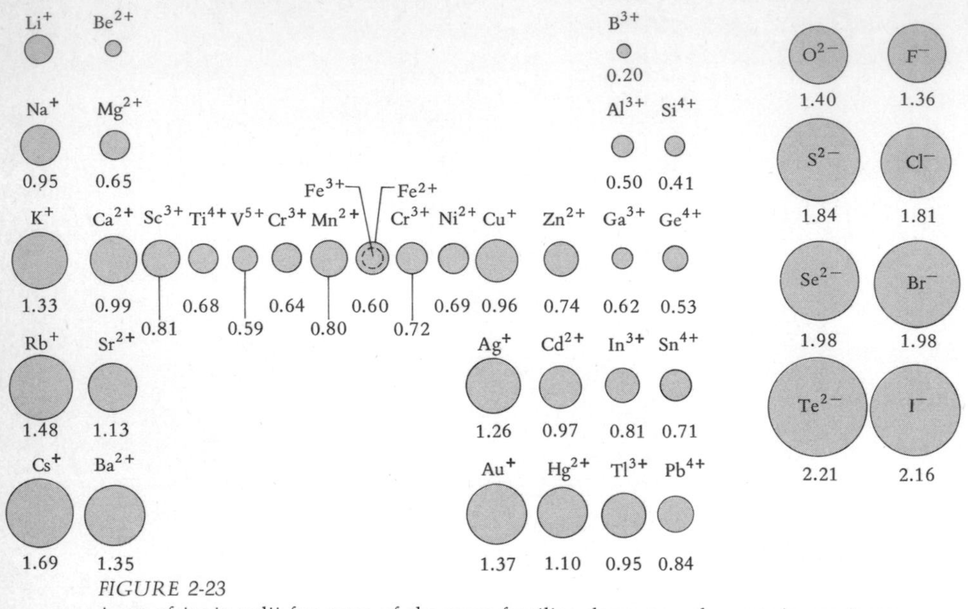

FIGURE 2-23
A set of ionic radii for some of the more familiar elements; values are in angstroms.

To find general structure features beyond this primary structure, attention must be restricted to compounds of a given type. Many protein structures have now been worked out and this class can be used to illustrate how generalities can be sought in the secondary structure of large molecules.

The overall structure of a protein chain can be treated in terms of the two planar units (Fig. 2-24) and their angular positions relative to the

TABLE 2-3
Van der Waals' Radii for Atoms of Molecules
Values deduced from the *b* parameter of van der Waals' equation are shown, as are values deduced from the packing of molecules in crystals.

atom	van der Waals' radii, Å	radii deduced from least contact distance of molecules in crystals, Å
H	1.1	0.95
C		1.5
N	1.5	1.3
O	1.4	1.3
F	1.3	
Cl	1.8	

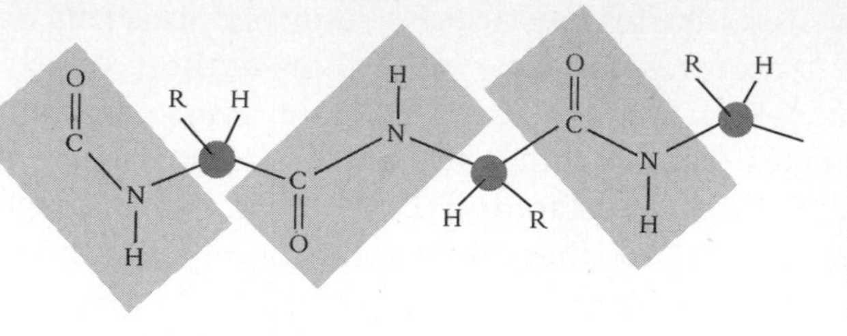

FIGURE 2-24
The planar units between connecting tetrahedral carbon atoms (here shown as dark spheres) of a protein chain.

carbon atom that connects them. The angular positions are usually indicated (Fig. 2-25*a*) by the angles φ, showing the rotation about the N—C bond, and ψ, showing the rotation about the C—C bond.

For a random protein chain the values of φ and ψ vary from one carbon atom to the next, but statistical tabulation of the values of these angles found in the random coils of globular proteins would show certain preferred positions. The values of φ usually adopted lie near 120°, corresponding to a rotation that leaves the sometimes bulky R group confronted directly only by a hydrogen atom. Then a range of values of ψ are taken on by the second planar unit, but ψ values between 0 and 90° are *not* adopted. From Fig. 2-25*b* (or better from models) you can see that once φ is given the value 120°, the carbonyls of the two planar units are in close proximity for ψ values from about 0 to 90°.

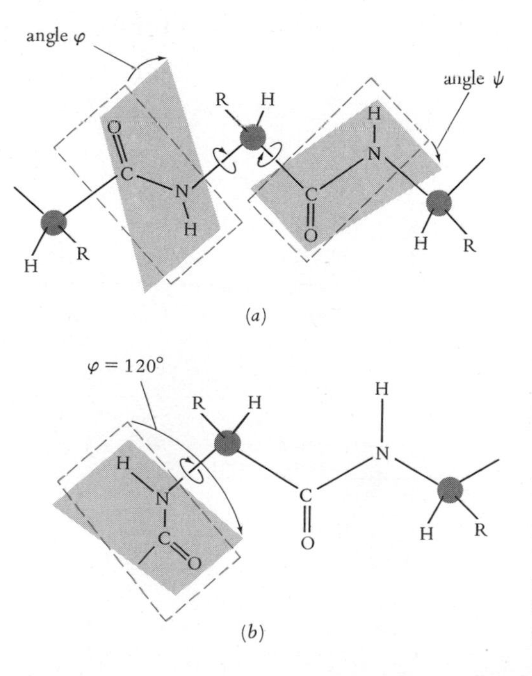

FIGURE 2-25
(*a*) Description of the rotation of the planar units about the N—C and C—C bonds by the angles φ and ψ, respectively. (*b*) Rotation to a value φ = 120° places the carbonyl group as far away from the often bulky R group as possible.

Regular, organized structures can also be described in terms of the φ and ψ values. In these structures, however, the values of these two angles are maintained at each carbon atom of the molecule. Again the general results from random shapes (that φ values near 120° are preferred and ψ values other than 0 to 90° are satisfactory) carry over. Now, however, organized structures are often those that allow good hydrogen bonding of the type

$$\rangle N-H---O=C\langle$$

Such bonding can be developed from *pleated sheet* structures, for example, as shown in Fig. 2-26. More often helical structures are adopted that allow hydrogen bonds to be formed within the helix. Figure 2-27 shows one such configuration, the biologically important *α helix*.

The structure of a fibrous protein can be indicated, in part, by locating a point on a plot of φ versus ψ to indicate the values for these two angles. The variety of φ and ψ values of a globular protein can be displayed by locating a point for each amino acid residue. The angles adopted by the amino acid residues of lysozyme are shown in Fig. 2-28. Such diagrams, known as *Ramachandran plots,* conveniently show the values of these two angles. That the angles indicated do in fact correspond to the deduced structure can be seen only with the help of models. Furthermore, models make clear several other features of the Ramachandran plot. One can see,

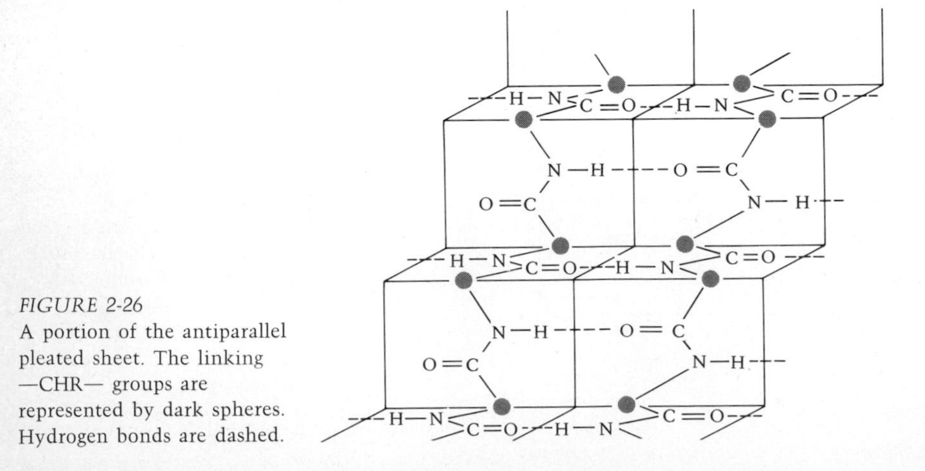

FIGURE 2-26
A portion of the antiparallel pleated sheet. The linking —CHR— groups are represented by dark spheres. Hydrogen bonds are dashed.

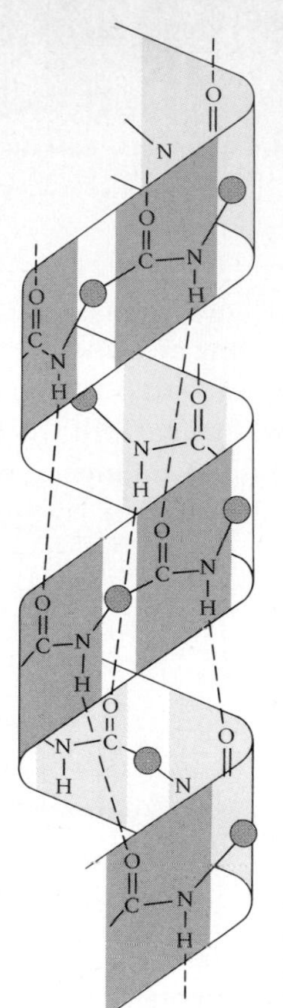

FIGURE 2-27
A portion of the α-helix protein structure. The —CHR— groups are represented by dark spheres.

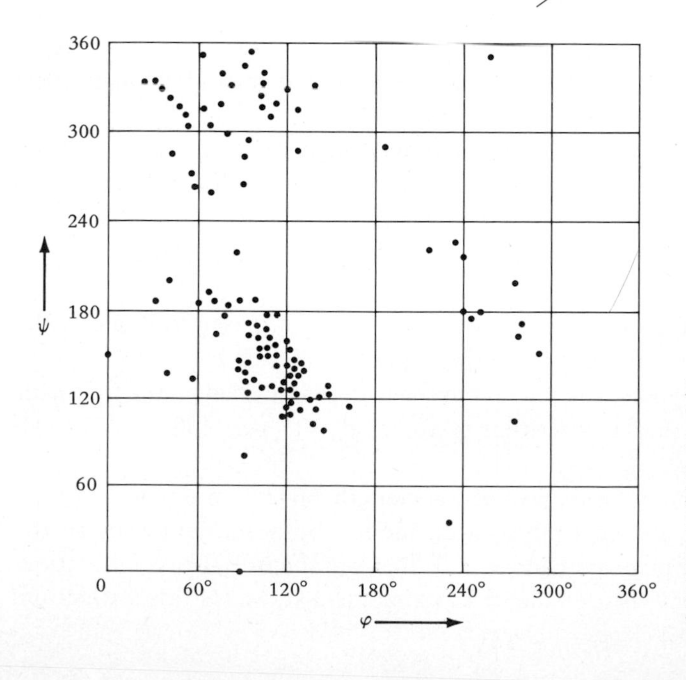

FIGURE 2-28
The values of the angles φ and ψ of Fig. 2-25 for the amino acid residues of lysozyme. [*After R. E. Dickerson and I. Geis, "The Structure and Action of Proteins," Harper & Row, Publishers, Incorporated, New York, 1969.*]

for example, positions that indicate helices of various types and regions of right- and left-handed helices.

PROBLEMS

1 Describe the number and position, as given by direction and distance, of the lattice points surrounding any one lattice point of the two tetragonal lattices shown in Fig. 2-4. Does your answer confirm that the lattices are different?

2 A fiber photograph was taken with Cu K_α x-rays of wavelength 1.542 Å and a film diameter of 30 mm. A millimeter scale placed next to the developed film gave the following values for the layer-line positions:

layer line	mm
3	37.4
2	18.7
1	8.2
0	0
−1	8.3
−2	18.6
−3	37.2

Calculate from these data the value of the repeat spacing along the fiber.

3 Besides the layer lines reported in Prob. 2, additional strong layer lines show up at about 44 mm above and below the equatorial line. These can be interpreted as being the first of the series of layer lines from an additional repeat spacing along the molecules of the fiber. What is the length of this additional repeat unit?

4 At what angles from the direction of the x-rays incident on a crystal will the first- and second-order diffraction be found if they are produced by x-rays of wavelength 1.542 Å being diffracted by planes with spacing of 2.500 Å? (Be careful to distinguish between the angles asked for here and the angle θ of Fig. 2-11.)

5 Write Miller indices for the planes that intercept the axes used to describe the crystal at (*a*) $\frac{1}{3}a, \frac{2}{3}b, c$; (*b*) $\frac{1}{2}a, b, \frac{1}{2}c$; (*c*) $a, 2b, 2c$; (*d*) $\frac{1}{2}a, 2b, c$.

6 In a diagram of a tetragonal unit cell sketch the planes described by the following Miller indices (note that $\bar{1}$ is written to indicate −1): (*a*) 100, (*b*) 010, (*c*) 120, (*d*) 111, (*e*) $1\bar{1}1$.

7 A diffraction experiment uses x-rays of wavelength 1.542 Å, produced with an x-ray tube containing a copper target; a diffraction beam makes a spot on the photographic detector plate at an angle of 5°27′ from the initial beam direction.

a What is the value of θ, as used in Fig. 2-11 and Eqs. **3** and **4,** for this diffraction?

b If the spot has not been indexed, what can be said about the spacing between the set of planes that produce this diffraction?

8 When x-rays with a wavelength of 2.29 Å, as produced by an x-ray tube with a chromium target, are diffracted by a crystal, a series of diffractions at values of $\sin\theta$ of 0.0722, 0.1440, 0.2164, and 0.2878 are produced. If the orientation of the crystal and the direction of these diffractions are such that this set of diffractions can be attributed to 100 planes, calculate a best value for the d_{100} spacing and provide indices for each of the observed diffractions.

9 The following data have been given for the protein ribonuclease-S by J. R. Knox [*J. Chem. Educ.*, **49**:476 (1972)]:

Unit cell volume	167,000 Å^3
Crystal density	1.282 g/cm^3
Protein fraction	0.68
Number of molecules per unit cell	6

What is the molecular mass of ribonuclease-S?

10 Consider a cubic unit cell that has atoms with unit scattering factor at the unit cell corners. The effect of these can be most simply described by placing one such atom at the origin corner with $x = y = z = 0$. (Alternatively you could place one-eighth of such an atom at each of the unit cell corners.)

a Introduce a constant factor (const) into the proportionality of Eq. **7** and express the intensity of the 100, 200, 110, and 220 diffractions.

b Now introduce an additional atom, also of unit scattering power, in the center of the unit cell. Since x, y, and z are coordinates expressed in terms of the unit cell dimensions, its coordinates will be $x = y = z = \frac{1}{2}$. Repeat the calculation of the intensities of the 100, 200, 110, 220 reflections for this *body-centered* unit cell. Draw the unit cell and the planes and rationalize the intensity results you obtained.

c Move the central atom so that $x = \frac{3}{4}$, with y and z remaining at $\frac{1}{2}$. What are the intensities of the four diffractions? This should illustrate, in a crude way, that the intensities of diffractions depend on the positions of atoms within the unit cell and can be used to determine them.

11 To what extent are the covalent radii of Table 2-2 consistent with the bond lengths of Fig. 2-22? Consider any justification there is for expecting that covalently bonded atoms can be given fixed radii.

12 Draw a diagram to represent the shape, as determined by the data of Tables 2-2 and 2-3 and Fig. 2-22, of H_2, HCl, and the linear CO_2 molecule.

13 For any element for which data are available, compare an ionic radius with a covalent radius and a van der Waals' radius.

REFERENCES

Crystal-Structure Determinations

BUERGER, M. J.: "Contemporary Crystallography," McGraw-Hill Book Company, New York, 1970. One of the best, for the nonspecialist, of the many books on the determination of crystal structures by means of x-ray diffraction. It is not aimed at the study of organic or biological substances, but it does emphasize the precession method.

TANFORD, C.: "Physical Chemistry of Macromolecules," John Wiley & Sons, Inc., New York, 1961. The opening chapter deals with crystal-structure studies.

KNOX, JAMES R.: Protein Molecular Weight by X-Ray Diffraction, *J. Chem. Educ.*, **49:**476 (1972).

Molecular-Structure Results

DICKERSON, R. E., and I. GEIS: "The Structure and Action of Proteins," Harper & Row, Publishers, Incorporated, New York, 1969. Beautifully presented structural diagrams of protein molecules are combined with clear and intriguing discussions of the relation of structure to action.

WOLD, F.: "Macromolecules: Structure and Function," Prentice-Hall, Inc., Englewood Cliffs, N.J., 1971. Structure, in its broadest sense, is one of the vehicles that leads into clear and penetrating analyses of chemical and biological actions.

RAMACHANDRAN, G. N., and V. SASISEKHARAN: Conformations of Polypeptides and Proteins, *Adv. Protein Chem.*, **23:**284 (1968). Considerations of favored structural arrangements and systemization of the treatment of conformations.

3 Liquids and the Motion and Structure of Molecules of Liquids

Chapters 1 and 2 have introduced some of the powerful methods for determining the motions of molecules of gases and the structures of molecules of crystals. It is an example of the perversity, or the challenge, of nature that these methods are inapplicable when we tackle the most interesting systems, those of substances in aqueous solutions. No comparable alternative methods are available. We proceed, in part, by carrying over and modifying the information from gas and solid-state studies. This chapter provides guides to the carryover and extension of such results to molecules in solution and presents some of the experimental approaches that can be applied to liquids. You will see that some valuable ideas about the molecular world of the liquid state can be obtained by bringing a variety of information together.

Most molecules in the liquid state show no evidence of free rotational motion, and their translational motion must be assumed to be transformed into a continually disturbed jostling with neighboring molecules. But there is evidence, provided primarily by the similarity of vibrational spectra of a molecule in different physical states, that the vibrations within the molecule, and thus the chemical bonds and primary structure of the molecule, are often little changed with change of state. This important observation allows us to take structural information from studies of gases and crystals and apply it, with certain precautions, to the less easily studied liquid state.

3-1 BACKGROUND TO MOLECULAR VIBRATIONS: VIBRATIONS OF BALL-AND-SPRING MODELS

Now let us investigate how particles joined together by flexible bonds or springs can vibrate. The net results of this study will provide further insight into the motions of atoms of molecules, the rigidity or flexibility of chemical bonds, and the persistence of the primary structure of molecules. This treatment of vibrations parallels that of rotation given in Secs. 1-6 to 1-8.

Consider, to begin with, the vibrational characteristic of a ball at the end of a spring, as in Fig. 3-1. In practice the spring would hang vertically, but we shall be led to diagrams that are the right way up if we draw

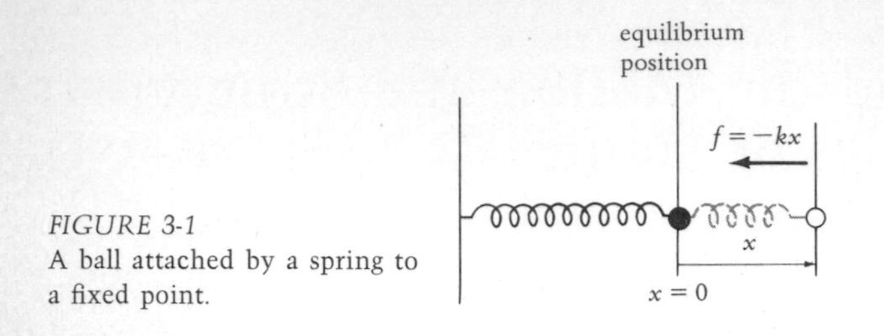

FIGURE 3-1
A ball attached by a spring to a fixed point.

the spring as in Fig. 3-1. You can, if you like, imagine it to be on a frictionless table.

The vibrational characteristics of such a ball or particle are determined by the mass of the particle and by the nature of the spring. The simplest assumption about the spring is that the particle experiences a restoring force pulling or pushing it back to its equilibrium position and that this force is proportional to the distance the particle has been displaced from its equilibrium position. This assumption turns out to be quite satisfactory for both ordinary-sized springs and for chemical bonds. Such a force-displacement relation is known as *Hooke's law.* Since displacing the particle in one direction brings about a force in the opposite direction, Hooke's law is written

$$f = -kx \tag{1}$$

where f is the restoring force and x is the displacement from the equilibrium position. The proportionality constant k, known as the *force constant,* is a measure of the stiffness of the spring. The force constant is equal to the restoring force operating for a unit displacement from the equilibrium position.

For many purposes it is helpful to recognize that Hooke's law implies that the potential energy U experienced by the bound particle rises parabolically as the particle is moved away from its equilibrium position. The potential energy is the stored work, and as the particle is moved a small distance dx away from its equilibrium position, a force equal and opposite to the spring force, $f = -kx$, must be applied. The net increase in potential energy as the particle is moved from its equilibrium position $x = 0$ to some other position x is obtained by the integral

$$\begin{aligned} U_{x=x} - U_{x=0} &= \int_0^x (+kx)\,dx \\ &= \tfrac{1}{2}kx^2 \end{aligned}$$

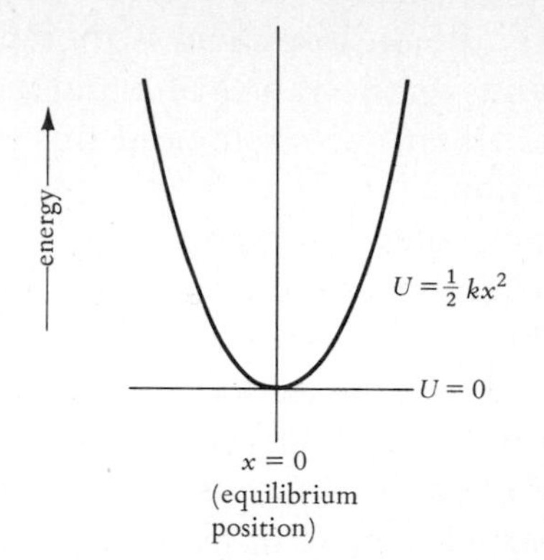

FIGURE 3-2
The potential-energy curve for a spring obeying Hooke's law.

If we set $U_{x=0} = 0$, we have the alternative expression for Hooke's law

$$U = \tfrac{1}{2}kx^2 \tag{2}$$

The parabolic curve for this function is shown in Fig. 3-2.

The motion of an ordinary-sized object attached to a Hooke's law spring can be deduced by returning to the force expression of Eq. **1** and substituting this along with the acceleration expression $a = d^2x/dt^2$ into Newton's $f = ma$ law. We obtain

$$-kx = m\frac{d^2x}{dt^2}$$

or

$$\frac{m}{k}\frac{d^2x}{dt^2} = -x \tag{3}$$

A solution to this equation can be seen by inspection (and verified by substitution) to be

$$x = A \sin \sqrt{\frac{k}{m}}t \tag{4}$$

A is a constant that is equal to the maximum value of x; that is, it is the vibrational amplitude. The position of the particle varies sinusoidally with time. Every time t increases by $2\pi\sqrt{m/k}$, the quantity $\sqrt{k/m}\,t$ increases by 2π and the particle traces out one complete cycle. The time corresponding to one oscillation, or vibration, is therefore $2\pi\sqrt{m/k}$. More directly useful is the reciprocal of this quantity, which is the frequency of vibration, i.e., the number of cycles performed per second. If this quantity is denoted by ν_{vib}, we have

$$\nu_{\text{vib}} = \frac{1}{2\pi}\sqrt{\frac{k}{m}} \tag{5}$$

For a system of ordinary dimension, there is therefore a natural frequency of oscillation that depends on the values of k and m. Any amount of energy can be imparted to the vibrating system, and this energy changes only the amplitude of the vibration.

Closer to the molecules in which we are interested than this single-particle system is an assembly made from a number of balls held in some flexible arrangement by a set of springs. With such an assembly in your hands so that you could investigate its vibrations, you would find that a general distortion and release would produce a jangled, disorganized vibrational motion. But for certain initial distortions, release leads to a special motion in which all particles move in phase and with the same frequency. These motions, known as *natural* or *normal* vibrations, provide the basis for the analysis of all types of vibrations of the system. Since a single vibrational frequency characterizes the motion in a normal vibration, each normal vibration can be looked on as the counterpart of the single vibration of a one-particle system. The details are more complicated in that many of the connecting springs and many of the particle masses are involved in the motion. Still, the roles of the force constants and masses are suggested by the single-particle equation, Eq. **5.**

Each normal vibration of an assembly of particles corresponds to a vibrational degree of freedom. How many vibrational degrees of freedom are there in an assembly with n particles? If the connecting springs are exceedingly weak, you would describe the system in terms of the three coordinates, or degrees of freedom, that would be needed to describe the positions or motions of each of the n particles. You would then ascribe $3n$ degrees of freedom to the entire system. If we imagine the connecting springs as gradually strengthening, at some stage it becomes more convenient to describe the motion of the system in terms of the translational motion of the entire assembly, the rotational motion of the entire assembly, and finally the internal or vibrational motion of the particles. Since the translational motion of the system accounts for 3 degrees of freedom and the rotational motion accounts for another 3, there remains $3n - 6$ to be accounted for by vibrational motions. Thus, an assembly of n particles has $3n - 6$ normal vibrations, each corresponding to a vibrational degree of freedom.

This assumes a generally shaped ball-and-spring model. A linear assembly has only 2 rotational degrees of freedom, and thus the expression for the number of vibrational degrees of freedom of a linear system of n particles is $3n - 5$.

This background, developed with regard to ordinary-sized ball-and-spring assemblies, can now be carried over to the molecular world.

3-2 INTRODUCTION TO MOLECULAR VIBRATIONS AND INFRARED SPECTRA

How stiff are chemical bonds, and with what frequency do they vibrate? We can answer this if we anticipate a result from the following sections. If a molecule or (for simplicity) an atom bound by a chemical bond to a wall is hit by another molecule so that the molecule is given $\frac{1}{2}kT$ energy of vibration, the bound atom will vibrate with an amplitude corresponding to something like 2 percent of its bond length. This information is enough to determine the stiffness of the representative bond and thus the frequency of its natural vibration.

If the bond has a length of 1.5 Å, a potential-energy increase of $\frac{1}{2}kT$, or 2.1×10^{-14} erg at 25°C, produces a distortion of

$$\frac{0.02}{100} \times 1.5 = 0.03 \text{ Å}$$

Thus, in $U = \frac{1}{2}kx^2$ we have

$$2.1 \times 10^{-14} = \tfrac{1}{2}k(0.03 \times 10^{-8})^2$$

and $\quad k \cong 5 \times 10^5$ dyn/cm

Further, from Eq. **5**, if a representative molar mass of 10 g or a molecular mass of $10/(6 \times 10^{23})$ g is used,

$$\nu = \frac{1}{2\pi}\sqrt{\frac{5 \times 10^5}{10/(6 \times 10^{23})}}$$
$$\cong 3 \times 10^{13} \text{ Hz}$$

Our estimate of the stiffness of chemical bonds thus leads us to expect that the natural vibrational frequencies of molecules are of the same order as the frequencies of infrared radiation (see Fig. 1-12). Again, as for molecular rotations, we are led to consider a possible coupling and energy exchange between radiation of this type and molecular motion.

Coupling with electromagnetic radiation can occur, as for rotating molecules treated in Sec. 1-7, if the vibrating molecule produces an oscillating dipole moment that can interact with the electric field of the radiation. It follows that homonuclear diatomic molecules, like H_2, N_2, and O_2, which necessarily have a zero dipole moment for any bond length, fail to interact. The dipole moment of molecules such as HCl, on the other hand, can be expected to be some function (usually unknown) of the internuclear distance. The vibration of such molecules leads to an oscillating dipole moment, and vibrational spectra can be expected.

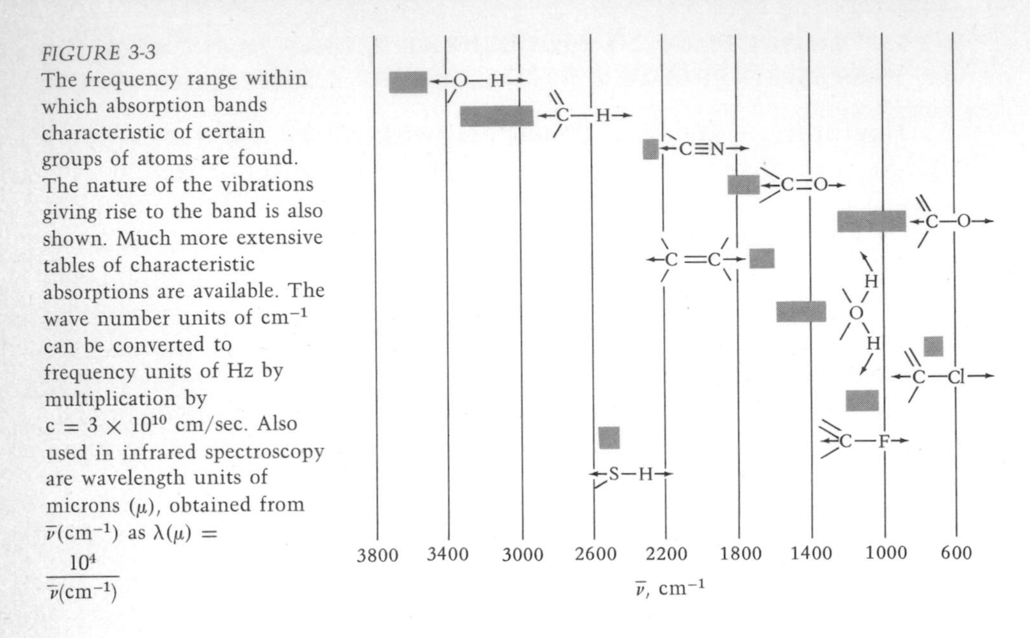

FIGURE 3-3
The frequency range within which absorption bands characteristic of certain groups of atoms are found. The nature of the vibrations giving rise to the band is also shown. Much more extensive tables of characteristic absorptions are available. The wave number units of cm^{-1} can be converted to frequency units of Hz by multiplication by $c = 3 \times 10^{10}$ cm/sec. Also used in infrared spectroscopy are wavelength units of microns (μ), obtained from $\bar{\nu}(\text{cm}^{-1})$ as $\lambda(\mu) = \dfrac{10^4}{\bar{\nu}(\text{cm}^{-1})}$

If we shine infrared radiation through a sample containing simple molecules (other than homonuclear diatomic molecules) or complex molecules, we generally find a rich display of absorption regions. These absorption regions are called *bands* rather than lines because they occur over an appreciable frequency range. The bands can be attributed to the transfer of energy from the infrared radiation to the vibrating molecules of the sample. Thus the general expectation developed above is borne out. Unexpected, however, is the absorption at particular frequencies rather than as a continuum throughout the infrared region. Apparently, again, the classical conclusion that a vibrating particle can have any amount of vibrational energy is not valid. Vibrating molecules are restricted to certain vibrational energies and thus can absorb only certain frequencies, or quanta, from incident infrared radiation.

Even without further development, infrared spectra can be used for identification and analysis. As Fig. 3-3 shows, certain chemical groups produce characteristic absorptions. The presence of this characteristic absorption in a spectrum means that this group occurs in the molecules of the sample. Furthermore, the exact spectral position of the band characteristic of the group often indicates something of the chemical environment of the group.

Also, at this stage, we can use the general similarity of the infrared spectra of materials obtained in different physical states or in solutions with different solvents to conclude that the chemical bonds and primary structure of molecules are little altered by such changes. This casual conclusion has extremely important consequences since it is the most direct basis for carrying the detailed structural results from x-ray studies of crystals over to the liquid state. However, it is only chemical bonds and angles between bonds that we know are preserved. The longer-range structure of large molecules usually has no noticeable effect on infrared spectra, and we therefore cannot conclude that the overall shape of large molecules is maintained.

Can we explain the presence of discrete bands in infrared spectra? To do so, we must see how the restriction to certain energies is imposed on vibrating molecules. In so doing, we come to the second way quantum restrictions are imposed to make ordinary expressions applicable to the molecular world.

3-3 THE ENERGY OF A PARTICLE CONFINED TO A SMALL REGION: AN INTRODUCTION TO THE WAVE NATURE OF PARTICLES

In 1923 Louis de Broglie suggested a way of deducing the special features displayed by particles whose movement is confined to small regions, e.g., the movement of the atoms bound together into a molecule by chemical-bond forces. These special features, he suggested, could be accounted for by associating waves with particles. He thus extended the wave-particle duality introduced for electromagnetic radiation. The observed behavior of confined particles can be deduced, he claimed, if the wavelength λ of this wave is related to the particle mass m and particle velocity v by the expression

$$\lambda = \frac{h}{mv} \tag{6}$$

where again h is Planck's constant. If the product mv is recognized as the momentum of the particle, this equation expresses the proportionality

$$\text{Wavelength} \propto \frac{1}{\text{particle momentum}} \tag{7}$$

The consequences of the idea that wave features can be considered even when we are dealing with a particle can be illustrated by studying a simple system that is artificial but revealing. Consider a particle that can move freely along a straight line over some length l but is prevented from moving beyond this segment of the line or off the line. Your picture should resemble a train running on a fixed length of straight track, but the train is replaced by a particle such as an atom and the section of track by a distance typical of the lengths of atoms or molecules.

How can the wave that is associated with this particle be accommodated in this length? First, the wave must be no longer than the length segment l. If the wave existed outside this length, it would imply that the particle could get outside this segment.

Waves can be fitted in, as in Fig. 3-4, so that they go to zero values at the ends of the segment and thus fit smoothly on the zero wave-amplitude regions outside the segment. Such special waves occur if some multiple of half their wavelength is equal to the length l. The ways in which the wave can be fitted into the length are given by the expression

$$n\frac{\lambda}{2} = l \qquad n = 1, 2, 3, \ldots \tag{8}$$

If the de Broglie relation $\lambda = h/mv$ is inserted and the result rearranged, we can find the kinetic energy $\frac{1}{2}mv^2$ the particle would have for each possible way the wave fits. Thus

$$n\frac{h/mv}{2} = l$$

or $$(mv)^2 = \frac{n^2h^2}{4l^2}$$

and $$\epsilon = \tfrac{1}{2}mv^2 = \frac{(mv)^2}{2m} = \frac{n^2h^2}{8ml^2} \tag{9}$$

where $n = 1, 2, 3, \ldots$

These *allowed energies,* which have been discovered by associating a wave nature with the confined particle, are shown graphically in Fig. 3-5. Such allowed energies are often the most important result when this procedure is applied to real problems.

What is the significance, if any, of the waves themselves? Max Born suggested that the probability of the particle's being at a particular position is proportional to the square of the wave function at that position. Thus, if the mathematical function that corresponds to each wave is represented by $\psi(n)$, where an integer must be inserted for n to specify a particular wave, the function $[\psi(n)]^2$ indicates the probability of the particle's being at various positions.

To investigate this, the wave function and the squares of the functions for these waves must be plotted. This is often done by placing them on a diagram (see Fig. 3-6) in accordance with energies to which they

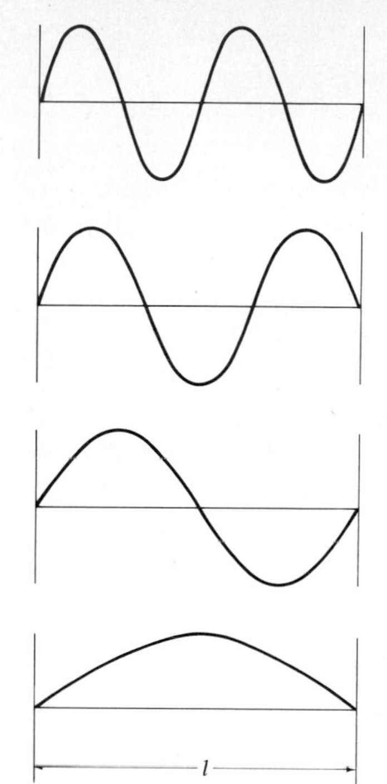

FIGURE 3-4
Some of the special ways in which a wave can be fitted into a region of length l.

correspond. Keep in mind that a one-dimensional system is being investigated. The amplitude of the wave and its square are plotted on an ordinate scale, but the particle does not go up and down the way these curves do. The dot-density diagrams of Fig. 3-6, which also show the values of the

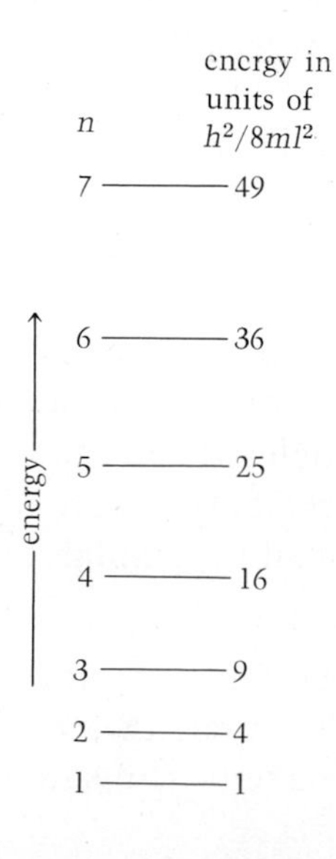

FIGURE 3-5
The energies allowed to a particle of mass m confined to a length l.

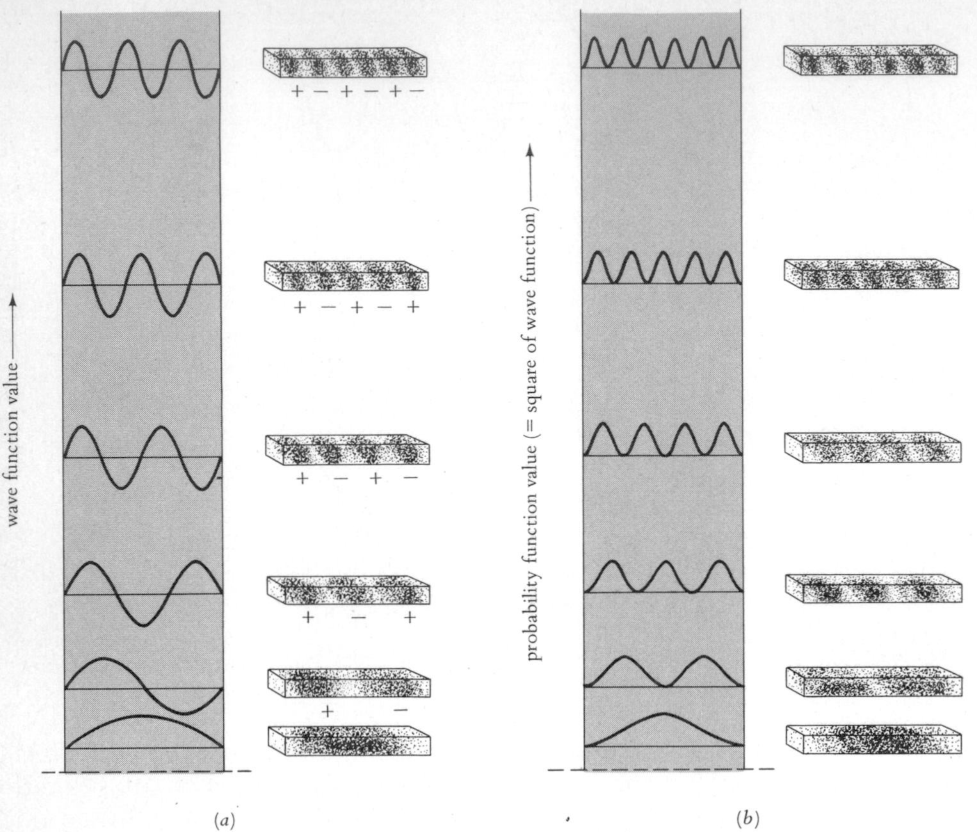

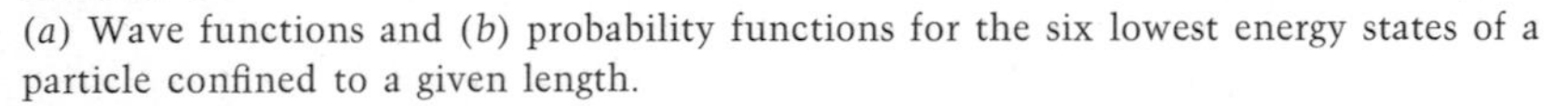

FIGURE 3-6

(*a*) Wave functions and (*b*) probability functions for the six lowest energy states of a particle confined to a given length.

amplitude and its square, avoid this implication. You should think in terms of these probability variations along the length *l* of the particle's region of freedom, but you should also see that the wave curves are equivalent and more quantitative.

The results of this simple but revealing exercise apply generally to confined particles and allow important general conclusions about the behavior of confined particles to be drawn. Such conclusions apply even to atomic and molecular systems, where simple wave fitting cannot be done and a more mathematically formidable approach must be used.

1. Particles confined to some region of space can have only certain energies, energies that are said to be *allowed*.
2. The *states of the particles,* i.e., their overall behavior, are further described by a wave or the square of a wave. These waves and the

corresponding energies are indexed by a number, often an integer, called a *quantum number.*

3 The more closely the particles are confined, i.e., the smaller the space in which they are free to move, the more widely the energies of the allowed states are spaced, as one sees from $\epsilon = n^2h^2/8ml^2$.

4 The smaller the mass of the particles, the more widely the energies of the allowed states are spaced, again as is seen from $\epsilon = n^2h^2/8ml^2$.

5 For a given particle confined to a given region, the waves for the states of higher energy have more nodes (more places where the amplitude is zero) than those for the lower energy states.

Fitting de Broglie waves in is readily done in some simple problems, such as the square-well problem. But generally this is not possible, and a mathematical rather than pictorial procedure is needed. If, for example, we ask about the allowed behaviors of particles held together by Hooke's law forces, we cannot simply fit waves into a clearly limited region.

A mathematical procedure that is a direct extension of the pictorial procedure of the preceding section is provided by an equation suggested by Erwin Schrödinger in 1926 and named for him. This equation involves a function for the wave, usually denoted, as before, by ψ. If, to begin with, only one dimension, the x direction, is considered, the function will involve x. Allowed behaviors for a particle of mass m experiencing a potential energy described by a function U are then revealed by solutions of the differential equation

$$-\frac{h^2}{8\pi^2m}\frac{d^2\psi}{dx^2} + U\psi = \epsilon\psi \tag{10}$$

where ϵ denotes the energy of the particle. If the fact that ψ and U are functions of x is shown explicitly, this is written

$$-\frac{h^2}{8\pi^2m}\frac{d^2\psi(x)}{dx^2} + U(x)\,\psi(x) = \epsilon\psi(x) \tag{11}$$

Satisfactory solution functions ψ can be found if the particle has some restriction on the region in which it can move, only for certain values of ϵ, and these are the allowed energies of the particle. The wave functions that then satisfy the equation reveal the position of the particle in the system. They do so, however, only through the related function ψ^2, which gives the probabilities of the particle's being at various positions.

All this can be illustrated by returning to the one-dimensional square-well problem and now treating it by means of the Schrödinger equation.

Outside the region between $x = 0$ and $x = l$ the potential energy is infinitely high, and there will be zero probability of finding the particle in these regions. (An electron-in-a-wire analogy is helpful. There the difference in energy of an electron in the wire and outside the wire is large, although not infinite.) It follows that since ψ^2 must be zero for $x < 0$ and $x > l$, so also must ψ be zero in these regions. In the region $0 < x < l$ the potential-energy function can be taken as $U(x) = 0$, and the Schrödinger equation in this region is

$$-\frac{h^2}{8\pi^2 m}\frac{d^2\psi}{dx^2} = \epsilon\psi \qquad \mathbf{12}$$

It is now necessary to find well-behaved solutions for this equation. To be well behaved and avoid a discontinuity at $x = 0$ and $x = l$, the function ψ in the region between 0 and l must be such that it equals zero at $x = 0$ and at $x = l$. Functions which solve the differential equation and also satisfy these boundary conditions can be seen by inspection to be

$$\psi = A \sin\frac{n\pi x}{l} \qquad \text{where } n = 1, 2, 3, \ldots \qquad \mathbf{13}$$

and A is some constant factor. The expression $n\pi x/l$ has been arranged, as can be checked, so that the function goes to zero at $x = 0$ and at $x = l$ for any integral value of n. That the function satisfies the Schrödinger equation can be tested by substitution in Eq. **12** to give

$$\text{Left side} = -\frac{h^2}{8\pi^2 m}\left(-\frac{n^2\pi^2}{l^2}\right) A \sin\frac{n\pi x}{l}$$

$$= \frac{n^2 h^2}{8ml^2}\left(A \sin\frac{n\pi x}{l}\right) \qquad \mathbf{14}$$

$$\text{Right side} = \epsilon\left(A \sin\frac{n\pi x}{l}\right) \qquad \mathbf{15}$$

The left and right sides of Eq. **12** are equal, and the expression $\psi = A \sin(nx/l)$ is therefore a solution if

$$\epsilon = \frac{n^2 h^2}{8ml^2} \qquad n = 1, 2, 3, \ldots \qquad \mathbf{16}$$

Thus we come again to the results obtained by the less generally applicable approach of fitting in the waves.

It should be clear that if the mass of the particle under study is known, and if the potential-energy function can be expressed mathematically, one need only find solutions to the Schrödinger equation to deduce the allowed behaviors of the particle. For complex systems this is often a formidable task. It is, however, a straightforward mathematical problem, and only mathematical ingenuity and the availability of high-speed computors are necessary to solve any problem.

Example 3-1

For some compounds, the absorption of visible light that characterizes colored compounds can be described in terms of the absorption of light by electrons that are free to move throughout the length of the molecule. If the molecule is treated as a box in which one of the electrons of the molecule is free to move, what length of molecule would be required to give an energy spacing between allowed states so that visible radiation could raise the electron from one state to the next?

Solution

Let us consider the energy difference between the lowest, $n = 1$, and the next higher, $n = 2$, states. We write, from Eq. **16,**

$$\Delta\epsilon = \epsilon_{n=2} - \epsilon_{n=1} = \frac{h^2}{8ml^2}(2^2 - 1^2)$$
$$= \frac{3h^2}{8ml^2}$$

A value of $\Delta\epsilon$ corresponding to visible-region quanta can be calculated from a representative visible wavelength of 5000 Å, or 0.5×10^{-4} cm, as

$$\Delta\epsilon = h\nu = h\frac{c}{\lambda}$$
$$= 6.62 \times 10^{-27} \frac{3 \times 10^{10}}{0.5 \times 10^{-4}}$$
$$= 4 \times 10^{-12} \text{ erg}$$

Now, with the electron mass of 0.91×10^{-27} g we can substitute in $\Delta\epsilon = 3h^2/8ml^2$ or $l = \sqrt{3h^2/8m\,\Delta\epsilon}$ to obtain

$$l = \sqrt{\frac{3(6.62 \times 10^{-27})^2}{8(0.91 \times 10^{-27})(4 \times 10^{-12})}}$$
$$= 6.7 \times 10^{-8} \text{ cm}$$
$$= 6.7 \text{ Å}$$

The result shows that a molecule must let one or more electrons move freely over a number of adjacent bonds if the electrons are to have allowed states with an energy separation as small as the quanta of visible light. When electrons are localized on atoms or in bonds, as they are in most molecules, higher energies would be required to excite the electrons. Such molecules absorb ultraviolet radiation but not visible radiation.

3-4 MOLECULAR VIBRATIONS AND FORCE CONSTANTS

The quantum-mechanical solution to the vibrating-particle problem provided by the Schrödinger equation shows, as required to explain the infrared spectra, that only certain amounts of vibrational energy are allowed. These can be deduced by entering the vibrational potential function $U = \frac{1}{2}kx^2$ into the Schrödinger equation and solving for the wave functions and energies of the allowed vibrational states. The procedure is thus analogous to that used for the square well. The parabolic potential function, however, makes determination of the solution wave functions rather more difficult, and these are shown, without derivation, in Fig. 3-7. Their similarity to the square-well functions is apparent.

The energies of the allowed vibrational states are also given by solution of the Schrödinger equation. Again, only the result is given. With v representing the integers that enter, the energies of the allowed states are

$$\epsilon_{\text{vib}} = (v + \tfrac{1}{2})\frac{h}{2\pi}\sqrt{\frac{k}{m}} \qquad v = 0, 1, 2, \ldots \tag{17}$$

This quantum-mechanical result therefore indicates a pattern of energy levels with a constant spacing $(h/2\pi)\sqrt{k/m}$, shown in Fig. 3-7.

The quantum-mechanical solution introduces the collection of terms $(1/2\pi)\sqrt{k/m}$ that correspond to the natural vibrational frequency of a classical oscillator. Equation **17** can therefore be written

$$\epsilon_{\text{vib}} = (v + \tfrac{1}{2})h\nu_{\text{vib}} \tag{18}$$

where the term ν_{vib} is interpreted according to $\nu_{\text{vib}} = (1/2\pi)\sqrt{k/m}$.

The *spacing* of the vibrational levels, according to the harmonic-oscillator or parabolic-potential approximation, is a constant energy

$$\Delta\epsilon_{\text{vib}} = \frac{h}{2\pi}\sqrt{\frac{k}{m}} \tag{19}$$

At room temperature the value of $\Delta\epsilon_{\text{vib}}$ is generally larger than the average thermal energy kT, so that most of the molecules are in the lowest allowed vibrational state. In a spectroscopic study, therefore, one investigates the absorption of radiation by these molecules in the $v = 0$ state.

Even when interaction between vibrating molecule and the radiation occurs, a further selection rule applies. This rule restricts transitions resulting from the absorption or emission of a quantum of radiation by the relation

$$\Delta v = \pm 1 \tag{20}$$

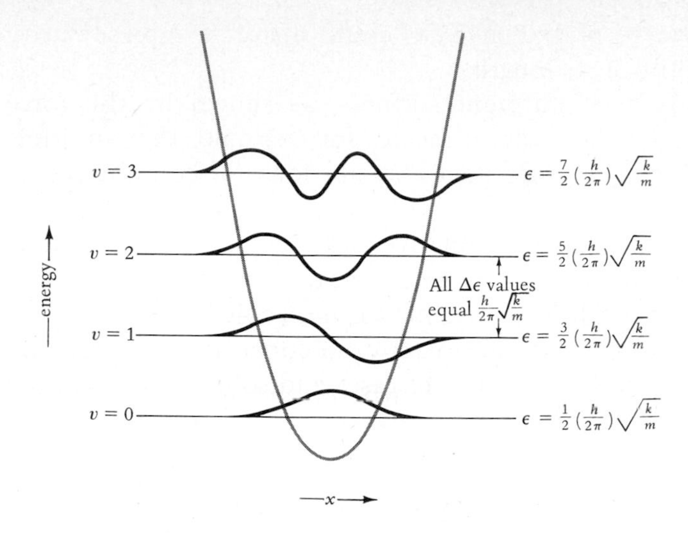

FIGURE 3-7
Wave functions and energies for the allowed states of a vibrating particle.

Vibrational spectra are usually determined by absorption spectroscopy, and then $\Delta v = +1$ is the only part of this selection rule which is pertinent.

For polyatomic molecules, many of the normal vibrations interact with incident infrared radiation, and the $\Delta v = +1$ transitions for each vibration produce an infrared absorption band. As mentioned in Sec. 3-2, the frequency, or quantum energy, of such bands can be used for many analytical purposes. Only with rather elaborate calculation, however, can the motions of the atoms and the distortion of the bonds of the molecule that are associated with the normal vibration be deduced. For the single-particle system, however, the relation $\Delta\epsilon_{\text{vib}} = (h/2\pi)\sqrt{k/m}$ has been obtained, and extension to diatomic molecules can be made simply by replacing m by the reduced mass μ, defined as $m_1m_2/(m_1 + m_2)$.

The use of vibrational spectra to obtain molecular properties can be illustrated by HCl. One observes in the infrared region an absorption band centered at a frequency of about 8.67×10^{13} Hz. This corresponds, with the relation $\Delta\epsilon = h\nu$, to the absorption of quanta of energy $(6.62 \times 10^{-27})(8.67 \times 10^{13}) = 5.74 \times 10^{-13}$ erg.

The observed HCl absorption can be assigned to the vibrational transition from $v = 0$ to $v = 1$ and can be equated, according to Eq. **19**, to $(h/2\pi)\sqrt{k/\mu}$. With $\mu = 1.628 \times 10^{-24}$ g, this yields

$$k = 4.83 \times 10^5 \text{ dyn/cm} \qquad \textbf{21}$$

The theory of vibrational spectra, together with the observed absorption, has now led to a value for the force constant of a chemical bond. Such measurements and calculations are, in fact, the basis for the estimate of Sec. 3-2, that an energy of $\frac{1}{2}kT$ at 25°C could distort a representative bond by about 2 percent of its length.

Different bonds have different stiffness, as shown by the force constants of Table 3-1. Such results indicate, for example, that multiple bonds are indeed stiffer than single bonds, as our ideas of bonding would lead us to expect.

One can also see now that the relative constancy of the frequencies of infrared absorptions with changes of physical state implies a corresponding constancy in bond force constants for molecules in gases, solids, and liquids. We can use force constants to draw our conclusion that primary structural features deduced from studies of gases and solids can be carried over to molecules in the liquid state.

Example 3-2

Replacement of the hydrogen by a deuterium in an O—H, N—H, or C—H group shifts the characteristic vibrational frequency of that group by a factor of about 1.35 to lower frequencies. Explain.

Solution

As you would recognize by considering a model with light and heavy balls connected by a spring, vibrational motion consists primarily of motion of

TABLE 3-1
Some Representative Bond Force Constants

molecule	bond	force constant, dyn/cm × 10^5
CH_4	C—H	5.6
NH_3	N—H	6.9
H_2O	O—H	7.8
HF	F—H	8.9
CH_3CH_3	C—H	5.1
	C—C	4.6
CH_2CH_2	C—H	5.6
	C=C	10.1
HCCH	C—H	6.1
	C≡C	16.1
C_6H_6	C—H	6.1
	C⩵C	7.9
H_2CO	C—H	4.8
	C=O	12.9
HCN	C—H	5.9
	C≡N	18.5

the lighter particle. Similarly, O—H, N—H, and C—H vibrations involve mostly hydrogen-atom motion.

The frequency of the vibrations of such groups can be treated to some extent like the vibration of a diatomic molecule indicated by X—H. Then we write

$$\nu = \frac{1}{2\pi}\sqrt{\frac{k}{\mu}}$$

where $\mu = \dfrac{m_H m_X}{m_H + m_X}$

Then, if $m_X \gg m_H$, we have

$$\mu \cong m_H$$

Thus, for the hydrogen- and the deuterium-containing bonds we write

$$\nu_{X-H} \cong \frac{1}{2\pi}\sqrt{\frac{k_{X-H}}{m_H}}$$

$$\nu_{X-D} \cong \frac{1}{2\pi}\sqrt{\frac{k_{X-D}}{m_D}}$$

Since isotopic substitution is not expected to affect the strength of the chemical bond, $k_{X-D} = k_{X-H}$ and

$$\frac{\nu_{X-H}}{\nu_{X-D}} = \sqrt{\frac{m_D}{m_H}} = 1.414$$

The observed ratio is somewhat less than this because the assumption that only the hydrogen or deuterium atom vibrates, or that $\mu_H = m_H$ and $\mu_D = m_D$, is only approximately valid.

3-5 INTERMOLECULAR INTERACTIONS: THE HYDROGEN BOND

The general deduction from infrared spectra of substances in gas, liquid, and solid states is that covalent chemical bonds are little affected by the physical state. A notable exception to this is found in studies of the spectral bands due to bonds involved in the specific interaction known as *hydrogen bonding*. This interaction is expected when any somewhat acidic hydrogen, e.g., those on oxygen or nitrogen atoms, ⟋O—H or ⟍N—H for example, is in the presence of any somewhat basic site, as :Ö— and —N̈—.

The interaction that occurs is generally indicated as

O—H-----O O—H-----O=C O—H-----:N N—H-----O=C

Vibrational spectral evidence for such interactions are illustrated by the examples of Figs. 3-8 and 3-9. The shift in the O—H band position indicates a modification of the vibrational motion and thus of the bonding. On this basis, without further analysis, infrared spectra are of great value in studies of hydrogen bonding since they allow the presence and extent of this bonding to be measured.

Measurements of the extent of hydrogen bonding in a given system as a function of temperature show that hydrogen-bond energies range up to about 10 kcal/mol (see Sec. 8-8). Thus, it takes somewhat more than the $\frac{3}{2}RT$, or 888 cal/mol, reference amount of thermal energy to disrupt hydrogen bonding. On the other hand, hydrogen bonds are much weaker than covalent bonds, which have strengths indicated by bond energies of the order of 100 kcal/mol. The hydrogen bond is thus remarkably well suited for its structure-determining role in the dynamic components of liquid biological systems.

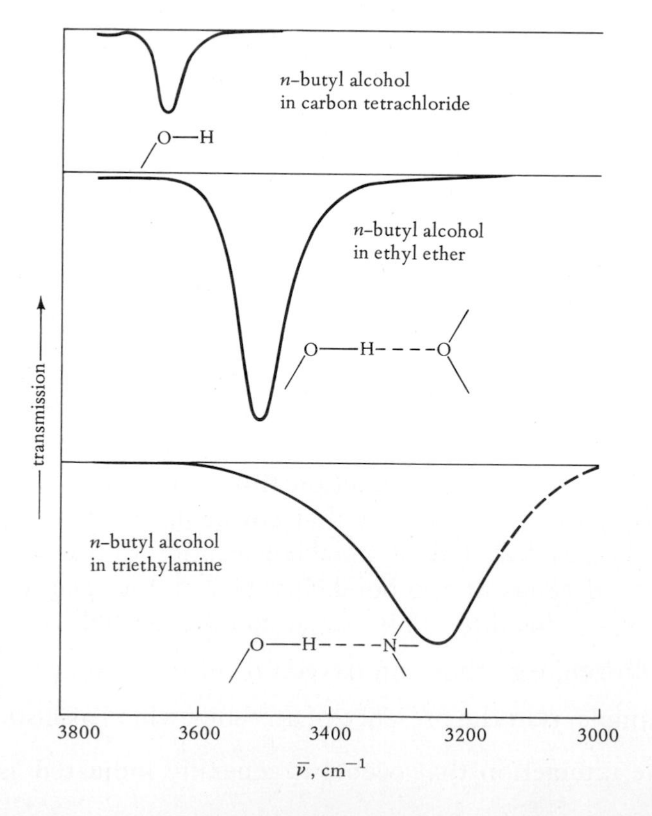

FIGURE 3-8
The shift in the frequency with which the O—H bond absorbs infrared radiation when it is involved in hydrogen bonding. Note that the intensity also increases and the absorption band broadens.

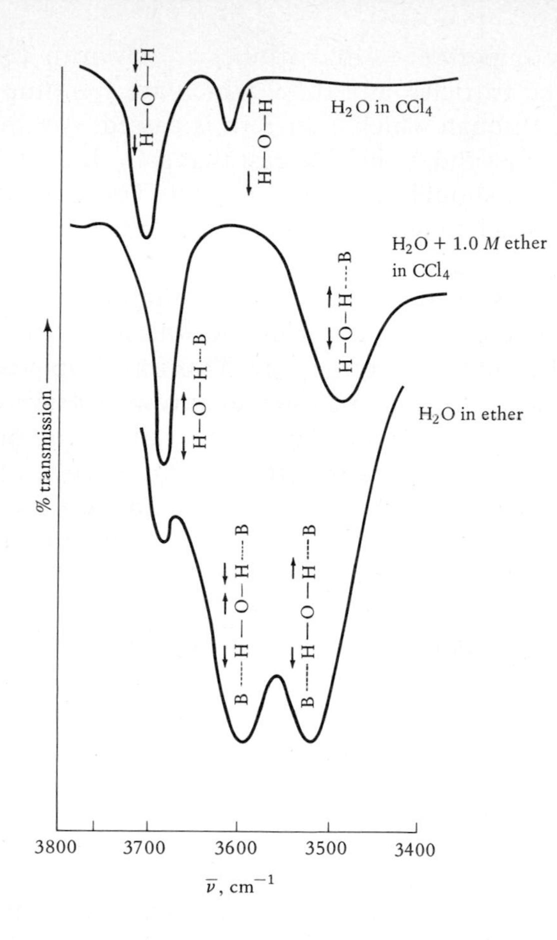

FIGURE 3-9
The effect of hydrogen bonding on the infrared absorption of water in the OH stretching region. The top and bottom spectra are said to show absorptions due to symmetric and antisymmetric vibrations. The middle spectrum would be said to show absorptions due to free and hydrogen-bonded O—H groups.

3-6 NUCLEAR MAGNETIC RESONANCE

A companion method to infrared spectroscopy reveals details of the structure of molecules in solution and is known as *nuclear-magnetic-resonance* (nmr) spectroscopy. Although it is used principally in deducing the primary structure of newly synthesized or isolated substances, it also offers some special insights into other structural features and processes of molecules in solution.

The nuclei of the atoms of molecules can be described as existing in certain rotational or *spin* states. These can be regarded as resulting from a rotation of the nucleus. In contrast to the variety of amounts of angular momentum that a rotating gas-phase molecule can acquire, each type of nucleus, i.e., each isotope, has a fixed, intrinsic nuclear angular momentum, or spin. The angular momentum of a nucleus with nuclear spin quantum number I has the value given by $\sqrt{I(I+1)}h/2\pi$. Examples of the nuclear spins of isotopes that are important in nmr studies are given in Table 3-2.

Nuclei are charged particles, and those which are spinning can be likened to a coil of wire through which a current is passed. Although the analogy is not quantitatively valid, it does suggest that spinning nuclei, like a current in a coil of wire, should act like a magnet. The magnitude of magnetic properties of a nucleus is indicated by the value of its *nuclear magnetic moment* μ.

If a magnetic field is applied to a sample containing nuclei with magnetic moments, to some extent, the nuclei will line up in the field in the direction corresponding to the lowest energy. This lining-up procedure, however, is subject to the angular-momentum quantum restriction (Sec. 1-8) that only certain components of angular momentum in the direction of the field are allowed. This rule means that only states with integer or half-integer spins along the direction of the magnetic field are allowed. The result, as illustrated in Fig. 3-10, is that the two or more states with different energies are drawn out of the original, single-energy set of states. The number of states depends on the nuclear spin. The energy spacing, as illustrated in Fig. 3-10, depends on the magnetic moment of the nucleus and on the value of the applied magnetic field.

The development of energy-spaced nuclear states provides the basis for a spectroscopic study. For example, irradiation of a sample containing protons with radiation of frequency ν and quantum energies $h\nu$ leads to absorption if the equation

$$h\nu = 2\mu_{\mathrm{H}}\mathcal{H} \qquad \mathbf{22}$$

is satisfied. In practice the absorption of energy is measured with transmitter and receiver coils arranged about the sample, as in Fig. 3-11. The principal magnetic field and the oscillator frequency are fixed at values that approximately satisfy Eq. **22** for the type of nucleus being studied. Then a small range of magnetic fields are swept through by the auxiliary sweep generator. At some stage in this field sweep, Eq. **22** will be satisfied, and the sample absorption and remission will produce a coupling between the transmitting

TABLE 3-2
Some Nuclear-Spin Quantum Numbers

isotope	spin	isotope	spin
^{1}H	$\frac{1}{2}$	^{14}N	1
^{2}H	1	^{16}O	0
^{13}C	$\frac{1}{2}$	^{17}O	$\frac{5}{2}$
^{12}C	0	^{31}P	$\frac{1}{2}$

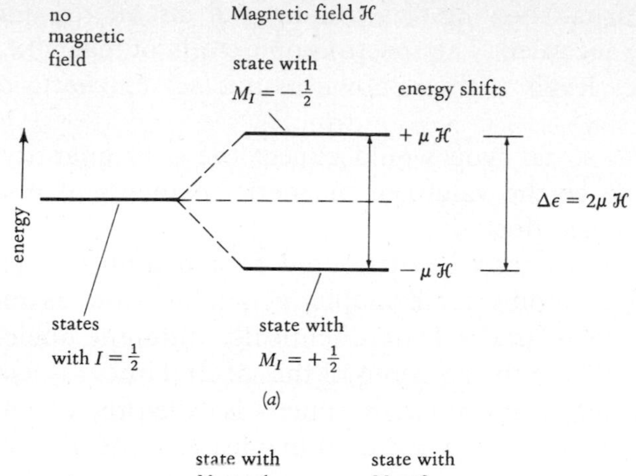

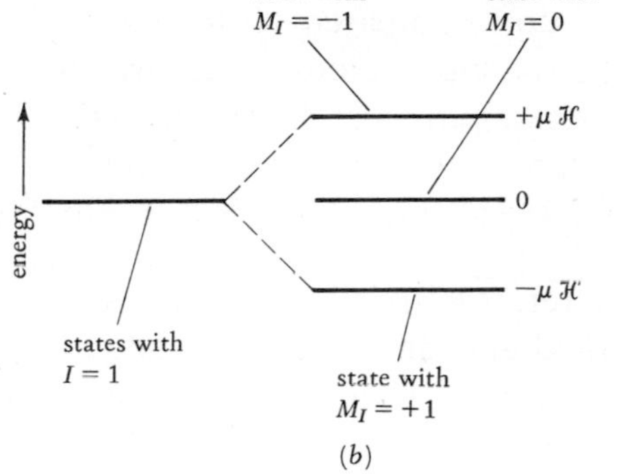

FIGURE 3-10
Effect of a magnetic field on the energies of nuclear states for nuclei with magnetic moments of μ. (*a*) Nuclear spin = $\frac{1}{2}$, as it does for protons. (*b*) Nuclear spin = 1, as it does for the ^{14}N nucleus.

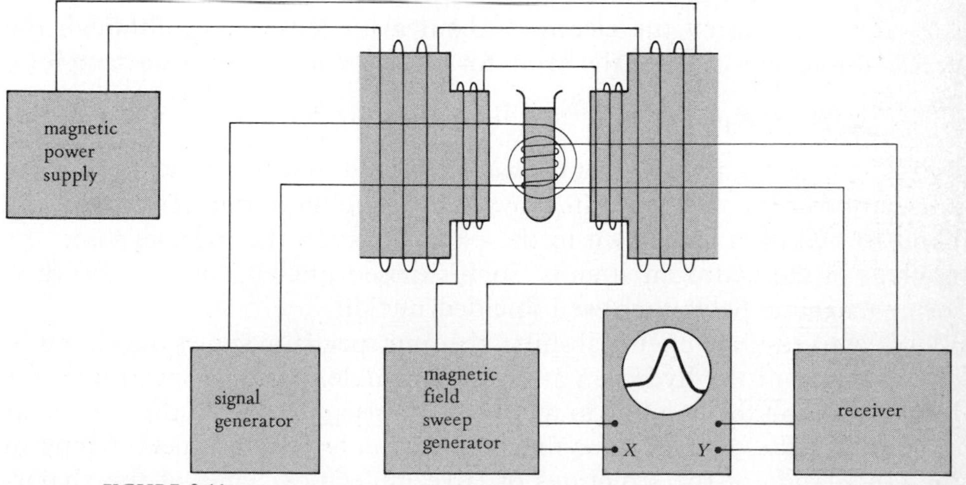

FIGURE 3-11
Principal components of a nmr spectrometer.

and receiving coils. A signal showing this coupling, or circuit resonance, can then be displayed or recorded. The spectroscopic study of magnetically produced nuclear energy levels is thus known as *nuclear magnetic resonance.*

From the analysis so far, you would expect the only quantity revealed by nmr studies to be the values of magnetic moments of nuclei. Two principal factors change this.

First, nmr spectra that can be attributed to a particular type of nucleus, that of hydrogen atoms for example, generally show as many spectral lines or sets of lines as there are chemically different nuclei in the molecule. The separation in the positions of the spectral lines associated with like atoms in different chemical environments is called the *chemical shift.* These shifts can be conveniently reported by means of the difference in magnetic field necessary for absorption compared with that necessary for absorption by the same type of atom in some reference compound. This difference is usually reported as the chemical shift δ, defined as

$$\delta = \frac{\mathcal{H}_{\text{sample}} - \mathcal{H}_{\text{ref}}}{\mathcal{H}_{\text{ref}}} \times 10^6 \qquad \textbf{23}$$

The reference chosen is now usually tetramethylsilane, $(CH_3)_4Si$, because of the conveniently located, well-defined absorption that it produces. Since hydrogen atoms in different samples show absorption, or resonance, at fields that differ by the order of milligauss when the magnetic field is tens of thousands of gauss, the values of δ are made of convenient size by the inclusion of the factor 10^6 in Eq. **23.**

The existence of the chemical shift can be attributed to the screening effect exerted by the electrons about a nucleus. Thus, although the external magnetic field is the same for all hydrogen atoms of a sample of CH_3OH, for example, the electron distribution in the $-\overset{|}{\underset{|}{C}}-H$ and $-O-H$ bonds screens the protons from the applied field to different extents. Some correlations have succeeded in showing that the more the electrons of the bond to hydrogen are drawn to the bonding atom, the more exposed the nucleus of the hydrogen atom is. Such exposed nuclei generally absorb at lower magnetic fields than well-shielded nuclei.

As a result of chemical shifts, the nmr spectrum shows the chemical environment of the hydrogen atoms of a sample. An nmr spectrum gives information on the bonding to hydrogen atoms and often to the molecular structure of the sample. Thus nmr complements infrared spectroscopy in the elucidation of the structures of large molecules. Some of the characteristic chemical shifts used in such analyses are shown in Table 3-3.

TABLE 3-3

Characteristic Values for the Chemical Shift δ for Hydrogen in Organic Compounds with Tetramethylsilane as the Reference Compound†

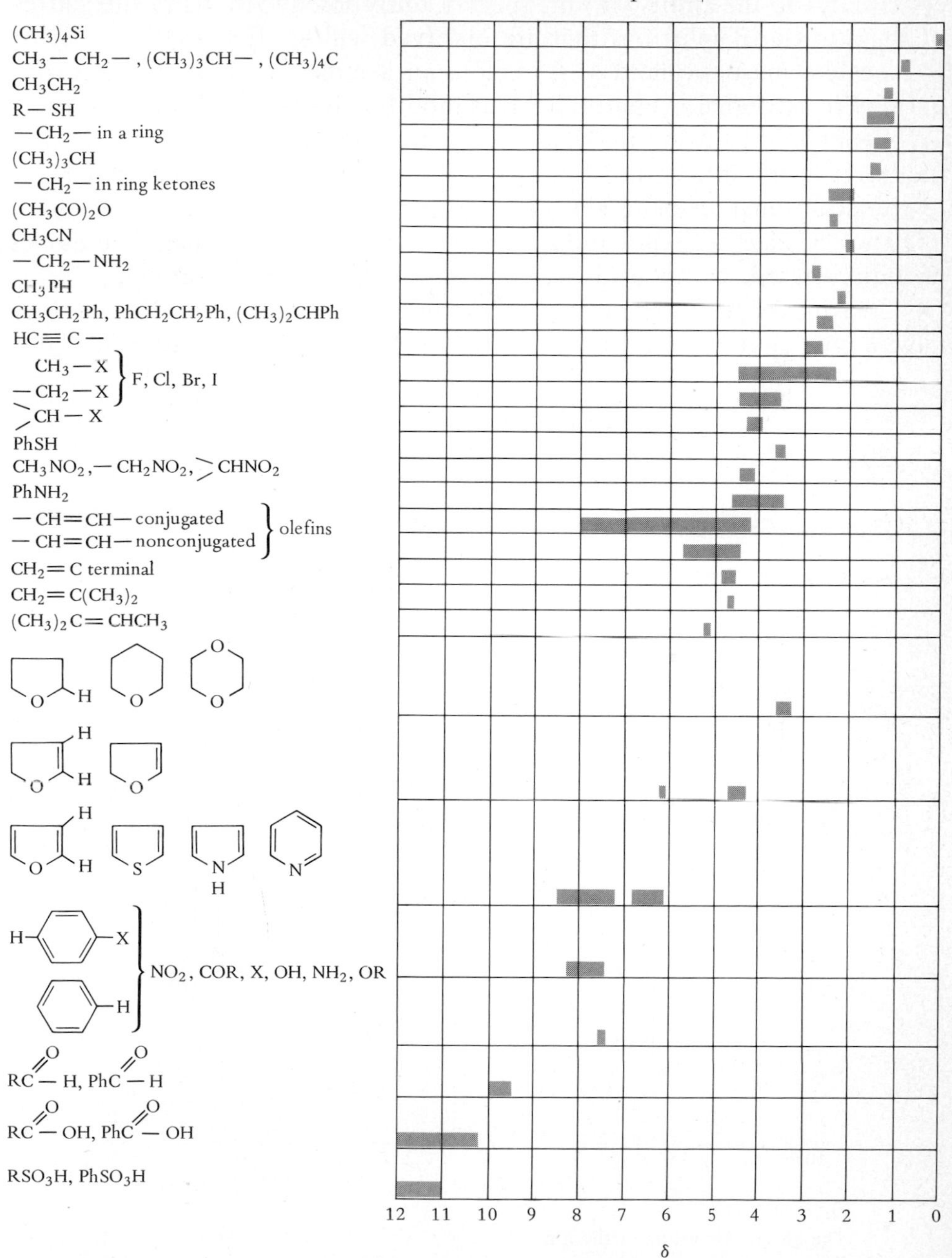

† From E. Mokacsi, *J. Chem. Educ.*, **41**:38 (1964).

The high-resolution detail, as illustrated by the example of acetaldehyde in Fig. 3-12, show the effects of a second factor that operates and contributes to the ability of nmr spectra. Only some features of the source of the additional splittings that are observed will be discussed.

The magnetic field at a nucleus in a molecule is determined not only by the external magnetic field, as modified by the shielding electrons. The other nuclei in the molecule act as magnets, i.e., have magnetic moments. It is the interaction between the nuclei of these atoms that leads to the additional splitting beyond that of the chemical shifts. Since both ^{16}O and ^{12}C have zero spin and zero magnetic moment, the magnetic nuclei of many organic compounds consist only of the hydrogen atoms.

An isotope of carbon, the ^{13}C isotope, does occur naturally and does have a spin. But since it is present naturally only to the extent of about

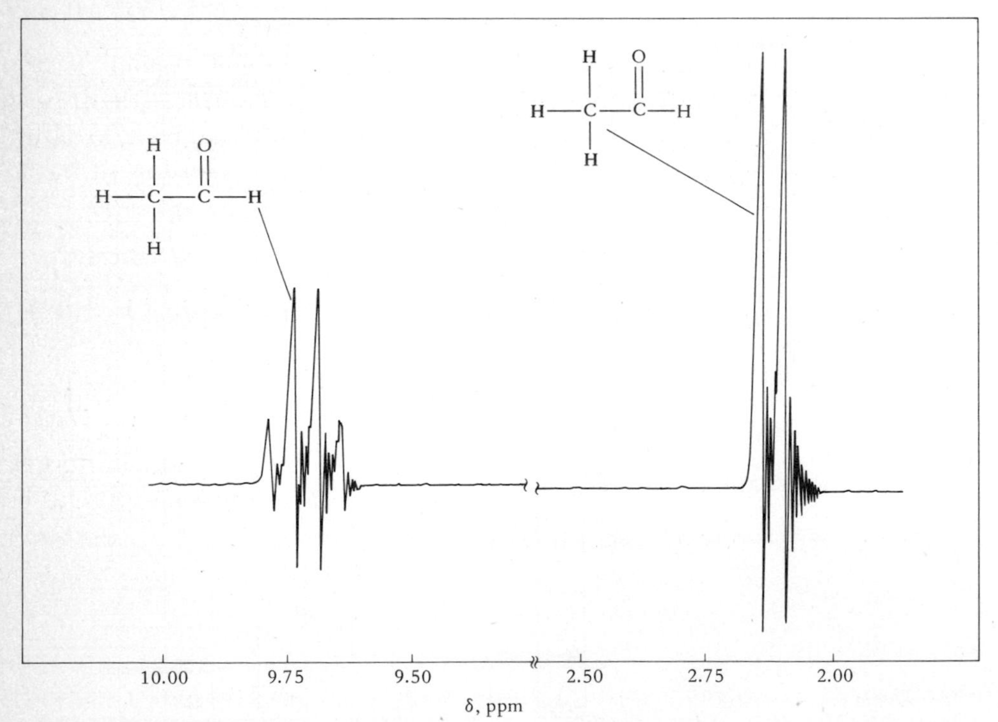

FIGURE 3-12
The observed splitting of the nmr lines due to the aldehyde hydrogen at about $\delta = 9.75$ ppm and to the methyl hydrogens at about $\delta = 2.1$ ppm. (*From F. A. Bovey, "Nuclear Magnetic Resonance Spectroscopy," Academic Press, New York, 1969.*)

1 percent, its effect on proton resonance signals is unnoticed. (Very sensitive magnetic-resonance instruments can detect the ^{13}C resonances themselves. As a result, the environment of the principal atomic species of most organic and biological substances can be investigated directly. An example is shown in Fig. 3-13.)

The nature and effect of the hydrogen atom interactions is illustrated by the spectrum of acetaldehyde, $\mathrm{CH_3{-}\overset{\overset{O}{\|}}{C}{-}H}$. The hydrogen atoms of the methyl group experience a magnetic field that depends on the applied field, on the chemical-shift effect of the shielding electrons, and on the influence of the magnetic field of the nucleus of the hydrogen atom adjacent to the carbonyl group. This nucleus, as Fig. 3-14 indicates, can line

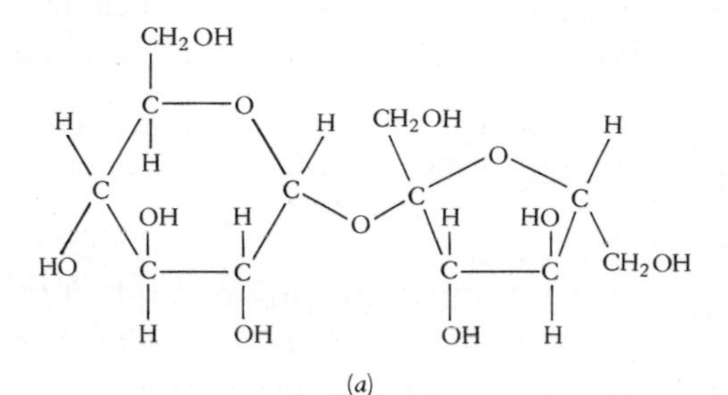

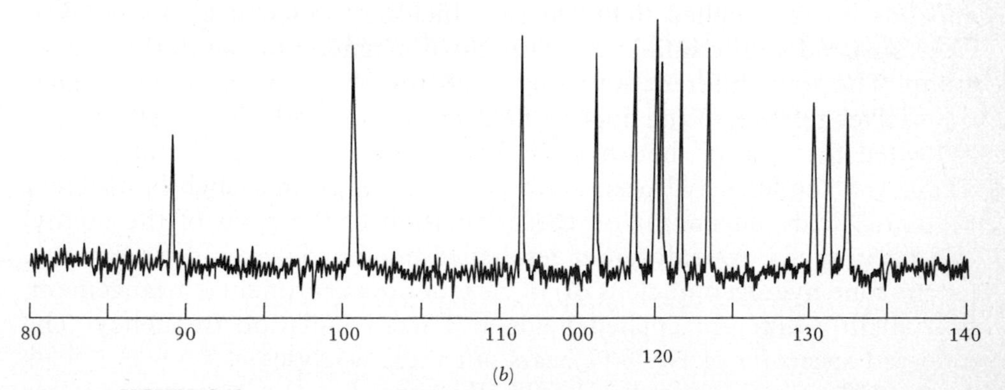

FIGURE 3-13

(*a*) Sucrose structure. (*b*) Natural-abundance ^{13}C Fourier transform nmr spectrum of aqueous 0.04 *M* sucrose with proton decoupling. The spectrum was obtained at 15.2 MHz in a sample tube 20 mm in diameter. The horizontal scale is parts per million upfield from $^{13}CS_2$. The 12 different carbons of sucrose are completely resolved. (*Courtesy of Prof. A. Allerhand, Dept. of Chemistry, Indiana University.*)

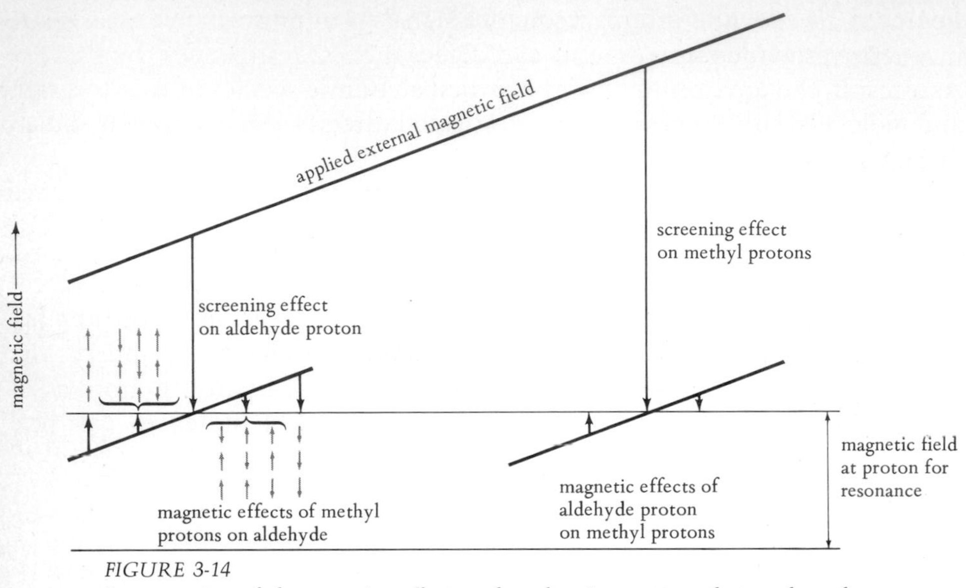

FIGURE 3-14
Interpretation of the screening effects and nuclear interactions that produce the nmr spectrum of acetaldehyde as shown in Fig. 3-12.

up with or against the principal magnetic field. Therefore the methyl hydrogen atoms will experience a slightly greater or lesser magnetic field depending on the orientation of the lone hydrogen atom. The methyl absorption will therefore be split into a doublet.

The single hydrogen atom also experiences a magnetic field that depends on the applied field, on the shielding provided by its bonding electrons, and on the influence of the three magnetic nuclei in the methyl group. The four different ways in which the three magnets can arrange themselves relative to the applied field are shown in Fig. 3-14, where it is indicated that two of the ways are 3 times as probable as the other two. Therefore, the lone hydrogen atom can experience four slightly different magnetic fields, depending on the orientation of the spins of the methyl hydrogen nuclei. Four different resonance frequencies would be expected for the lone hydrogen nucleus or, in view of the experimental arrangement, four slightly different applied fields at a fixed radiation frequency. The observed spectrum of Fig. 3-12 bears out these analyses.

This simple example should illustrate that the magnetic nuclear interactions give information about the type of neighbors of any hydrogen atom or group of atoms in the molecule. Such intimate details can be obtained even for quite large molecules, and it is this feature which makes the fine splittings of nmr spectra of great value in molecular-structure studies.

Our analysis of nmr spectroscopy has proceeded as if molecules were completely fixed, rigid structures. This is permitted because the rapid vibrational motion of the atoms of a molecule and the constant jostling of molecules by one another in the liquid sample do no more than develop average environments for the nuclei that are studied. However, some atomic-position changes do affect the nmr spectrum, and in such cases, the spectral changes can be used to gain information on the rate of the disrupting process.

Nmr spectral lines are extremely narrow and often very close to one another. If a spectrum is shown with a hertz scale, line widths and line spacings can be as little as a fraction of a hertz. (A comparable quantity in infrared spectra is, say 10 cm^{-1}, or a frequency of $10 \times 3 \times 10^{10} = 3 \times 10^{11}$ Hz.) The narrowness of nmr spectral lines implies that the states involved in the transition have sharply defined energies. Anything that broadens the energy level of a state will broaden any spectral line that involves the state.

A guide to the broadening of energy levels, and thus of spectral lines, is provided by one form of the *uncertainty principle*. It says that the well-defined energy of a quantum state is lost as the lifetime of the state becomes short. The energy of the state broadens out, or becomes uncertain, by an amount $\Delta\epsilon$ when the lifetime of the state is reduced to Δt according to the relation

$$\Delta\epsilon \, \Delta t = \frac{h}{2\pi} \qquad \textbf{24}$$

Insertion of numerical values would show that a noticeable broadening effect might be expected to show up in nmr spectra when lifetimes are reduced to something like 1 sec. (By contrast, corresponding infrared broadening from lifetime effects do not appear until lifetimes as short as 10^{-12} sec are reached.)

The lifetime we deal with in nmr studies is the length of time, on the average, that a nucleus remains on a site with a given average environment. Two processes that affect such lifetimes are the internal rotation, or other motion, that moves an atom from one isomer position to another and the exchange of an atom between two sites. In such processes we deal with two sites and the broadening effects that occur because of the short lifetime of both. If the average lifetime on each site is of the order of 1 sec, the nmr lines characteristic of the two sites will show some broadening. As the lifetime on the two sites becomes extremely short, the two separate nmr signals will merge and what is observed will be due to the average environment of the two sites. For fast enough interchange a single sharp nmr line will be observed.

An example of the spectral changes that accompany such stages is shown in Fig. 3-15 by the temperature dependence of the nmr spectrum of *N,N*-dimethylformamide

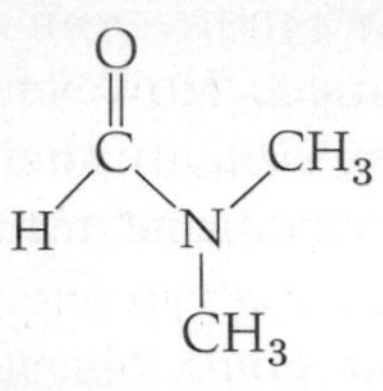

At low temperatures the methyl groups are distinguishable, as Fig. 3-15 shows, and this we attribute to a fixed planar geometry for the molecule. The resonance-structure rationale for this planarity is provided by the bonding diagrams

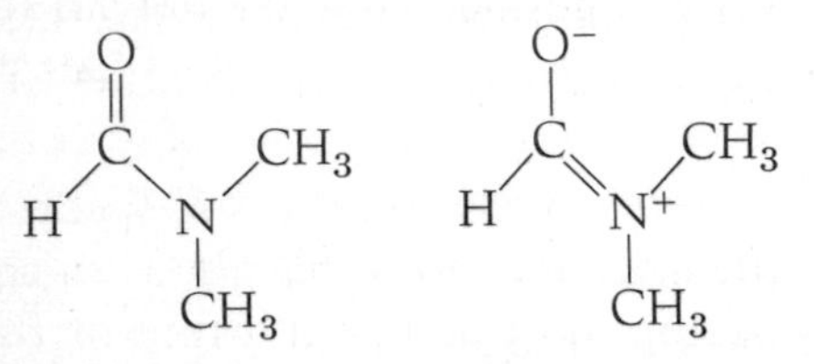

At higher temperatures the barrier to rotation about the C—N bond is apparently overcome, and, as Fig. 3-15 shows, a single —CH_3 resonance results.

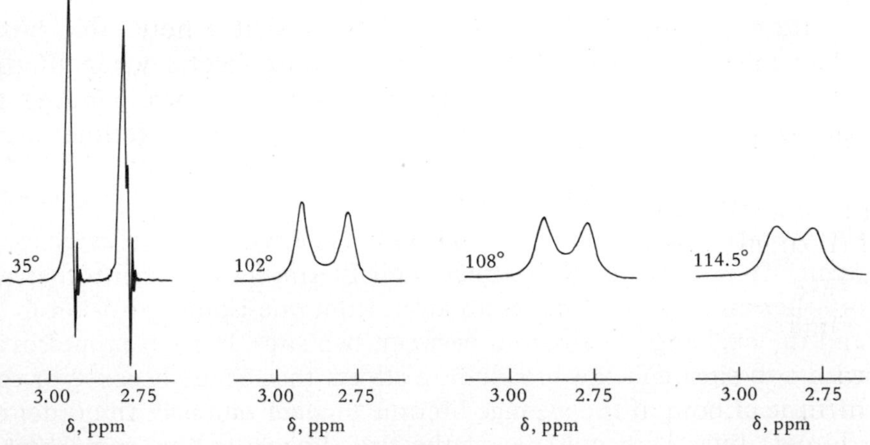

FIGURE 3-15

The temperature dependence, temperatures in °C, of the 60-MHz nmr signal due to the CH_3 groups of *N,N*-dimethylformamide. At 60 MHz a δ interval of 0.25 ppm corresponds to 15 Hz. The separation of the doublet at lower temperatures is therefore about 9 Hz. (*Spectrum from F. A. Bovey, "Nuclear Magnetic Resonance Spectroscopy" Academic Press, New York, 1969.*)

The relation of line shapes to state lifetimes can be worked out so that all stages, like those of Fig. 3-15, can be used to deduce lifetime values. A convenient simplification of such results allows the lifetime of the two states to be deduced for that stage at which the two separate lines have just merged. Suppose that the two states a and b connected by the exchange process produce nmr lines at frequencies ν_a amd ν_b when they are not appreciably disturbed by the exchange process. To broaden and merge these lines so that a single broad line is formed, the lifetime τ of the states must be short enough to satisfy the relation

$$(\nu_a - \nu_b)\tau = \frac{\sqrt{2}}{\pi} \qquad \mathbf{25}$$

If both sides are multiplied by Planck's constant h, and if $\Delta\epsilon$ is written for $h\,\Delta\nu = h(\nu_a - \nu_b)$ and Δt for τ, this equation becomes

$$\Delta\epsilon\,\Delta t = \frac{\sqrt{2}h}{\pi}$$

This shows its relationship to the general uncertainty relation, Eq. **24,** for the broadening of an energy level.

Application of Eq. **25** to the example of Fig. 3-15 shows that at 118°C the lifetime of the proton on each of the sites is

$$\tau = \frac{\sqrt{2}}{\pi(9)} = 0.04 \text{ sec}$$

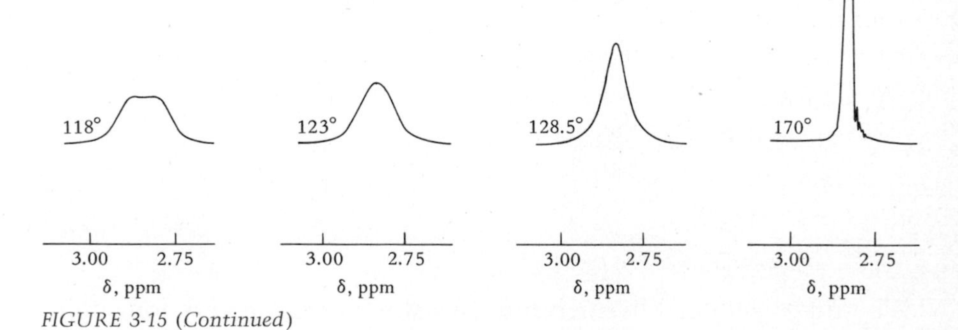

FIGURE 3-15 (Continued)

At lower temperatures the lifetime of the methyl groups in the two positions is longer than this amount.

Another illustration is provided by the —OH protons of ethyl alcohol and the protons of water in the spectra of Fig. 3-16. If only a little water is present in the alcohol, the exchange between the R—OH and the H_2O protons is slow enough for the two proton sites to be distinguished, as in Fig. 3-16*b*. But the addition of a small amount of acid accelerates the exchange, and a single line corresponding to the average of the H_3O^+, H_2O, and ROH environments results.

This exchange also affects the spectral lines of the adjacent $—CH_2$ group. In the absence of rapid R—OH proton exchange, the $—CH_2$ group would be split into a quartet by the action of the $—CH_3$ protons. Since, in addition, it is doubled by the action of the —OH protons, the result is an octet. Full resolution is not obtained in Fig. 3-16*a*, but structure beyond a quartet is apparent. In contrast, a rapidly exchanging —OH proton will produce no distinguishable spin-up and spin-down effects on the $—CH_2—$ protons. Thus, in the presence of rapid —OH exchange, the $—CH_2—$ resonance is the well-defined quartet that results from the uncomplicated $—CH_3$ influence.

A similar distinction between slow and fast exchanges is provided by the comparison of the spectra of water solutions of $CH_3NH_3^+$ with and without the addition of acid. Spectra are shown in Fig. 3-17. The $—CH_3$ resonance most clearly reflects the lifetime of the attachment of the protons on the nitrogen. Only in the acidified form $CH_3—NH_3^+$ are they present long enough to split the $—CH_3$ resonance into a quartet. In the absence of acid no $—CH_3$ splitting is observed, and we must conclude that there is a rapid coming and going of protons on the methylamine nitrogen, of the small amount of CH_3NH_2 that is present. This is suggested by

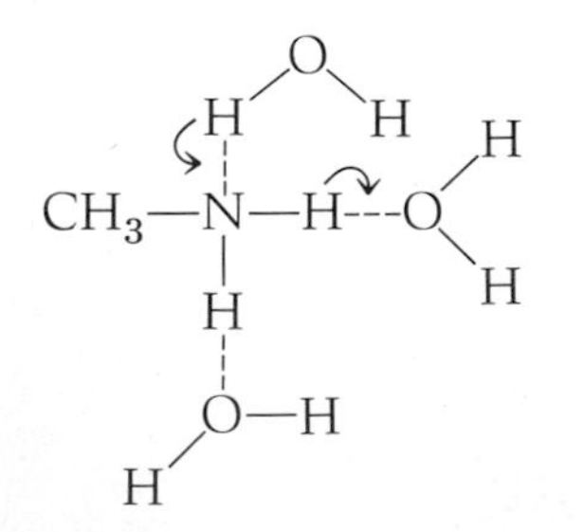

The times deduced by such nmr measurements are very long compared with such rapid molecular events as molecular vibrations or molecular collisions in liquids. Species with a long lifetime in relation to such phenomena as molecular vibrations, molecular rotations, and their liquid-phase

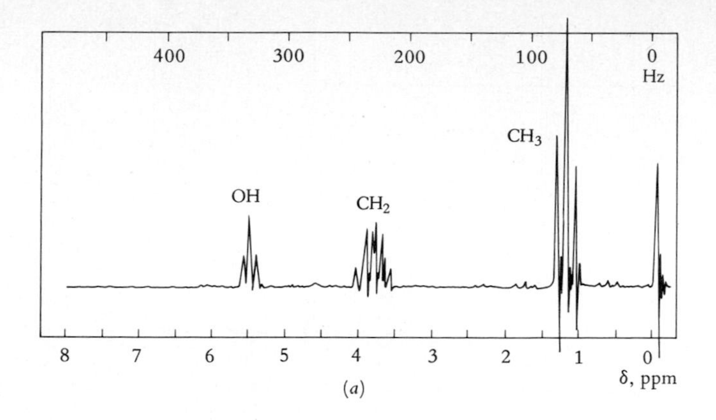

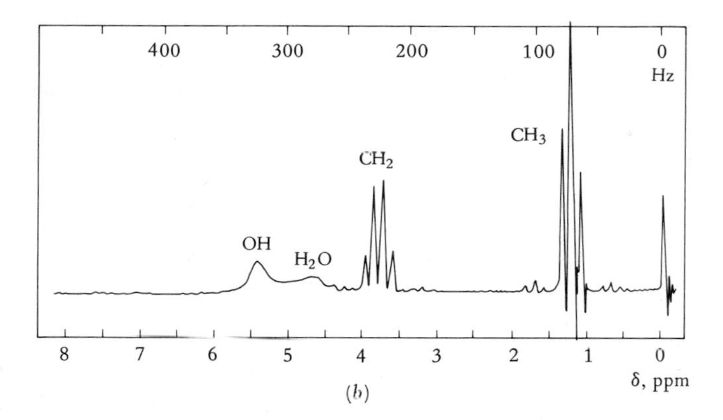

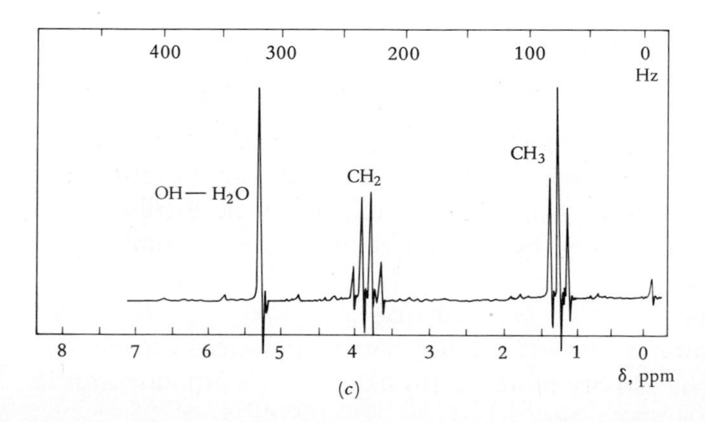

FIGURE 3-16
The 60-MHz nmr spectra of (*a*) 100% ethanol; (*b*) ethanol with one drop of water added; and (*c*) ethanol with one drop of dilute hydrochloric acid. [*From G. Glaros and Norman H. Cromwell, J. Chem. Educ.,* **48:**202 (1971) *by permission.*]

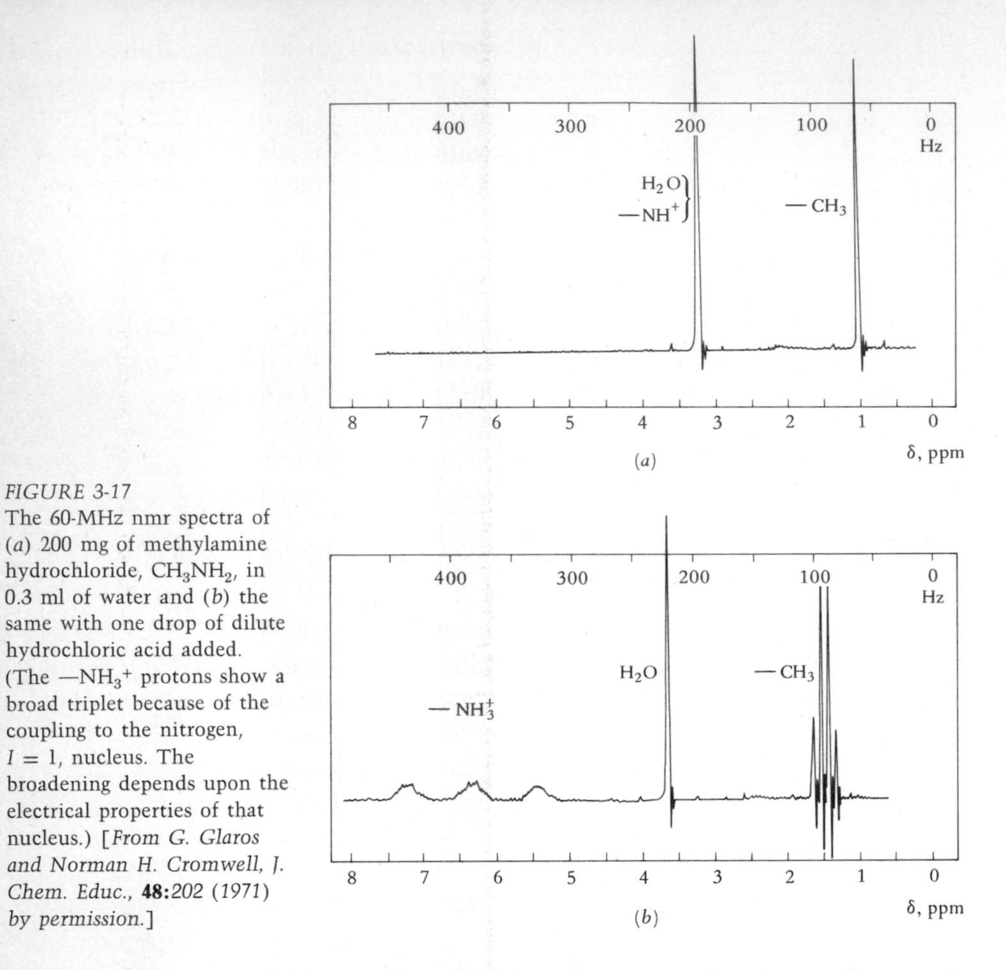

FIGURE 3-17
The 60-MHz nmr spectra of (*a*) 200 mg of methylamine hydrochloride, CH_3NH_2, in 0.3 ml of water and (*b*) the same with one drop of dilute hydrochloric acid added. (The $—NH_3^+$ protons show a broad triplet because of the coupling to the nitrogen, $I = 1$, nucleus. The broadening depends upon the electrical properties of that nucleus.) [*From G. Glaros and Norman H. Cromwell, J. Chem. Educ.,* **48:***202 (1971) by permission.*]

counterpart can still be too short-lived to have an identity well enough established to qualify as an nmr species.

3-7 EFFECT OF THE DIELECTRIC OF THE SOLVENT ON THE STRUCTURE OF SOLUTE MOLECULES

The most dramatic and general structural effect of solvents, water in particular, is the formation of relatively independently moving ions when acids, bases, or salts are added. Much of the effect can be attributed to the *dielectric* of the solvent, a property that reduces the coulomb interaction between charges.

Even when ions cannot be released from a crystal or molecule by this effect, the weakening of the interaction between charges can change the shape of the molecule. Many proteins, for example, incorporate amino acids that have acidic or basic side chains. The basic side chains take on positive charges in sufficiently acidic solutions and the acidic side chains ionize to leave negative charges in sufficiently basic solutions. Then the same factors that act to free or to associate the ions of simple inorganic substances also act to open up or condense the structure of a protein molecule carrying charged sites. Thus the overall structure of a protein in solution can be expected to depend on the pH of the solution. Similar effects must be expected with nucleic acids.

Of particular importance is the *isoelectric point,* the pH at which the protein carries no net charge. At that point the solute molecules carry equal numbers of positive and negative charges. Thus at both higher and lower pH values, greater chain-opening repulsive forces, due to the greater number of like charges, will set in.

The basic law that expresses the interaction of charged particles, known as *Coulomb's law,* states that the force of interaction between particles in a vacuum (or approximately in air) with charges q_1 and q_2 separated by a distance r is given by the proportionality

$$f \propto \frac{q_1 q_2}{r^2} \qquad \textbf{26}$$

If electrostatic units (esu) are used for the charges and centimeters for the distance, the equality corresponding to Eq. **26** has a proportionality constant of unity and is

$$f = \frac{q_1 q_2}{r^2} \qquad \textbf{27}$$

where f has the units of dynes.

Often it is convenient to deal with the potential energy associated with electrostatic interactions rather than with the corresponding force. The relation of potential energy to force can again be deduced from the elementary ideas that work is force times distance and that potential energy is stored work. Thus, for a small increase dr in the distance between particles between which a force f acts, the potential-energy change dU is given by

$$dU = -f\,dr \qquad \textbf{28}$$

The negative sign leads to an increase in U when particles that attract each other, and thus have a force that carries a negative value, are further separated.

Insertion of the inverse-square law of force into Eq. **28,** and integration, lead to the potential-energy expression

$$U = \frac{q_1 q_2}{r} \tag{29}$$

For charges in electrostatic units and distance in centimeters, the energy has units of ergs.

These expressions apply to the force and potential energy for interactions between charges in vacuum or (approximately) in air. When the charges are in a medium other than a vacuum, the force of interaction is reduced. This effect can be added by introducing the *dielectric constant* ϵ, and writing

$$f = \frac{q_1 q_2}{\epsilon r^2} \tag{30}$$

and thus

$$U = \frac{q_1 q_2}{\epsilon r} \tag{31}$$

For water the bulk dielectric constant has a value of about 80. Thus charges in water interact with a force that is less than that between charges similarly placed in air by this factor. As a result, attraction between opposite charges and repulsion between like charges is sometimes not an overwhelming factor in an aqueous environment.

As an illustration, consider the energy that would be required to separate ions carrying unit positive and negative charges from a distance of 5 Å to complete separation. If the ions are in air, the potential energy at 5 Å is calculated, with the electronic-charge magnitude of 4.80×10^{-10} esu, as

$$\begin{aligned} U_{5\,\text{Å}} &= \frac{(+4.80 \times 10^{-10})(-4.80 \times 10^{-10})}{5 \times 10^{-8}} \\ &= -460 \times 10^{-14} \text{ erg} \end{aligned} \tag{32}$$

At infinite separation, the reference state for the $U = q_1 q_2 / r$ expression, the potential energy is zero. Thus the separation process entails a potential-energy increase of 460×10^{-14} erg.

By contrast, if the ions are immersed in water, we calculate

$$\begin{aligned} U_{5\,\text{Å}} &= \frac{(+4.80 \times 10^{-10})(-4.80 \times 10^{-10})}{80(5 \times 10^{-8})} \\ &= -5.8 \times 10^{-14} \text{ erg} \end{aligned} \tag{33}$$

and, again with $U_\infty = 0$, the increase in potential energy for the separation is $+5.8 \times 10^{-14}$ erg.

These results can be compared with the reference thermal-energy amount $\frac{1}{2}kT$ of about 2×10^{-14} erg at 25°C. Comparison with the energy values for the separation of opposite charges obtained above suggests the relative ease with which such charges can be separated in aqueous solution.

You can, from this example, see that a molecule carrying *like* charges at various positions might still adopt a compact structure in a water environment but it would be driven to an extended structure with the charges well separated in a solvent of low dielectric.

3-8 SHAPES OF MACROMOLECULES IN SOLUTION: VISCOSITY MEASUREMENTS

The two preceding sections have dealt with factors that influence the shape of a macromolecule in solution—to the extent that the shape is not fixed by the relatively unperturbed primary chemical bonds. Now we turn to an experimental technique that gives some information about the overall shape a macromolecule adopts. This technique, the measurement of the viscosity of the solution, can be carried out quickly, easily, and with inexpensive equipment. In some situations, moreover, the results can be used to estimate the molecular weight of the dissolved material. As a result, it is a commonly used technique in studies of both natural and synthetic macromolecular materials.

Viscosity studies are included in this chapter because they provide an approach to the structures of macromolecules in solution. Customarily, however, these studies are included with the transport processes taken up in Chap. 12. Without creating any difficulties in the intervening material, study of this section can be postponed until Chap. 12 is reached.

The viscosity of a fluid, i.e., a gas, pure liquid, or solution, is an index of its resistance to flow. It is most simply defined, if not easily measured, with an arrangement like that of Fig. 3-18. The force f required to move a segment of the fluid of area A with a velocity v compared to fluid a distance y away for a given fluid is expressed by the proportionality

$$f \propto \frac{Av}{y} \qquad \mathbf{34}$$

Introduction of a proportionality constant η, known as the *coefficient of viscosity*, permits us to write this as the equality

$$f = \eta \frac{Av}{y} \qquad \mathbf{35}$$

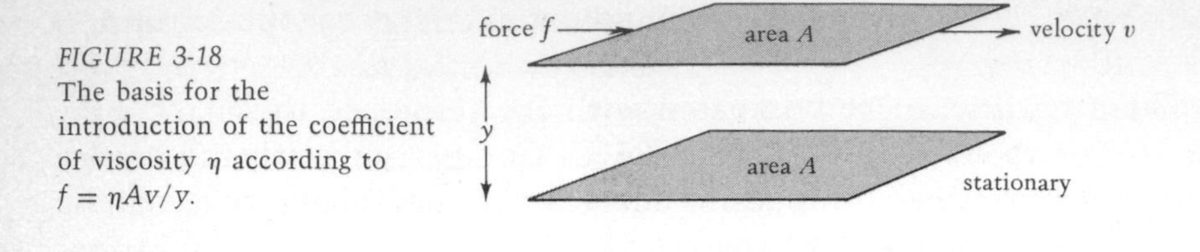

FIGURE 3-18
The basis for the introduction of the coefficient of viscosity η according to $f = \eta Av/y$.

The units of η can be seen from Eq. **35** to be dyne seconds per square centimeter. Or with the replacement of dynes by gram centimeters per second per second, as suggested by $f = ma$, the units of η become grams per centimeter second. These units are given the name *poise* (P). Furthermore, many liquids have viscosities of the order of 0.01 P, and as a result, viscosities are often reported in centipoise (cP) units; 1 cP = 0.01 P. At 20°C, the viscosity of water is 1.005 cP. At 25°C it has fallen to 0.894 cP.

The viscosities of liquids are usually determined by measuring the rate of flow through a tube with circular cross section. A common device for this is shown in Fig. 3-19. The rate of flow through a tube of such a device depends on the viscosity of the liquid because the film of liquid adjacent to the glass surface is stationary. Flow of the liquid through the tube thus requires movement relative to this stationary layer.

The velocity profile that results when a pressure P is applied to make a fluid flow through a tube of length l and radius R, as in Fig. 3-20, can be deduced by applying Eq. **35.** If the expression $f = \eta\, Av/y$ is turned around to $v = fy/\eta A$, the form of the cylindrical-tube result

$$v = \frac{P}{4\eta l}(R^2 - r^2) \qquad \mathbf{36}$$

can be appreciated. For such flow v is seen to be greatest in the center of the tube, when $r = 0$, and to fall off to a zero value at $r = R$. The velocity profile this gives is shown in Fig. 3-21*a*. (At very high flow rates a different type of flow, known as *turbulent flow,* sets in. Such flow, which has a different velocity profile, is not reached in viscosity-measuring devices.)

The net flow rate, which can be obtained by using Eq. **36** and integrating over the entire tube cross section, is given by

$$\text{Rate of flow} = \frac{\pi P R^4}{8l\eta} \qquad \mathbf{37}$$

Usually viscosities are determined by comparing the time it takes for a given volume of solution to pass through an apparatus like that of Fig. 3-19 with the time for a reference solution to pass through. The comparison is made as

$$\frac{t_1}{t_2} = \frac{\eta_1/\rho_1}{\eta_2/\rho_2} \quad \text{or} \quad \frac{\eta_1}{\eta_2} = \frac{\rho_1 t_1}{\rho_2 t_2} \qquad \mathbf{38}$$

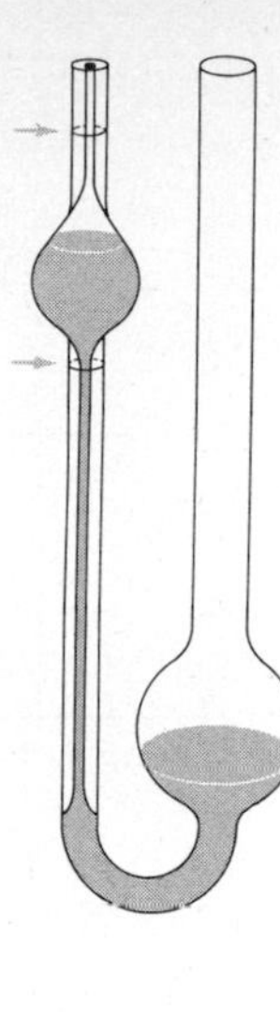

FIGURE 3-19
An Ostwald viscometer; the time for the liquid level to move from the upper mark to the lower mark is measured.

where t_1 and t_2 are the two flow times and ρ_1 and ρ_2 are the two densities. Densities enter because the pressures driving the solutions are proportional to them.

When the viscosities of solutions are studied, so that the properties of the solute molecules can be deduced, viscosities are obtained for the

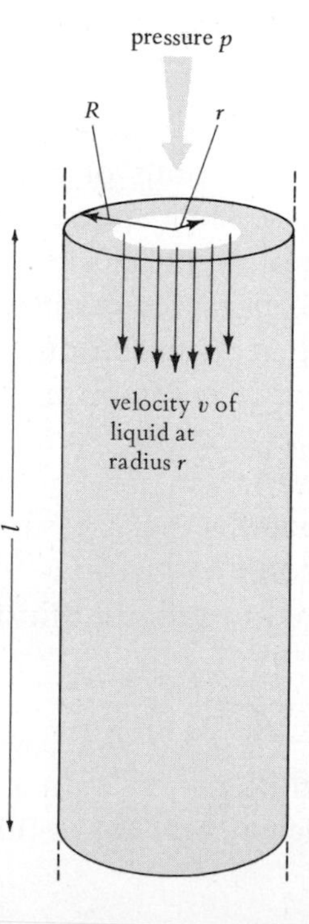

FIGURE 3-20
The velocity of liquid at a radius r in a tube of radius R.

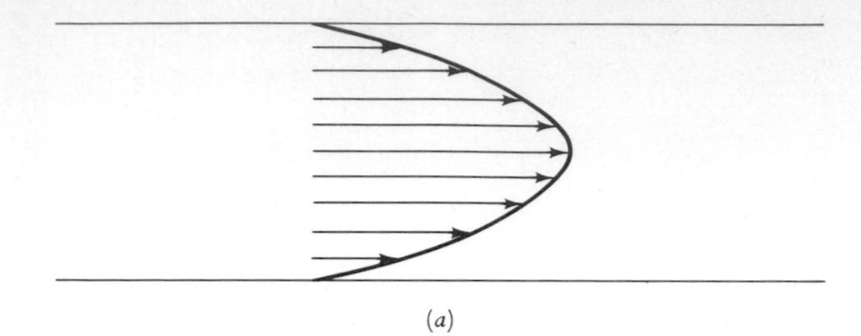

(*a*)

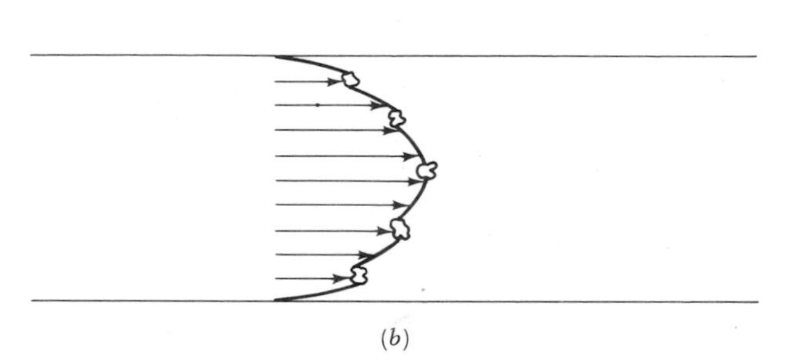

(*b*)

FIGURE 3-21
(*a*) The velocity profile for flow in a cylindrical tube according to Eq. **36.** (*b*) The disruption caused by macromolecules. If the molecules are assumed to be moving without rotation, the velocity is the same on both sides of the molecule.

pure solvent and for solutions of various concentrations. At each concentration the effect of the solute can be conveniently shown by calculating

$$\frac{\eta - \eta_0}{\eta_0} \quad \text{or} \quad \frac{\eta}{\eta_0} - 1 \qquad \mathbf{39}$$

where η and η_0 are the viscosity of the solution and the solvent, respectively.

The properties of solutions of macromolecules are often affected by the interaction of these large solute particles. Results that reflect the properties of the individual particles rather than their interactions are obtained by extrapolating measured quantities to infinite dilution. Here this is done by introducing the concentration c as *mass per unit volume* rather than as the chemically more common molar concentration. The viscosity effect per unit concentration of the solute can now be expressed as

$$\frac{\eta/\eta_0 - 1}{c} \qquad \mathbf{40}$$

and the value extrapolated to infinite dilution is represented by $[\eta]$, where

$$[\eta] = \lim_{c \to 0} \frac{\eta/\eta_0 - 1}{c} \qquad \mathbf{41}$$

If c is in grams per milliliter, the values of $[\eta]$ are known as *intrinsic viscosities*. Notice, however, that $[\eta]$ is not really a viscosity. It depends on the ratio η/η_0 and thus does not contain viscosity units. The intrinsic viscosity of a solution of a given macromolecule is used to deduce the shape and sometimes the molecular mass of the molecule.

The addition of a macromolecular solute to a solvent invariably increases the viscosity. This qualitative result seems reasonable in view of the familiar thickening that accompanies the formation of such solutions. This thickening, however, is usually the result of molecular interactions and is not related to the intrinsic-viscosity effect.

The molecular explanation of the increase in viscosity that results from the addition of macromolecules is based on the disruption of the flow pattern or velocity gradient they produce. As Fig. 3-21*b* attempts to suggest, the fluid in contact with a macromolecule moves with the same velocity as that molecule. The smooth velocity contour of Fig. 3-21*a* is destroyed, and less efficient flow lines are followed.

When this approach is developed quantitatively, the fraction of the volume of the solution attributable to the solute is most directly related to the viscosity effect. If v_{eff} is introduced as the effective specific volume of the macromolecular material in solution, i.e., the volume per unit mass, then since c is the mass of solute per unit volume of solution, cv_{eff} is the effective volume of solute per unit volume of solution. Thus if c is the concentration in grams per milliliter and v_{eff} is in units of milliliters per gram, cv_{eff} is the volume of solute in milliliters per milliliter of solution.

According to derivations by Einstein and by Simha, the extrapolation of viscosities of solutions of molecules that have spherical or ellipsoidal shapes should conform to

$$\lim_{c\to 0} \frac{\eta/\eta_0 - 1}{cv_{\text{eff}}} = \nu \tag{42}$$

where ν is a shape-dependent parameter shown in Fig. 3-22. If v_{eff} is removed from the extrapolation term and treated separately, we can write

$$\lim_{c\to 0} \frac{\eta/\eta_0 - 1}{c} = \nu v_{\text{eff}}$$

or

$$[\eta] = \nu v_{\text{eff}} \tag{43}$$

Thus, the intrinsic viscosity of a macromolecular substance depends on the shape of its molecules and on its specific volume. (Additional factors must be taken into account if viscosities deduced from high flow-rate studies are considered. In such cases the flow is said to be nonnewtonian, implying that the proportionality between driving force and flow rate does not hold. The molecular interpretation of viscosities obtained under such

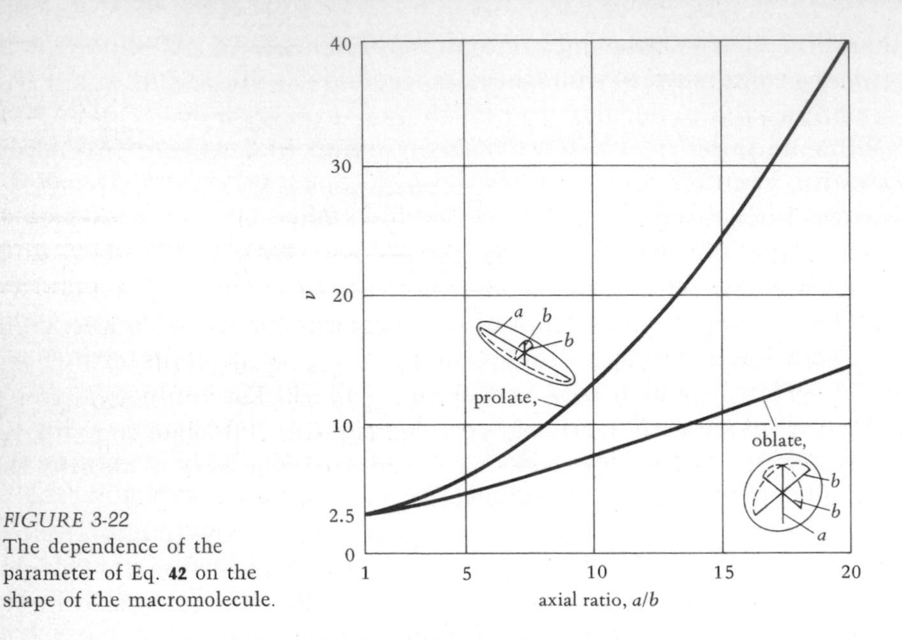

FIGURE 3-22
The dependence of the parameter of Eq. **42** on the shape of the macromolecule.

conditions must allow for an alignment of molecules with the flow lines. Here such effects will not be considered.)

The effective specific-volume term v_{eff} raises some difficulties. The specific volume of the dry solute, which we denote by v_2, is nothing more than the reciprocal of the density of the dry solute and can be measured. But in solution many types of macromolecules hold solvent molecules in their cavities and on their surface. If we introduce δ to represent the mass of solvent that is bound in this way per unit mass of solute and v_1 to represent the specific volume of the solvent, we can write

$$v_{eff} = v_2 + \delta v_1 \qquad \textbf{44}$$

Then Eq. **43** can be written as

$$[\eta] = \nu(v_2 + \delta v_1) \qquad \textbf{45}$$

Application of this result is awkward because of the two unknowns ν and δ. A variety of procedures can be followed.

For example, we can consider a particular class of compounds, say proteins. Then we can proceed by using representative values of $v_2 = 0.75$ ml/g, $\delta = 0.2$, and $v_1 = 1$ ml/g. Then we have

$$[\eta] = \nu(0.95) \cong \nu$$

The values for the intrinsic viscosities of the globular proteins in Table 3-4 conform to this expectation in showing values that are independent of their molecular masses and, moreover, correspond to ν values near the left, the spherical-shape side, of Fig. 3-22. If we press on to the deduction of molecular shape, we see that values of $[\eta]$, and thus values of ν, of about 3.5 lead, according to Fig. 3-22, to ellipsoidal shapes with axial ratios of about 3:1.

Some refinements to this deduction are possible. These sustain the general conclusion reached here that the species listed in Table 3-4 as "globular" do, in fact, have a balled-up more or less spherical shape when dissolved in water.

For macromolecules that are to be pictured as extended chains both ν, according to Fig. 3-22, and the effective specific volume will be expected to increase with the size of the molecule. Higher intrinsic viscosities are indeed found for the species of Table 3-4, for which either a random coil or a more rigid extended rod shape is expected.

For many synthetic polymers the dependence of the intrinsic viscosity on molecular mass is represented for a given molecular type in a given solvent by an empirical equation of the form

$$[\eta] = KM^a \tag{46}$$

where M is the molecular mass and K and a are empirical constants. Once these parameters have been determined by measurements on polymer

TABLE 3-4
Intrinsic Viscosities of Biologically Important Macromolecules

shape	substance	molecular weight	$[\eta]$, ml/g
Globular	Ribonuclease	13,680	3.4
	Serum albumin	67,500	3.7
	Ribosomes (*E. Coli*)	900,000	8.1
	Bushy stunt virus	10,700,000	3.4
Random coils	Insulin	2,970	6.1
	Ribonuclease	13,680	16.0
	Serum albumin	68,000	52
	Myosin subunit	197,000	93
Rods	Fibrinogen	330,000	27
	Myosin	440,000	217
	Poly-α-benzyl-L-glutamate	340,000	720

fractions with known molecular masses, the equation can be used to obtain such mass data from the easily obtained intrinsic-viscosity results.

A similar application to proteins and polypeptides can be made by using a solvent that opens up the molecules so that they form *random coils.* The term can be taken to mean the result when a flexible long molecule adopts an overall configuration as a matter of chance rather than as the result of forces producing a tightly balled-up or rigidly extended species. Suitable solvents to produce such random coils can be made by adding high concentrations of polar species, e.g., detergents, urea, or guanidine hydrochloride, to a water solution of the protein or polypeptide. If the parent protein contains disulfide cross links, they can be ruptured by the addition of a small amount of a reducing agent such as β-mercaptoethanol.

For a variety of polypeptides and proteins in such solvents, Tanford has shown that an empirical correlation between intrinsic viscosity, as listed in Table 3-5, and number of amino acid residues can be written. A suitable equation is

$$[\eta] = 0.732n^{0.656} \tag{47}$$

where n is the number of amino acid residues. This expression, apparently

TABLE 3-5
Intrinsic Viscosities for Proteins in 6 *M* Guanidine Hydrochloride and 0.1 *M* β-Mercaptoethanol†

	molecular mass	**residues per chain**	$[\eta]$, **ml/g**
Insulin	2,970	26	6.1
Ribonuclease	13,680	124	16.6
Hemoglobin	15,500	144	18.9
Myoglobin	17,200	153	20.9
β-Lactoglobulin	18,400	162	22.8
Chymotrypsinogen	25,700	245	26.8
Glyceraldehyde 3-phosphate dehydrogenase	36,300	331	34.5
Pepsinogen	40,000	365	31.5
Aldolase	40,000	365	35.3
Serum albumin	69,000	627	52.2
Thyroglobulin	165,000	1500	82
Myosin	197,000	1790	92.6

†From C. Tanford, K. Kawahara, and S. Lapanje, *J. Am. Chem. Soc.*, **89:**729 (1967).

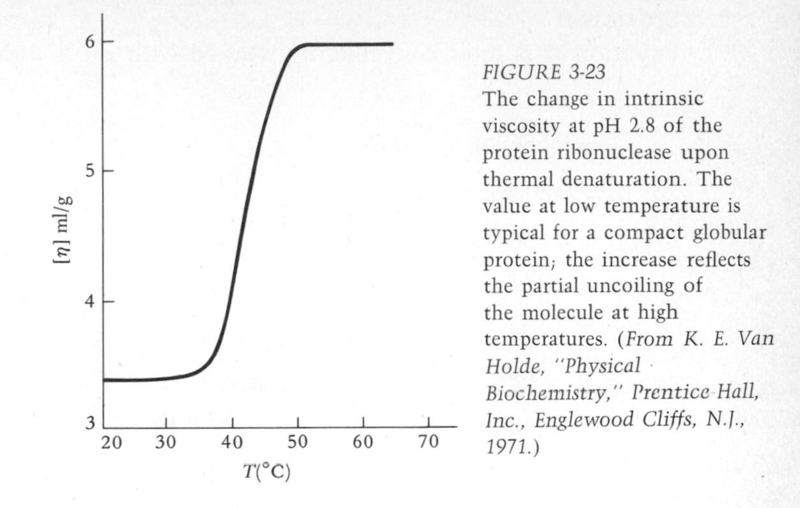

FIGURE 3-23
The change in intrinsic viscosity at pH 2.8 of the protein ribonuclease upon thermal denaturation. The value at low temperature is typical for a compact globular protein; the increase reflects the partial uncoiling of the molecule at high temperatures. (*From K. E. Van Holde, "Physical Biochemistry," Prentice-Hall, Inc., Englewood Cliffs, N.J., 1971.*)

applicable to proteins with random-coil shapes, appears to be valid for polypeptides in a great range of sizes up to large fragments of myosin with nearly 3000 residues. Thus viscosity measurements of solutions that convert proteins to the random-coil configuration can be simply related to the size of the polypeptide or protein.

A change in shape of protein molecules generally accompanies denaturation of the protein. For example, the shape change might be from a globular form to a random-coil form. Then we would expect the intrinsic viscosity to change from some small value, as in the 3 to 4 range, to larger values such as listed in Table 3-5. Measurement of viscosity, as this suggests, is a convenient way of studying protein denaturation. The effect of heat on ribonuclease provides an example. The abrupt viscosity change at about 40°C, as Fig. 3-23 shows, indicates a partial uncoiling of the compact globular form that exists at lower temperatures. The process is reversible. As the temperature is lowered, the viscosity curve is retraced and the original globular form is recovered.

Example 3-3

A protein isolated from *Paramecium* [A. H. Reisner and J. Rowe, *Nature,* **222:**558 (1969)] was shown to have 2930 amino acid residues and a molecular mass of about 300,000. Measurements of the viscosity of solution of this protein in 6 *M* GuHCl led to an intrinsic viscosity of 133 ml/g. Is this consistent with the empirical relation of Eq. **47?**

Solution

Equation **47** is most conveniently worked with in its logarithmic form

$$\log [\eta] = \log 0.732 + 0.656 \log n$$
$$= 0.656 \log n - 0.1355$$

For $n = 2930$ we obtain

$$\log [\eta] = 0.656(3.467) - 0.1355$$
$$= 2.274 - 0.1355$$
$$= 2.139$$

and $[\eta] = 138$

The result is in satisfactory agreement with the observed value of 133 ml/g.

PROBLEMS

1 Describe, in words, the effect of the mass of the vibrating object and the stiffness of the spring to which it is attached on the frequency of the natural vibration.

2 How many normal vibrations will there be for the molecules HCl, N_2, HCN, CH_4, and NH_2CH_2COOH?

3 Imagine a ball-and-spring model that is the counterpart of the CO_2 molecule.
  a Draw pictures to represent the initial distortions of the model that, on release, would lead to normal vibrations of the system.
  b How many of the normal vibrations can you depict?
  c Which ones could couple with electromagnetic radiation?

4 Calculate the wavelength of the de Broglie wave associated with:
  a An electron accelerated by a voltage of 10,000 V. The speed acquired by the electron is then about 6×10^9 cm/sec.
  b A molecule of nitrogen moving with the speed corresponding to the average translational energy at 25°C.
  c A car weighing 1 ton traveling with a speed of 60 mi/h (1 ton = 2000 lb, 1 lb = 454 g, 1 mi/h = 44.7 cm/sec).

5 Calculate the energies of the two lowest quantum states for an electron confined to an infinitely walled potential well 2 Å wide. If radiation could be absorbed to raise the electron from the lowest energy state to the next higher, what electromagnetic region would present quanta of suitable energy?

6 Obtain an expression for the allowed energies of a baseball, of mass 5 oz, confined to a baseball park of length 350 ft. What velocity jumps are implied? (1 oz = 28 g, 1 ft = 30 cm.)

7 The average one-dimensional translational energy of an N_2 molecule was shown in Chap. 1 to have the value $\frac{1}{2}kT$.

a What would be the quantum number, according to the square-well treatment, of an average N_2 molecule at 25°C in a container 10 cm long?

b What would be the energy separation between successive allowed quantum states at this energy?

c How does this energy separation compare with the total average energy of $\frac{1}{2}kT$?

8 Make qualitative statements about the effect of (*a*) stiffness of a chemical bond and (*b*) mass of the vibrating atoms on the frequency at which a vibrational band due to the stretching motion of that bond will occur.

9 The fundamental absorption band, corresponding to the $v = 0$ to $v = 1$ transition, of carbon monoxide is found in the infrared at a frequency of 6.50×10^{13} Hz.

a What is the reduced mass of the CO molecule?

b What is the force constant of the CO bond in this molecule?

10 In the vapor phase two absorption bands are observed for H_2O that can be assigned to symmetric and antisymmetric O—H stretching vibrations. They have frequencies of 10.96×10^{13} and 11.27×10^{13} Hz. Corresponding frequencies for D_2O are found at 8.00×10^{13} and 8.35×10^{13} Hz.

a Using averages of the symmetric and antisymmetric vibration frequencies, obtain values that can be associated with the frequencies of simple O—H and O—D stretching vibrations.

b Calculate the reduced mass of O—H and O—D.

c Calculate the force constants of the O—H and O—D bonds. Comment on the results.

11 The nmr spectrum of methanol which contains a small amount of water shows two sharp singlets. If, however, methanol is dissolved in acetone, the lines attributed to the methanol consist of a doublet and a quartet, the total intensity of the quartet being approximately one-third that of the doublet. Explain all these observations.

12 The relation of Eq. **22**, $h\nu = 2\mu_H \mathcal{H}$ shows that for protons there is a proportionality between the magnetic field to which they are exposed and the frequency of the radiation that will cause transitions between the nuclear states. For protons the proportionality is approximately $\nu/\mathcal{H} = 4.26 \times 10^3$ Hz/gauss = 4.26 MHz/kilogauss. A common nmr frequency is 60 MHz. What magnetic field would be necessary to produce proton resonances? (Ignore, here, the small chemical shifts and spin-spin splitting that make nmr a chemically useful tool.)

13 A 60-MHz study of absolute methanol reports that the nmr lines due to the —OH and the —CH_3 protons are separated by a δ value of 1.5 ppm.

a To what magnetic-field difference does this correspond? (Use the information obtained from the preceding problem.)

b To what frequency difference, assuming the magnetic field is fixed, in hertz, does this δ value correspond?

c Suppose the study were repeated with an nmr instrument operating at 100 MHz. What would be the δ value, the magnetic-field difference, and the frequency difference?

14 Compare the energy spacing of the nuclear states produced in a typical nmr study with the value of kT at room temperature. Qualitatively, what do this and the Boltzmann distribution imply about the relative populations of such states?

15 β-Casein is a protein with no disulfide links, a molecular mass of 24,000, and 213 residues. The intrinsic viscosity of aqueous solutions at pH 7 is 23 ml/g. In guanidine hydrochloride solutions the intrinsic viscosity is virtually unchanged, the value 22 ml/g being reported. What can you conclude from these results and from their comparison with Eq. **47**?

16 The viscosity of solutions of native DNA and of fractions of DNA produced by sonic breakup of the native material have been reported [J. Eigner and P. Doty, *J. Mol. Biol.*, **12**:549 (1965)]. For molecular masses in the range 2×10^6 to about 130×10^6, the intrinsic viscosities can be represented by the empirical relation $[\eta] = 0.069M^{0.70}$, where M is the molecular mass.

a What value of $[\eta]$ does this equation yield for the high-molecular-mass fractions, with M of about 130×10^6, corresponding to unfragmented DNA?

b What structure do this value and the dependence of $[\eta]$ on mass suggest for the DNA molecule in solution?

REFERENCES

Infrared Spectroscopy

TANFORD, CHARLES: "Physical Chemistry of Macromolecules," pp. 73–89, John Wiley & Sons, Inc., New York, 1961.

BARROW, G. M.: "The Structure of Molecules," chaps. 3 and 4, W. A. Benjamin, Inc., New York, 1964.

FRASER, R. D. B., and E. SUZUKI: Infrared Methods, chap. 13 in S. J. Leach (ed.), "Physical Principles and the Techniques of Protein Chemistry," pt. B, Academic Press, Inc., New York, 1970. Extensive introductions to instrumental techniques followed by applications to protein systems.

Hydrogen Bonding

PIMENTEL, G. C., and A. L. MCCLELLAN: "The Hydrogen Bond," W. H. Freeman and Company, San Francisco, 1960.

Nuclear Magnetic Resonance

ROBERTS, J. D.: "Nuclear Magnetic Resonance," McGraw-Hill Book Company, New York, 1959. An elementary introduction to the subject.

METCALFE, J. C.: NMR Spectroscopy, chap. 14 in S. J. Leach (ed.), "Physical Principles and the Techniques of Protein Chemistry," pt. B, Academic Press, Inc., New York, 1970. The basic principles of nmr spectroscopy followed by many protein-system applications.

ROBERTS, G. C. K., and O. JARDETZKY: Nuclear Magnetic Resonance Spectroscopy of Amino Acids, Peptides, and Proteins, *Adv. Protein Chem.*, **24:**448 (1970).

Viscosity

TANFORD, CHARLES: "Physical Chemistry of Macromolecules," pp. 390–391, John Wiley & Sons, Inc., New York, 1961.

VAN HOLDE, K. A.: "Physical Biochemistry," chap. 7, Prentice-Hall, Inc., Englewood Cliffs, N.J., 1971.

TANFORD, C., K. KAWAHARA, and S. LAPANJE: Proteins as Random Coils, *J. Am. Chem. Soc.*, **89:**729 (1967) and related papers by C. Tanford and various co-authors referred to therein.

BRADBURY, J.: Viscosity, chap. 11 in S. J. Leach (ed.), "Physical Principles and the Techniques of Protein Chemistry," pt. B, Academic Press, Inc., New York, 1970. Experimental methods, molecular models, and applications, with many references to original work.

KRAGH, A. M.: Viscosity, chap. 5 in P. Alexander and R. J. Block (eds.), "Laboratory Methods of Protein Chemistry," vol. 3, Pergamon Press, New York, 1961.

4 Molecular Species in Solution: Acid-Base Reactions

The molecular species present in many solutions of chemical or biological interest are the equilibrium products that result from a variety of chemical reactions. Most widespread are acid-base reactions, which lead to the addition or removal of protons from basic or acidic reagents. A more generalized view of acids and bases can also be taken, and attention can then be focused on additional types of equilibria. All such reactions determine the species that are present in solution. Studies of the nature and structure of species in liquid solution should therefore include consideration of these reactions and their products.

We shall focus chiefly on acid-base equilibria and the various molecular species that result from these reactions. As we proceed from simple acid-base systems to more complex ones—polyprotic acids, amino acids, proteins and DNA—it will be apparent that the association of reagents other than H^+ can be treated in a similar way. It will be found, for example, that the treatment used to decide what species are present when hemoglobin is exposed to oxygen is similar to that used to deduce the degree of protonation of a protein when it is in a solution of given pH.

In this chapter we shall assume that the equilibria we are dealing with are not appreciably disturbed by molecules and ions in the solution that are not directly involved in the reaction. With this assumption we can focus on the often challenging problems of equilibrium calculations. A more complete treatment, and one that is required for most laboratory and natural solutions of biological interest, would have to take into account the disrupting effects of all the species, particularly ions, that are present. Often these disrupting effects are treated in terms of the *ionic strength*, this being a measure of the total ionic content of the solution. In this chapter references to experimental studies will occasionally mention the ionic strength of the solution on which the studies were made. We shall take this to imply a characterization of the solution, as is the specification of the solvent or the temperature. More attention will be paid to ionic strength effects in the thermodynamic treatments of Chaps. 7 and 8.

4-1 SPECIES PRESENT WHEN SIMPLE ACIDS AND BASES ARE DISSOLVED IN WATER

Chapter 3 considered the structure and motion of molecular species in solution. Then we made reference to molecules that retain their identity and much of their structure when they are brought from their pure state to the solution state. But many substances change markedly when brought into solution, aqueous solution in particular. Familiar examples are strong acids, strong bases, and salts derived from strong acids and bases. For not too concentrated solutions the aqueous environment leads to independently moving anions and cations in place of the parent molecules or ionic crystals.

The species that result from the solution of weak acids, weak bases, or their salts also differ in part from the parent species. Here, to introduce simple acid-base equilibrium calculations, we shall see to what extent the molecular and ionic species derived from weak acids and weak bases are present.

Measurements of how solutions of weak acids conduct an electric current and many other types of studies show that a weak acid, HA, in solution sets up the equilibrium system that can be written

$$HA \rightleftarrows H^+ + A^- \tag{1}$$

or

$$HA + H_2O \rightleftarrows H_3O^+ + A^- \tag{2}$$

The second of these reflects the *Brønsted* theory that acid-base reactions involve the *transfer* of a proton from an acid, here HA, to a base, here H_2O. This description avoids the unsatisfactory implication that bare protons exist as chemical species in solution. However, we should expect all ions to be more or less strongly associated with water molecules in aqueous solution. Then H^+ and H_3O^+ can be said to carry similar implications, the latter being more explicit about one of the associated water molecules. Thus, although you should picture acid-base reactions as involving proton-transfer processes, you can work with equations that include or omit the constant-concentration water term and the explicit hydration of the proton.

Equilibrium-constant expressions for the above equations can then be written, using brackets to indicate the molar concentration of the enclosed species. From Eq. **1** we write

$$K_a = \frac{[H^+][A^-]}{[HA]} \tag{3}$$

From Eq. **2**

$$K'_a = \frac{[H_3O^+][A^-]}{[HA][H_2O]}$$

or

$$K_a = [H_2O]K'_a = \frac{[H_3O^+][A^-]}{[HA]} \tag{4}$$

A convenient procedure for using such equilibrium expressions to calculate the concentrations of the H^+, A, and HA species in a solution of a weak acid consists of first writing the equation that connects these species. Then the initial and the equilibrium concentration are described, a procedure that usually requires the introduction of an unknown quantity. Thus

	HA	$\rightleftarrows$ H$^+$	+ A$^-$
Initial concentration:	c	0	0
Equilibrium concentration:	$c - x$	x	x

where x is the molar concentration of H^+ and of A^- resulting from the dissociation. Then the equilibrium-constant expression allows the unknown x to be deduced from

$$K_a = \frac{[H^+][A^-]}{[HA]} = \frac{(x)(x)}{c - x} \quad 5$$

Given a value of K_a for the acid and a value of c for the solution under consideration, a value of x can be calculated. Rearrangement of Eq. **5** gives the quadratic expression

$$x^2 + K_a x - cK_a = 0 \quad 6$$

We recall that for a general quadratic expression

$$ax^2 + bx + c = 0 \quad 7$$

the solution values for x are given by

$$x = \frac{-b \pm \sqrt{b^2 - 4ac}}{2a} \quad 8$$

Applied to Eq. **6,** this general formula gives

$$x = \frac{-K_a \pm \sqrt{K_a{}^2 + 4c}}{2} \quad 9$$

Frequently this cumbersome solution form can be avoided by trying the assumption $x \ll c$ in Eq. **5.** Then from Eq. **5** or **6**

$$x^2 - cK_a \cong 0 \qquad \text{and} \qquad x \cong \sqrt{cK_a} \quad 10$$

With the inevitable example of acetic acid, HAc, with $K_a = 1.8 \times 10^{-5}$ at 25°C, the species present in an 0.1 *M* acetic acid solution can be calculated as an illustration. We write

	HAc	$\rightleftarrows$ H$^+$	+ Ac$^-$
Initially:	0.10	0	0
At equilibrium:	0.10 − x	x	x

Then

$$1.8 \times 10^{-5} = \frac{[H^+][Ac^-]}{[HAc]}$$

$$= \frac{x^2}{0.10 - x}$$

If $x \ll 0.10$, then

$$1.8 \times 10^{-5} = \frac{x^2}{0.10}$$

and $x = 0.0013$

Since this is considerably less than 0.10, the $x \ll 0.10$ assumption is valid. We thus have the result

$$[H^+] = 0.0013$$
$$[Ac^-] = 0.0013$$
$$[HAc] = 0.099$$

In this calculation the self-dissociation of water has been ignored. We have not recognized the fact that the solution contains H^+ and OH^- ions from the reaction

$$H_2O \rightleftarrows H^+ + OH^-$$

If we assume that the concentration of H_2O is essentially constant in the solutions we deal with, this process leads to the equilibrium-constant expression

$$K_w = [H^+][OH^-] \qquad \textbf{11}$$

with $K_w = 1.0 \times 10^{-14}$ at 25°C **12**

For a solution of a weak acid in water, the water itself contributes some of the reactive species H^+. In many situations, as in the above example, the concentrations that stem from water dissociation are negligible compared with that from the added acid. In such cases the water-dissociation equilibrium can be ignored.

But often the concentration of OH^-, as well as that of H^+, is of interest. Then the ion product of water *is* used. In an 0.10 *M* acetic acid solution, for example, we calculate from $[H^+] = 0.0013$

$$[OH^-] = \frac{1.0 \times 10^{-14}}{0.0013} = 7.7 \times 10^{-12}$$

The deduction of the species present when a weak base is added to water follows a similar pattern. Thus pyridine, N, or C_5H_5N, is a weak base that in water enters into the reaction

$$C_6H_5N + H_2O \rightleftarrows C_6H_5NH^+ + OH^-$$

The equilibrium constant for such reactions is usually indicated by K_b, and here

$$K_b = \frac{[C_6H_5NH^+][OH^-]}{[C_6H_5N]}$$

where

$$K_b = 1.5 \times 10^{-9} \qquad \text{at } 25°\text{C}$$

The species present in an 0.10 *M* pyridine solution can be calculated as

$$C_6H_5N + H_2O \rightleftarrows C_6H_5NH^+ + OH^-$$

	C_6H_5N	$C_6H_5NH^+$	OH^-
Initially:	0.10	0	0
At equilibrium:	0.10 − x	x	x

Then

$$1.5 \times 10^{-9} = \frac{[C_6H_5NH^+][OH^-]}{[C_6H_5N]}$$

$$= \frac{x^2}{0.10 - x}$$

If $x \ll 0.10$,

$$1.5 \times 10^{-9} = \frac{x^2}{0.10}$$

and $\quad x = 1.2 \times 10^{-5}$

The assumption $x \ll 0.10$ is valid, and we have

$$[C_6H_5NH^+] = 1.2 \times 10^{-5}$$

$$[OH^-] = 1.2 \times 10^{-5}$$

$$[C_6H_5N] = 0.10$$

and $\quad [H^+] = \frac{1.0 \times 10^{-14}}{[OH^-]}$

$$= \frac{1.0 \times 10^{-14}}{1.2 \times 10^{-5}} = 0.8 \times 10^{-9}$$

Pyridine is such a weak base that the hydroxide concentration we have calculated is not overwhelming compared with the 10^{-7} value of water itself. The example is therefore close to the situations where account of the water equilibrium must be taken.

The strength of a base can be thought of in terms of its ability to extract a proton from water. This is indicated by the value of K_b. The greater the value of K_b, the stronger the base. Thus ammonia, with a K_b value of 1.8×10^{-5}, is a stronger base than pyridine, whose K_b value is only 1.5×10^{-9}.

It is often more convenient to think of the strength of a base in terms of the reluctance of its *conjugate* acid to lose a proton. Pyridine, C_6H_5N, is a base; its conjugate acid is $C_6H_5NH^+$. The strength of this conjugate acid is indicated by K_a of the reaction

$$C_6H_5NH^+ \rightleftarrows C_6H_5N + H^+$$

The stronger the base the more tightly it holds an added proton and the weaker its conjugate acid. Pyridine is a rather weak base, and we expect its conjugate acid to be relatively strong.

The relation between K_b of a base and K_a of its conjugate acid can be seen by first writing the equilibrium expression for the ionization of the conjugate acid (in the pyridine example) as

$$K_a = \frac{[C_6H_5N][H^+]}{[C_6H_5NH^+]}$$

The value of the expression is not changed by inserting an $[OH^-]$ term in the numerator and denominator. This leads us to

$$K_a = \frac{[C_6H_5N][H^+][OH^-]}{[C_6H_5NH^+][OH^-]}$$

The value of this manipulation is that it lets us recognize K_w and K_b expressions on the right. Dividing numerator and denominator by $[C_6H_5N]$ leads us to

$$K_a = \frac{[H^+][OH^-]}{[C_6H_5NH^+][OH^-]/[C_6H_5N]}$$

$$= \frac{K_w}{K_b}$$

Thus we can interconvert K_b values and conjugate-acid K_a values by using the relation

$$K_aK_b = K_w \qquad \textbf{13}$$

For pyridine and ammonia, for example, the conjugate acids have K_a values of 6.6×10^{-6} and 5.6×10^{-10}, respectively. The conjugate acid of pyridine is stronger than the conjugate acid of ammonia. Some values of K_a of acids and the conjugate acids of bases are given in Table 4-1.

The comparison of the concentrations of the species A^- and HA that result from the partial dissociation of an acid HA or of the species BH^+ and B that result from the partial reaction of a base B with water can be conveniently represented by introducing α, the degree of dissociation. For a weak acid, α is taken to mean the fraction of the initial acid that dissociates. Thus, one writes

	HA	$\rightleftharpoons H^+$	$+ A^-$
Initially:	c	0	0
At equilibrium:	$c - c\alpha = c(1 - \alpha)$	$c\alpha$	$c\alpha$

The equilibrium-constant expression is then

$$K_a = \frac{[H^+][A^-]}{[HA]} = \frac{c\alpha^2}{1 - \alpha}$$

TABLE 4-1
Dissociation Constants K_a of Weak Acids and the Conjugate Acids of Weak Bases in Water at 25°C and Values of $pK_a = -\log K_a$

acid	**formula**	K_a	pK_a
Monochloroacetic	$ClCH_2COOH$	1.3×10^{-3}	2.88
Nitrous	HNO_2	5.1×10^{-4}	3.29
Formic	$HCOOH$	1.8×10^{-4}	3.75
Lactic	$CH_3CHOHCOOH$	1.4×10^{-4}	3.85
Benzoic	C_6H_5COOH	6.2×10^{-5}	4.21
Anilinium ion	$C_6H_5NH_3^+$	2.5×10^{-5}	4.60
Acetic	CH_3COOH	1.8×10^{-5}	4.75
Butyric	$CH_3CH_2CH_2COOH$	1.5×10^{-5}	4.82
Pyridinium ion	$C_5H_5NH^+$	6.6×10^{-6}	5.18
Ammonium ion	NH_4^+	5.6×10^{-10}	9.25
Hydrocyanic	HCN	4.9×10^{-10}	9.32
Trimethylammonium ion	$(CH_3)_3NH^+$	1.2×10^{-10}	9.91
Phenol	C_6H_5OH	1.0×10^{-10}	10.00
Methylammonium ion	$CH_3NH_3^+$	2.5×10^{-11}	10.60

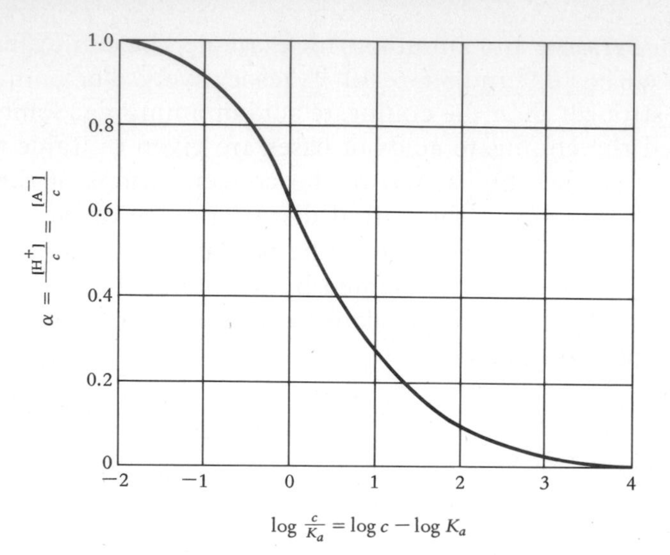

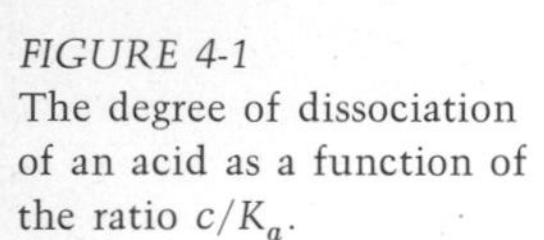
FIGURE 4-1
The degree of dissociation of an acid as a function of the ratio c/K_a.

By rearranging this to

$$\frac{\alpha^2}{1-\alpha} = \frac{K_a}{c}$$

or $$\alpha^2 + \frac{K_a}{c}\alpha - \frac{K_a}{c} = 0 \qquad \textbf{14}$$

the degree of dissociation α is seen to be a function of K_a/c. For small amounts of dissociation, i.e., when $\alpha \ll 1$, this dependence is simply

$$\alpha = \sqrt{\frac{K_a}{c}} \qquad \textbf{15}$$

It follows from Eq. **14,** or, for small degrees of dissociation from Eq. **15,** that the dependence of the degree of ionization on acid strength and acid concentration can be represented by a single α-versus-K_a/c curve. However, to cover a wide enough range of possible K_a and c values it is necessary to use an α-versus-$\log(K_a/c)$ curve, as in Fig. 4-1.

The dependence of α on concentration for given K_a values is shown explicitly for several K_a values in Fig. 4-2. Since

$$\log \frac{K_a}{c} = \log K_a - \log c \qquad \textbf{16}$$

the curves of Fig. 4-2 are seen to be like that of Fig. 4-1 and, for various values of K_a, are simply displaced from one another.

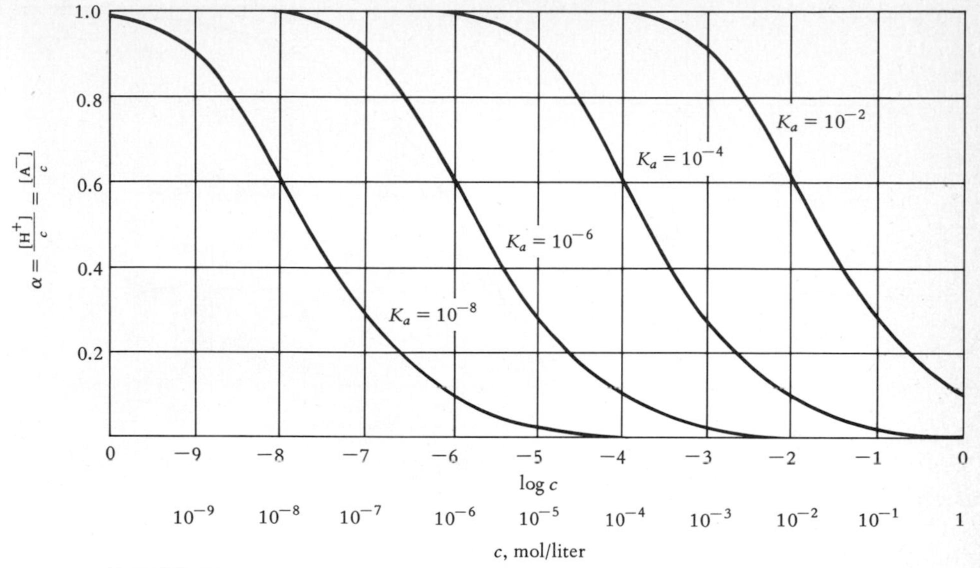

FIGURE 4-2
The degree of dissociation of acids as a function of concentration for various values of the acid dissociation constant K_a.

In anticipation of species-concentration displays that will be used in more complex situations, the fraction of the total acid that is present as HA and as A^- is given in Fig. 4-3 for the example of $K_a = 10^{-5}$. The symbol ν is used to express this extent of protonation, i.e., $\nu = \dfrac{[HA]}{c}$. Here since $[HA] = c - [A] = c - c\alpha$, we have the relation $\nu = (c - c\alpha)/c = 1 - \alpha$.

A similar treatment of weak bases follows from writing

	B	$+ H_2O \rightleftarrows BH^+$	$+ OH^-$
Initially:	c	0	0
At equilibrium:	$c - c\alpha$ $= c(1 - \alpha)$	$c\alpha$	$c\alpha$

Development, as for acids, then leads to the display of species present, as in Fig. 4-4.

All this introductory material (with the exception of the abscissa scales of Figs. 4-1 to 4-4) has been presented in terms of concentrations. Often more convenient is the use of logarithms of concentrations and of equilibrium constants. It is customary, in particular, to deal not with the hydrogen-ion concentration or even the logarithm of this concentration but with the negative of this logarithm. The term pH is used for this quantity,

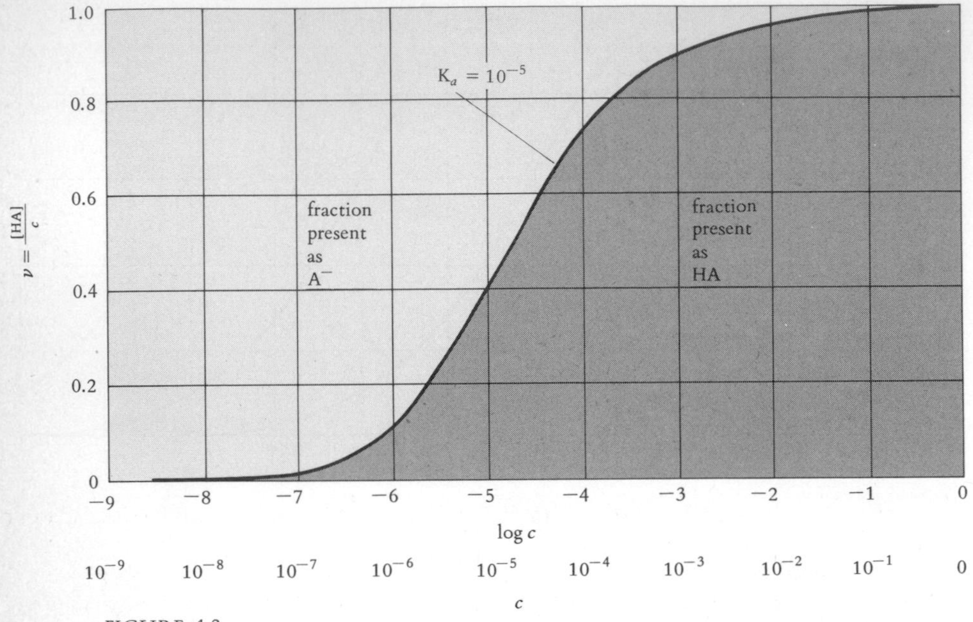

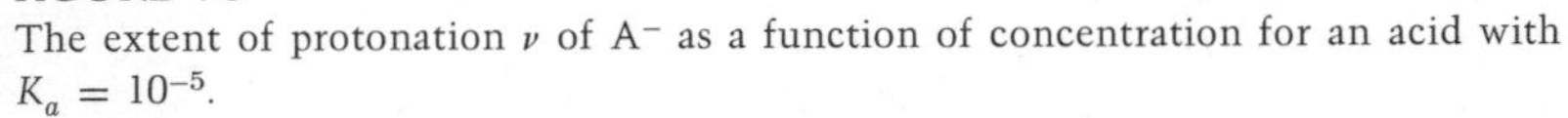

FIGURE 4-3
The extent of protonation ν of A^- as a function of concentration for an acid with $K_a = 10^{-5}$.

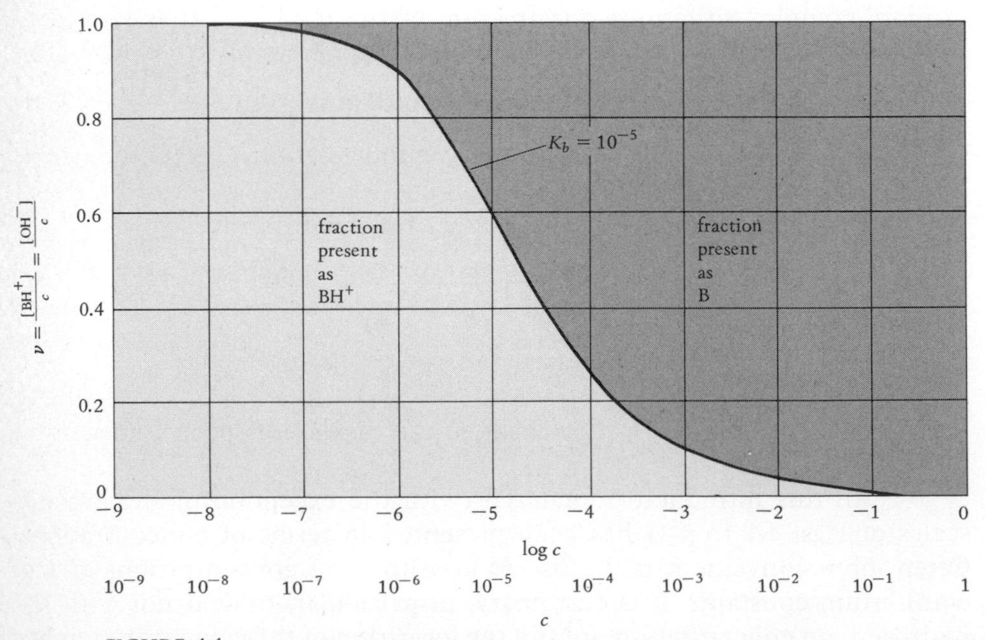

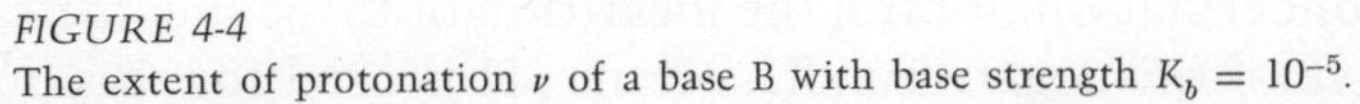

FIGURE 4-4
The extent of protonation ν of a base B with base strength $K_b = 10^{-5}$.

and thus we have the defining equation

$$pH = -\log [H^+] \quad \textbf{17}$$

In a similar but somewhat less familiar way we write

$$pOH = -\log [OH^-] \quad \textbf{18}$$

$$pK_a = -\log K_a \quad \textbf{19}$$

$$pK_w = -\log K_w \quad \textbf{20}$$

and so forth.

Much convenience results from these logarithmic quantities. Here this is only anticipated by noting that the relation

$$K_w = [H^+][OH^-]$$

becomes with this p notation

$$pK_w = pH + pOH$$

or at 25°C

$$pH + pOH = 14 \quad \textbf{21}$$

In this chapter, we ignore the modifying effects of the ionic content of a solution on each species in the solution. In particular, we treat the involvement of the hydrogen ion in equilibria simply in terms of the hydrogen-ion concentration. In solutions with any appreciable ionic content the "effective" concentration of an ion, the hydrogen ion for example, will be somewhat different from the actual concentration. This complication is mentioned here because pH values of solutions are often deduced from the reading given by a pH meter and not by a measurement of the hydrogen-ion concentration. In view of this, pH is now usually defined in terms of the electrochemical cell upon which operation of the pH meter is based. This definition differs from that of Eq. **17.** It is, however, in the spirit of this chapter to ignore the complications that lead one to define pH other than by Eq. **17.**

4-2 SOLUTIONS OF SALTS OF WEAK ACIDS AND SALTS OF WEAK BASES

Little needs to be added to the ideas of the preceding section to develop expressions for the species present when a salt of a weak acid or a weak base is added to water. Examples of such salts are sodium acetate and ammonium chloride. Reactions with water, such as these salts undergo, are called *hydrolysis reactions.*

Consider, to be specific, sodium acetate, NaAc. No evidence for NaAc salt molecules, or "ion pairs" in solution exists—unless very concentrated solutions are considered. We thus can treat separately the consequences of the Na^+ and the Ac^- interactions with water. The first, the sodium ion, enters into no water-disrupting reaction. The acetate ion, Ac^-, does react. This ion can be recognized as the conjugate base of the acid, acetic acid, HAc. Since acetic acid is not a strong acid, the Ac^- ion will have some tendency to accept a proton, i.e., to act as a base. The reaction is described by the equation

$$Ac^- + H_2O \rightleftarrows HAc + OH^-$$

Determination of the species present at equilibrium as a result of this reaction can be carried out using the same format as in the preceding section. We write

	Ac^-	$+ H_2O \rightleftarrows$	HAc	$+ OH^-$
Initially:	c		0	0
At equilibrium:	$c - x$		x	x

The equilibrium expression for this reaction is

$$K_{hyd} = \frac{[HAc][OH^-]}{[Ac^-]} \qquad 22$$

where the subscript indicates that the reaction is a hydrolysis reaction. As you recognize, this is not an expression for which tabulated values of the constant are available. It can be converted to recognizable expressions, as in the development that led to Eq. **13,** by inserting an $[H^+]$ factor in the numerator and denominator. This leads to

$$K_{hyd} = \frac{[HAc][OH^-][H^+]}{[Ac^-][H^+]} = \frac{[OH^-][H^+]}{[H^+][Ac^-]/[HAc]} = \frac{K_w}{K_a}$$

or $\quad K_a K_{hyd} = K_w$

Nothing distinguishes this treatment from that which led to Eq. **13** except the point of view that Ac^- is the anion of a salt and that B is a base.

For the acetate-ion example

$$K_{hyd} = \frac{1.0 \times 10^{-14}}{1.8 \times 10^{-5}} = 5.5 \times 10^{-10}$$

This result can be inserted in Eq. **22** to give

$$K_{hyd} = 5.5 \times 10^{-10} = \frac{[HAc][OH^-]}{[Ac^-]}$$

$$= \frac{x^2}{c - x} \qquad 23$$

For a given value of c, a value of x can be calculated.

One could also proceed in terms of the extent to which A^- becomes protonated as a result of the hydrolysis reaction. If ν is used again, here as [HAc]/c we have $c\nu$ in place of [HAc], or x, and, in general,

$$K_{\text{hyd}} = \frac{K_w}{K_a} = \frac{c\nu^2}{1 - \nu} \qquad 24$$

or

$$\frac{K_w/K_a}{c} = \frac{\nu^2}{1 - \nu} \qquad 25$$

In the region where the hydrolysis is slight, where $\nu \ll 1$, this simplifies to

$$\nu \cong \sqrt{\frac{K_w/K_a}{c}} \qquad 26$$

Thus you can see that the extent of hydrolysis increases as we go to smaller concentrations and as we go to anions that are derived from weaker acids, i.e., those with small values of K_a.

The effect of such hydrolyses on the pH of the solution can be illustrated by considering an 0.10 M solution of sodium acetate. Insertion of this value for c in Eq. **23** or in equations of the form of Eq. **25** or **26** yields

$$[\text{HAc}] = [\text{OH}^-] = 0.74 \times 10^{-5}$$

and $$[\text{Ac}^-] = 0.10 - 0.74 \times 10^{-5} = 0.10$$

Also $$[\text{H}^+] = \frac{1 \times 10^{-14}}{0.74 \times 10^{-5}} = 1.3 \times 10^{-9}$$

and $$\text{pH} = 8.9$$

Thus a somewhat basic solution is produced by the hydrolysis of a salt of a weak acid. You come to this qualitative conclusion directly by recognizing that the acetate ion is a base, the conjugate base of acetic acid.

Likewise, pyridine hydrochloride would hydrolyze to produce a somewhat acidic solution, a consequence of the acidity of the pyridinium ion, the conjugate acid of the weak base pyridine.

4-3 THE pH DEPENDENCE OF SPECIES DERIVED FROM SIMPLE ACIDS AND BASES: TITRATION CURVES

Systems of principal interest are those in which the species that develop from a parent substance do so as a result of their interaction with water and with other reagents in the system. The preceding sections treated the equilibrium products resulting when a single acid, base, or salt is dissolved in water. Now we must see what species are present when in addition the H^+ or OH^- concentrations, usually expressed by the pH of the solution, are adjusted by the addition of strong acid or base. Here, focusing attention on simple reagents, we begin the study of the major area of the pH dependence of the species present in solution of substances that react with acids and bases.

The principal experimental entry into the subject is through *titrations* and *titration curves*. We start, for example, with a solution of an acid; then a solution of a strong base can be added in small increments, and the change in $[H^+]$ or pH of the solution can be measured and the results plotted to yield curves like those of Fig. 4-5. In a similar way typical titration curves for the titration of a base by means of an acid can be obtained, as in Fig. 4-6.

Titrations and titration curves are put to a variety of uses. Analytical application depends on recognizing that the rather abrupt change in the pH exhibited in Figs. 4-5 and 4-6 occurs when an *equivalent amount* of base has been added to the acid solution or an equivalent amount of acid to the base solution. The titration then allows us to deduce the amount or the concentration of the acid or base equivalent to that in the original solution.

Our principal interest is in the use of titration curves to give information on the relative amounts of the species HA and A^- from a weak acid or B and HB^+ from a weak base present in solutions with various pH values. Such information can be obtained directly from titration curves. A better route, however, leads from titration curves to K_a and K_b, or pK_a and pK_b, values and then to calculation of the species present.

Consider the equilibrium set up when a weak acid is titrated. The equation that describes this is

$$HA \rightleftarrows H^+ + A^- \qquad \textbf{27}$$

Now the special case of $[H^+] = [A^-]$, as used in Sec. 4-1, is not applicable and a more general development is necessary. Still valid is the equilibrium expression

$$K_a = \frac{[H^+][A^-]}{[HA]} \qquad \textbf{28}$$

Some rearrangement leads to a more convenient form. First

$$[H^+] = K_a \times \frac{[HA]}{[A^-]}$$

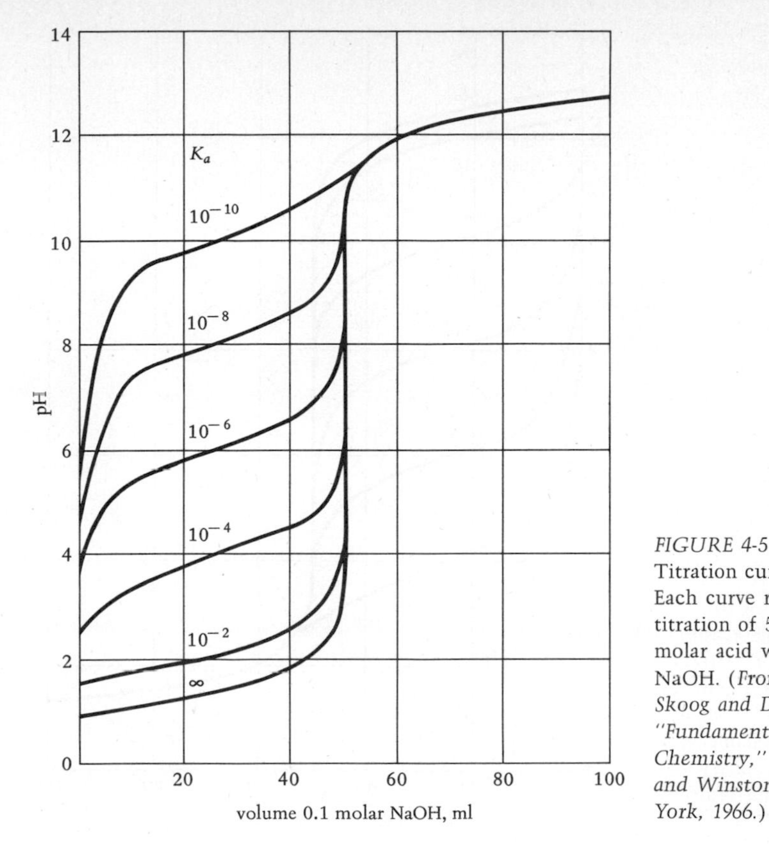

FIGURE 4-5
Titration curves for acids. Each curve represents the titration of 50.0 ml of 0.100 molar acid with 0.100 molar NaOH. (*From D. A. Skoog and D. M. West, "Fundamentals of Analytical Chemistry," Holt, Rinehart and Winston, Inc., New York, 1966.*)

Taking logarithms gives

$$\log [H^+] = \log K_a + \log \frac{[HA]}{[A^-]}$$

or

$$-\log [H^+] = -\log K_a + \log \frac{[A^-]}{[HA]}$$

Finally, with the p standing for $-\log$,

$$pH = pK_a + \log \frac{[A^-]}{[HA]} \tag{29}$$

Clearly, if we now know the pK_a of an acid, we can calculate the $[A^-]/[HA]$ ratio at any pH. This useful expression is one form of the relation known in biology and biochemical literature as the *Henderson-Hasselbalch equa-*

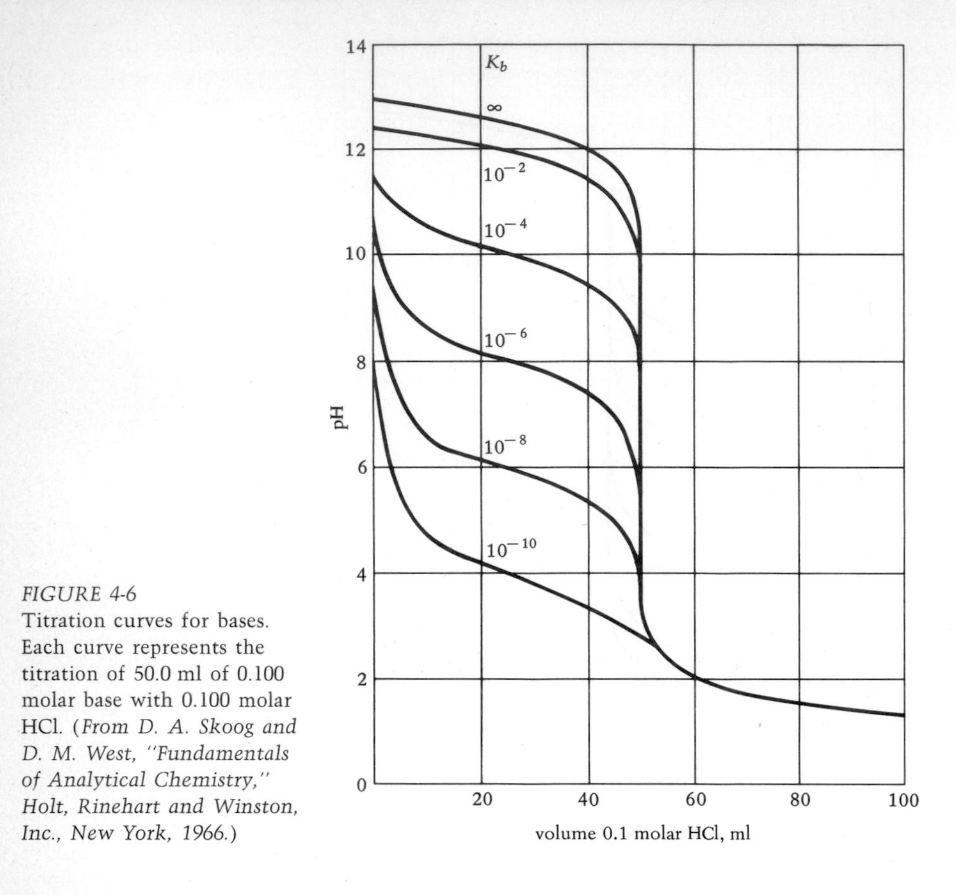

FIGURE 4-6
Titration curves for bases. Each curve represents the titration of 50.0 ml of 0.100 molar base with 0.100 molar HCl. (*From D. A. Skoog and D. M. West, "Fundamentals of Analytical Chemistry," Holt, Rinehart and Winston, Inc., New York, 1966.*)

tion. A similar development for a weak base B leads to

$$pH = pK_a + \log \frac{[B]}{[BH^+]} \qquad \textbf{30}$$

where $K_a = K_w/K_b$ is the ionization constant for the conjugate acid of the base and $pK_a = pK_w - pK_b = 14 - pK_b$.

Values for the pK_a terms of Eq. **29** or **30** are usually deduced from titration curves. A convenient method depends on recognizing that when half the acid or base has been titrated, $[A^-] = [HA]$ or $[B] = [BH^+]$. Then since the final terms become $\log 1 = 0$, Eqs. **29** and **30** reduce simply to $pH = pK_a$. Thus, as can be verified in Figs. 4-5 and 4-6, the value of pK_a of a weak acid or the conjugate acid of a weak base is given directly by the pH when the titration has been carried halfway to the equivalence point. It is not even necessary to know the concentrations of the solutions, except that solutions so dilute that the self-ionization of water would produce appreciable H^+ or OH^- must be avoided.

To display the relative amounts of the species A^- and HA or B and BH^+ present as a function of pH we usually plot pH along the abscissa rather than along the ordinate, as for titration curves. We are still left with a variety of options for showing the concentrations or relative concentrations of the species present. A procedure that is convenient when more complex systems are dealt with consists of showing the extent of protonation of the substance involved in the titration. If c_A represents the total concentration of acid, that is, $[HA] + [A^-]$, and if c_B represents the total concentration of base, that is, $[B] + [BH^+]$, the degree of protonation, again represented by ν, can be shown by plotting

$$\nu = \frac{[HA]}{c_A} = \frac{[HA]}{[HA] + [A^-]} \quad \textbf{31}$$

or

$$\nu = \frac{[BH^+]}{c_B} = \frac{[BH^+]}{[B] + [BH^+]} \quad \textbf{32}$$

against pH. These fractions protonated can be calculated from the ratios $[HA]/[A^-]$ or $[BH^+]/[B]$, which are given as a function of pH and pK_a or pK_b by Eqs. **29** and **30.** The fraction of acid or base protonated is then given by

$$\nu = \frac{[HA]}{c_A} = \frac{1}{1 + [A^-]/[HA]} \quad \textbf{33}$$

and

$$\nu = \frac{[BH^+]}{c_B} = \frac{1}{1 + [B]/[BH^+]} \quad \textbf{34}$$

Although the ratios $[HA]/[A^-]$ and $[BH^+]/[B]$ can take on all values between zero and infinity, the fraction ν of acid or base protonated varies only from 0 to 1.

Examples of the extent of protonation of an acid and a base as a function of pH are shown in Figs. 4-7 and 4-8.

The ordinates of Figs. 4-7 and 4-8 are expressed as the ratio of the concentrations of the protonated species to total acid or base. A more specific molecular interpretation can be given. The ordinate can also be said to show the ratio of the number of protonated molecules to the total number of acid or base molecules; or it shows the fraction of a proton that would be possessed by each acid or base molecule if an averaging process were envisaged. This final interpretation has special value when the proton acceptor has many acceptor sites and the average number of protons accepted is not limited to values up to unity.

When titration curves for more complex species are studied, pK_a values cannot easily be deduced. Then one must go directly from the

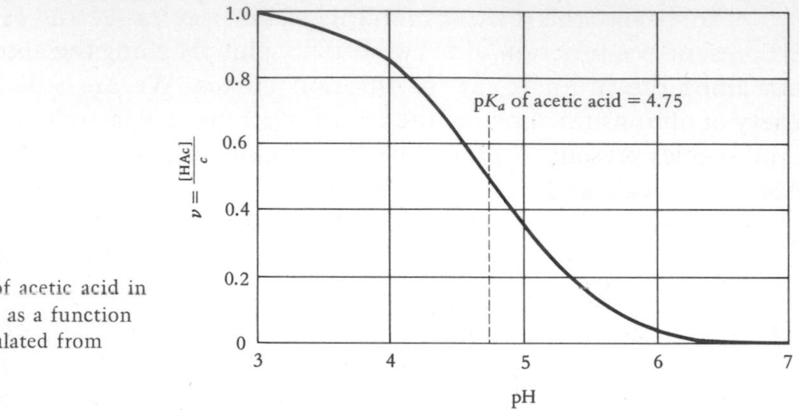

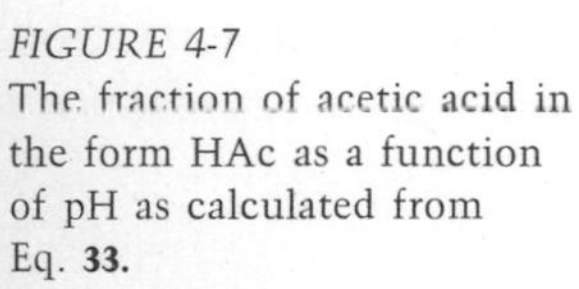
FIGURE 4-7
The fraction of acetic acid in the form HAc as a function of pH as calculated from Eq. **33.**

titration curve to results like those in Figs. 4-7 and 4-8. The fraction of the acid present as the protonated species can be calculated as follows. Throughout the titration the number of moles of HA plus the numbers of moles of A^- is a fixed, constant amount. At any stage in the titration the balance of positive and negative charges allows us to write

$$[Na^+] + [H^+] = [A^-] + [OH^-] \tag{35}$$

Except for solutions of rather high or low pH, the values of $[H^+]$ and $[OH^-]$ will be negligible compared to those of $[Na^+]$ and $[A^-]$. Then we have simply $[A^-] = [Na^+]$. Further, the number of moles of A^- is equal to the number of moles of the Na^+ added, and this can be calculated from the volume and concentration of the base added at that stage of the titration. Subtraction of this number from the total moles of A^- plus HA gives the number of moles of HA in the solution. Then ν, the ratio of moles of HA to total moles, can be calculated and associated with the pH at this stage in the titration. For very acidic and very basic solutions or for solutions very dilute in the acid being titrated, the concentration of H^+ or of OH^- must also be calculated and inserted, along with $[Na^+]$, in Eq. **35.**

4-4 BUFFERS AND BUFFER ACTION

In some situations it is the control the species in solution have on the pH of the solution rather than the control the pH has on the species, that is of importance and interest. This is true of synthetic and natural solutions that resist changes in pH when more acid or base is added to the solution. Solutions that perform in this way are called *buffer solutions.*

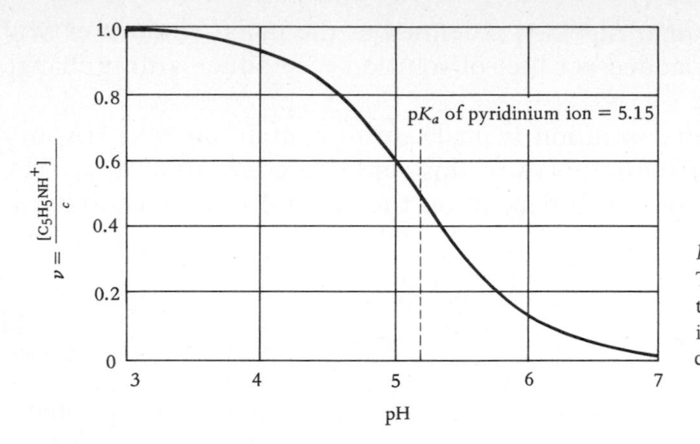

FIGURE 4-8
The fraction of pyridine in the form of the pyridinium ion as a function of pH as calculated from Eq. **33.**

Inspection of the titration curves of Figs. 4-6 and 4-7 shows particularly flat portions of the titration curves corresponding to little pH change with added acid or base. Such regions occur when the titration has been carried halfway to the equivalence point. The solutions corresponding to these regions contain about equal amounts of the parent acid HA and the ion A^- or, for a base, equal amounts of the parent base B and the product BH^+. It follows that buffers can be made up by forming solutions that contain equal concentrations of a weak acid and the salt of the weak acid or a weak base and the salt of the weak base. Furthermore, the pH of such buffer solutions will be equal to the pK_a of the weak acid or the pK_a of the conjugate acid of the base. Thus, an equimolar mixture of acetic acid and sodium acetate will form a buffer solution that will hold the pH near the value 4.75, the value of the pK_a of acetic acid. A solution containing equimolar amounts of ammonia and ammonium chloride would have buffering action that tends to hold the pH at 9.25, the pK_a value of NH_4^+, the conjugate acid of ammonia.

Even for a given buffering system the pH of the buffer solution can be adjusted somewhat by moving away from the equimolar ratio of the two buffer-solution ingredients. The pH values then reached can be calculated by applying the Henderson-Hasselbalch equation, Eq. **29** or **30.** The concentrations of HA and A^-, or of B and BH^+, must not be shifted too far from their initial values. If the imbalance between the two reagents is too great, the system does not correspond to the best buffer region and the buffering action is weakened. Some useful buffer solutions for various pH ranges are shown in Table 4-2.

The ability of a buffer solution to resist pH change is reported as the *buffer value* or *buffer index*. It is defined as the molar amount of acid or base that must be added per liter of solution to produce a unit change in pH.

Suppose a buffer solution is made up to contain an acid HA at a concentration c_a and the salt NaA of this acid at a concentration c_s. The pH and the value of $[H^+]$ will depend on the A^- and HA concentrations through the relation

$$\frac{[H^+][A^-]}{[HA]} = K_a \quad (36)$$

Initially $[A^-] = c_s$ and $[HA] = c_a$, but these relations will not be maintained as acid or base are added to test the buffering action of the buffer. Still valid, as long as the volume of the buffer is not appreciably changed, will be the relation

$$[A^-] + [HA] = c_s + c_a$$

or
$$[HA] = (c_s + c_a) - [A^-] \quad (37)$$

TABLE 4-2
Buffer Solutions and the pH about Which Their Effective Ranges Are Centered
All but the acetic acid example are known as *Lubs buffer mixtures.*

reagents used in buffer preparation	**buffering equilibrium**	pK_a **of acid** = pH **of buffer for equimolar amounts**
Boric acid and sodium hydroxide	$H_3BO_3 \rightleftarrows H^+ + H_2BO_3^-$	9.2
Potassium dihydrogen phosphate and sodium hydroxide	$H_2PO_4^- \rightleftarrows H^+ + HPO_4^{2-}$	7.2
Potassium acid phthalate and sodium hydroxide	$C_6H_4(COOH)(COO^-) \rightleftarrows H^+ + C_6H_4(COO^-)_2$	5.4
Acetic acid and sodium acetate	$CH_3COOH \rightleftarrows H^+ + CH_3COO^-$	4.8
Potassium acid phthalate and hydrochloric acid	$C_6H_4(COOH)_2 \rightleftarrows H^+ + C_6H_4(COOH)(COO^-)$	2.9

Insertion of this relation in Eq. **36** gives

$$\frac{[H^+][A^-]}{(c_s + c_a) - [A^-]} = K_a$$

and $$[A^-] = \frac{(c_a + c_s)K_a}{[H^+] + K_a} \qquad 38$$

Addition of base to the buffer solution produces a corresponding increase in A^- as a result of the reaction $HA + OH^- \rightleftarrows A^- + H_2O$. Thus, the buffer index is given by the derivative

$$\frac{d[OH^-]}{d(pH)} = \frac{d[A^-]}{d(pH)} \qquad 39$$

This can also be expressed in terms of a derivative with respect to $[H^+]$ as

$$\frac{d[OH^-]}{d(pH)} = \frac{d[A^-]}{d[H^+]}\frac{d[H^+]}{d(pH)} \qquad 40$$

The first term is obtained from Eq. **38** as

$$\frac{d[A^-]}{d[H^+]} = -\frac{(c_a + c_s)K_a}{([H^+] + K_a)^2}$$

The second term of Eq. **40** is evaluated with the aid of the derivative relation

$$d \ln x = \frac{1}{x}\,dx$$

or $$-2.303d(-\log x) = \frac{1}{x}\,dx$$

If this general expression is interpreted in terms of $x = [H^+]$, then $-\log x = pH$ and

$$-2.303d(pH) = \frac{d[H^+]}{[H^+]}$$

or $$\frac{d[H^+]}{d(pH)} = -2.303[H^+] \qquad 41$$

Finally, we arrive at

$$\text{Buffer index} = \frac{2.303(c_a + c_s)K_a\,[H^+]}{([H^+] + K_a)^2} \qquad 42$$

You see that the greater the concentration of the buffering reagents the greater the buffering index. Using the calculus procedure to find the value

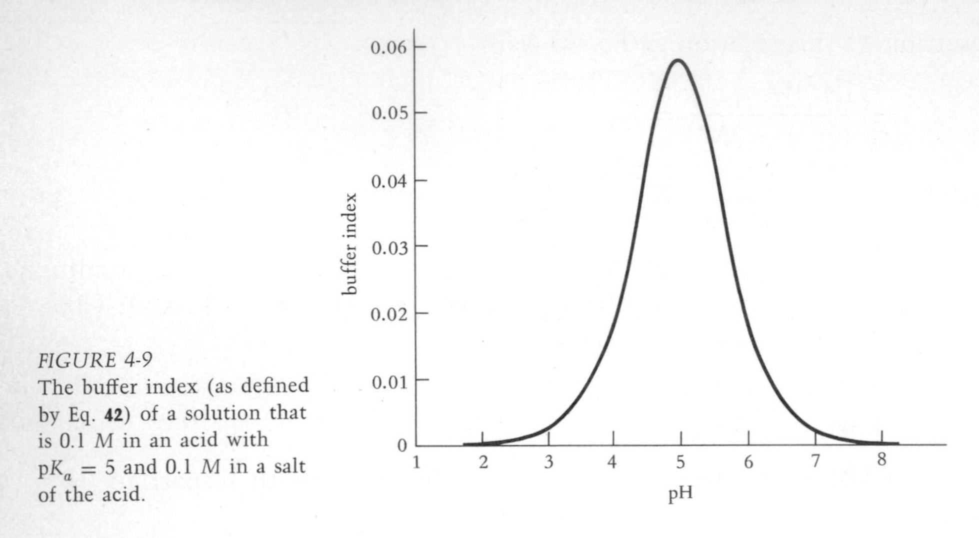

FIGURE 4-9
The buffer index (as defined by Eq. **42**) of a solution that is 0.1 M in an acid with $pK_a = 5$ and 0.1 M in a salt of the acid.

of $[H^+]$ for a maximum buffer index shows that this occurs when $[H^+] = K_a$, that is, when $pH = pK_a$. A display of the buffer-index–pH variation is graphed in Fig. 4-9, from which you can see that a buffering action is really effective only within about one pH unit of the pK_a value.

4-5 POLYPROTIC ACIDS

Now let us consider the deduction of pK_a values and their use to calculate species present in a solution of given pH for polyprotic acids. An example of such an acid is phosphoric acid, H_3PO_4, for which three ionization steps can be recognized:

$$\begin{aligned} &H_3PO_4 \rightleftarrows H_2PO_4^- + H^+ \\ &H_2PO_4^- \rightleftarrows HPO_4^{2-} + H^+ \\ &HPO_4^{2-} \rightleftarrows PO_4^{3-} + H^+ \end{aligned} \qquad \textbf{43}$$

Phosphoric acid is particularly suitable here as an example because many important biological molecules contain phosphate-related groups. These groups enter into multiple-step acid-base processes closely analogous to those of phosphoric acid itself.

Phosphoric acid is an example of a *polyprotic*, or *polybasic*, acid, whose ionization steps can be treated one at a time. A titration curve for phosphoric acid solutions, Fig. 4-10, shows the stepwise neutralizations of the three acid hydrogens. As a result, it is possible to carry over the procedures developed for simple acids in the preceding sections. For example, the pK_a values for the three ionization steps can be deduced, as shown

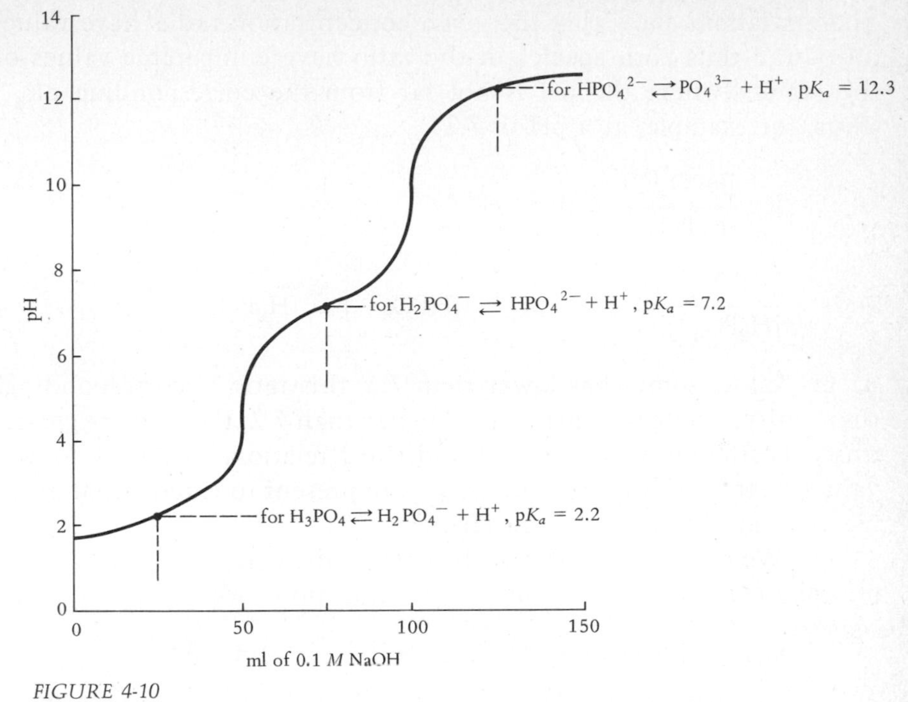

FIGURE 4-10
The titration of 50 ml of 0.1 *M* H_3PO_4 solution by an 0.1 *M* NaOH solution.

in Fig. 4-10. We thus have

$$\begin{aligned} H_3PO_4 &\rightleftarrows H_2PO_4^- + H^+ && pK_a = 2.2 \\ H_2PO_4^- &\rightleftarrows HPO_4^{2} + H^+ && pK_a = 7.2 \\ HPO_4^{2-} &\rightleftarrows PO_4^{3-} + H^+ && pK_a = 12.3 \end{aligned} \tag{44}$$

We can proceed to quantitative results for the relative amounts of the species present by recognizing that at any pH there will be, at most, two major phosphate species. For example, at pH values equal to the pK_a values there are equal concentrations of the species related by the equation corresponding to that pK_a. Additional concentration ratios are obtained from the Henderson-Hasselbalch equations, which can be rearranged to

$$\begin{aligned} \log \frac{[H_2PO_4^-]}{[H_3PO_4]} &= pH - 2.2 \\ \log \frac{[HPO_4^{2-}]}{[H_2PO_4^-]} &= pH - 7.2 \\ \log \frac{[PO_4^{3-}]}{[HPO_4^{2-}]} &= pH - 12.3 \end{aligned} \tag{45}$$

These relations show that the given concentration ratios have values near unity and thus both species in the ratio have comparable values only in the regions where the pH is not far from the corresponding pK_a value. Thus, for example, at a pH of 7.2,

$$\log \frac{[HPO_4^{2-}]}{[H_2PO_4^-]} = 0$$

and $$\frac{[HPO_4^{2-}]}{[H_2PO_4^-]} = 1 \qquad \text{or} \qquad [HPO_4^{2-}] = [H_2PO_4^-]$$

At pH values somewhat lower than 7.2, this ratio is correspondingly less than unity, whereas at pH values higher than 7.2, this ratio is greater than unity. Furthermore, as the first and third relations of Eq. **45** show, at pH values near 7, both H_3PO_4 and PO_4^{3-} are present to extents that are negligible compared to $H_2PO_4^-$ and HPO_4^{2-}.

We can proceed also to displays of the extent of protonation of the phosphate ion as a function of concentration. We need only introduce

$$c_{total} = [H_3PO_4] + [H_2PO_4^-] + [HPO_4^{2-}] + [PO_4^{3-}] \qquad \textbf{46}$$

The fact that at a given pH at most two of the four species have appreciable concentrations allows the system to be treated in the same way as simple acids were in Sec. 4-3. Thus we have the display of the species as a function of pH shown in Fig. 4-11. The extent of protonation can also be calculated. It is given by

$$\nu = \frac{3[H_3PO_4] + 2[H_2PO_4^-] + [HPO_4^{2-}]}{c}$$

where c is the total phosphate concentration. As Fig. 4-12 shows, the curve for this average number of protons has steep portions in regions where the dominant species is changing over to a less protonated one and flat portions where no appreciable change in species is occurring.

In addition to phosphate-containing compounds, amino acids and proteins represent large classes of compounds that can accept a number of protons. The simplest examples are the amino acids whose side chains are neither acidic nor basic. The simplest of these, glycine, illustrates the acid-base behavior of these compounds. The three stages of proton donation to the proton-stripped species that would be present in a basic solution are:

Basic form:

$$H_2NCH_2COO^-$$

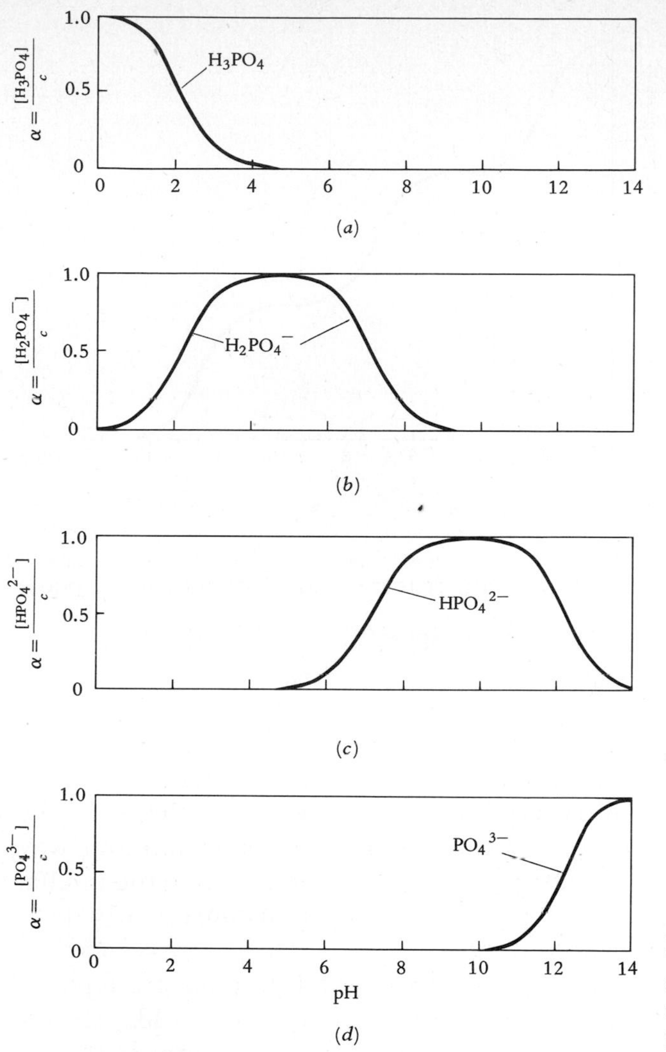

FIGURE 4-11
The fractions of phosphoric acid present at each of the four species as a function of pH.

Intermediate:

$$H_3\overset{+}{N}CH_2COO^- \text{ or } H_2NCH_2COOH$$

Acid form:

$$H_3\overset{+}{N}CH_2COOH$$

The equilibrium between the two intermediate forms, the uncharged H_2NCH_2COOH and the *zwitterion* $H_3\overset{+}{N}CH_2COO^-$, involves no net

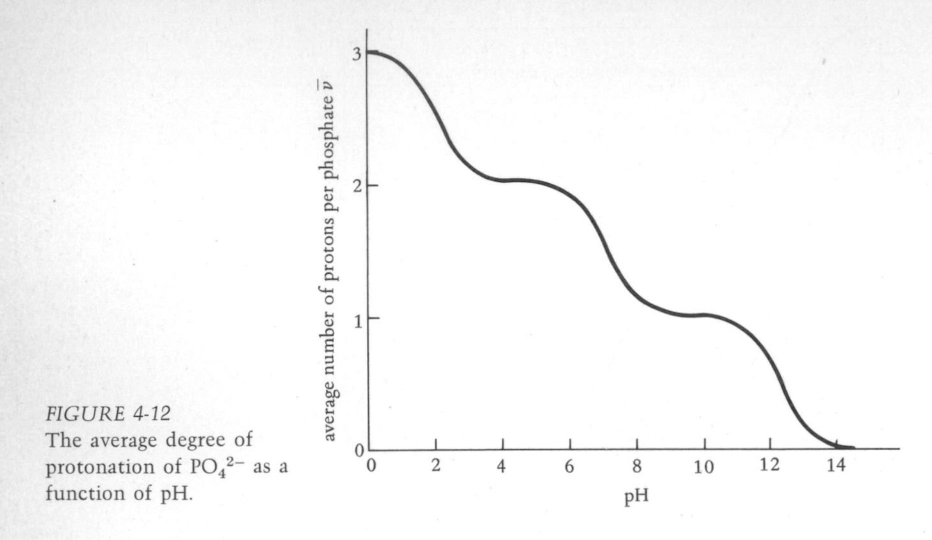

FIGURE 4-12
The average degree of protonation of PO_4^{2-} as a function of pH.

addition or removal of protons. The reaction connecting these species is

$$H_2NCH_2COOH \rightleftharpoons H_3\overset{+}{N}CH_2COO^-$$

and $$K = \frac{[H_3\overset{+}{N}CH_2COO^-]}{[H_2NCH_2COOH]} \qquad \textbf{47}$$

Thus the ratio of these species is independent of the pH. As a result, in treating acid-base equilibria we could describe the species that exists in the intermediate pH range in either way, or in some noncommittal way, as $H_5NC_2O_2$. In fact, at 25°C the ratio of the zwitterion form to the uncharged form is about 200,000:1, and on the basis it is customary to depict the species present by the zwitterion form.

The titration curves for amino acids and proteins are reported in three different ways. One, illustrated for glycine in Fig. 4-13*a*, shows the pH changes that occur when acid is added to the basic form of the amino acid, i.e., to the amino acid that would be present in the form it adopts in a strongly basic solution. For glycine this species is $H_2NCH_2COO^-Na^+$, and the titration curve is thus like that for a base with two proton sites. Alternatively, you will see titration curves, as in Fig. 4-13*b*, for the acid form of the amino acid, which for glycine is $Cl^-H_3\overset{+}{N}CH_2COOH$. Finally, the titration curve is often shown for the separate additions of acid and of base to a solution of the amino acid itself. For glycine the presentation would correspond to starting at the middle of Fig. 4-13 and working to the right and the left.

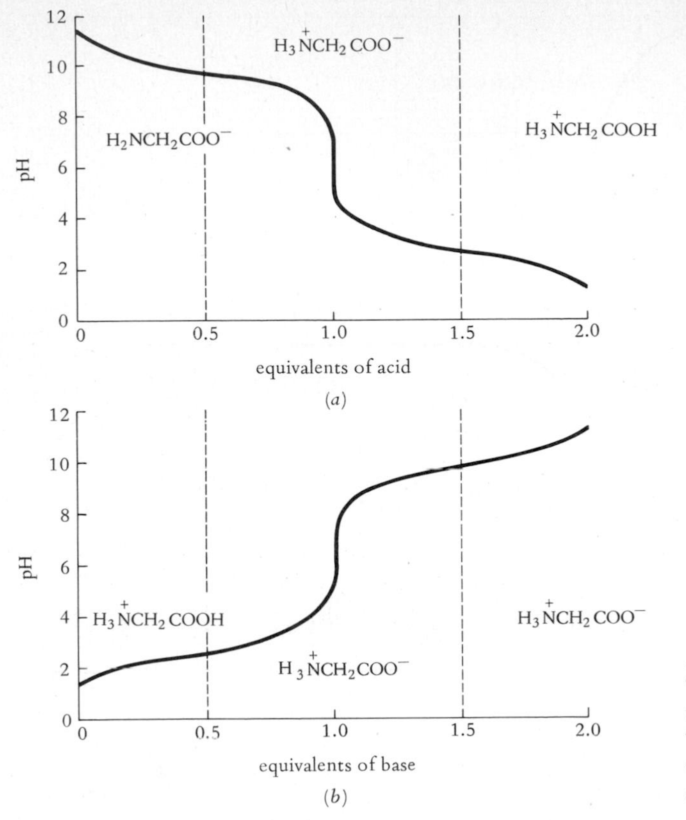

FIGURE 4-13
Two ways of presenting the titration curve for glycine. The principal species in the three titration regions are shown.

From any of the presentations in Fig. 4-13 the dissociation pK_a values that are deduced are

$$H_3\overset{+}{N}CH_2COOH \rightleftharpoons H_3\overset{+}{N}CH_2COO^- + H^+ \qquad pK_a = 2.6$$

$$H_3\overset{+}{N}CH_2COO^- \rightleftharpoons H_2NCH_2COO^- + H^+ \qquad pK_a = 9.8 \qquad \textbf{48}$$

The species present in a glycine solution as a function of pH can now be deduced, e.g., by using Eq. **29,** and displayed as in Fig. 4-14.

Often reported for amino acids and proteins is the pH at which the average species present as a result of the net dissociation and association of protons carries no net charge. For glycine the species that qualify are the zwitterion $H_3\overset{+}{N}CH_2COO^-$ and the uncharged form H_2NCH_2COOH. But solutions in which these are the predominant forms also contain minor amounts of the positive species $H_3\overset{+}{N}CH_2COOH$ and the negative species $H_2NCH_2COO^-$. The pH at which the ratio of each of these to the neutral

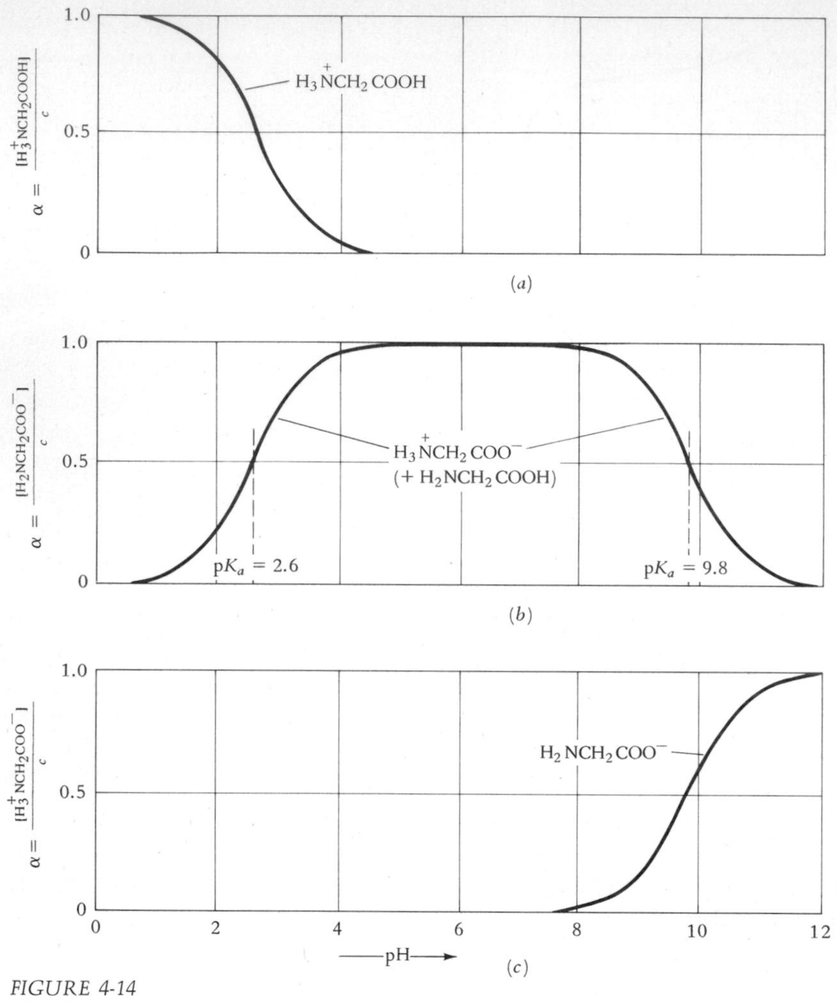

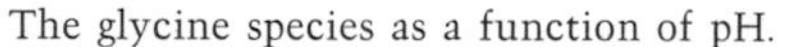

FIGURE 4-14
The glycine species as a function of pH.

species is the same is seen from Fig. 4-14 to be that which occurs halfway between the titration-curve breaks on either side of the neutral species. For glycine this pH, often known as the *isoionic point,* is 6.2. (The isoionic point is usually nearly equal to the *isoelectric point,* which was defined as the pH at which the solute species has no tendency to migrate in an applied electric field. This lack of movement means that the species and any ions that are associated with it constitute a neutral entity. Since some of the associated ions can be other than H^+ ions, the isoelectric point can differ somewhat from the isoionic point.)

Some of the important amino acids carry acidic or basic side chains, as shown in Fig. 4-15. Titration curves for two examples, lysine and glutamic acid, are shown in Figs. 4-16 and 4-17. Breaks in the curves corresponding to the titration of the various acid and basic groups do not appear in a clear stepwise sequence. As a result, pK_a values cannot be assigned to the various reactive groups in the molecule. A similar situation exists with protein titrations where the many amino acid side chains lead to a titration curve with no clear steps. In such cases two procedures can be followed.

One is to deduce the extent of protonation from the data of the titration curve. The titration curve shows the amount of acid or base that must be added to a solution containing some known amount of protein to produce various pH values. Except in solutions of very high or low pH values, each molecule of acid added to the solution transfers a proton to the protein molecule. Each molecule of base removes a proton. Thus the change in degree of protonation of the protein that corresponds to a change in the pH of the solution can be deduced from the amount of acid or base required to produce that pH change. The results that are obtained are illustrated by those for ribonuclease shown in Fig. 4-18. No direct information is given on which groups are protonated at any given pH, but the net protonation or average number of protons accepted at any pH is revealed.

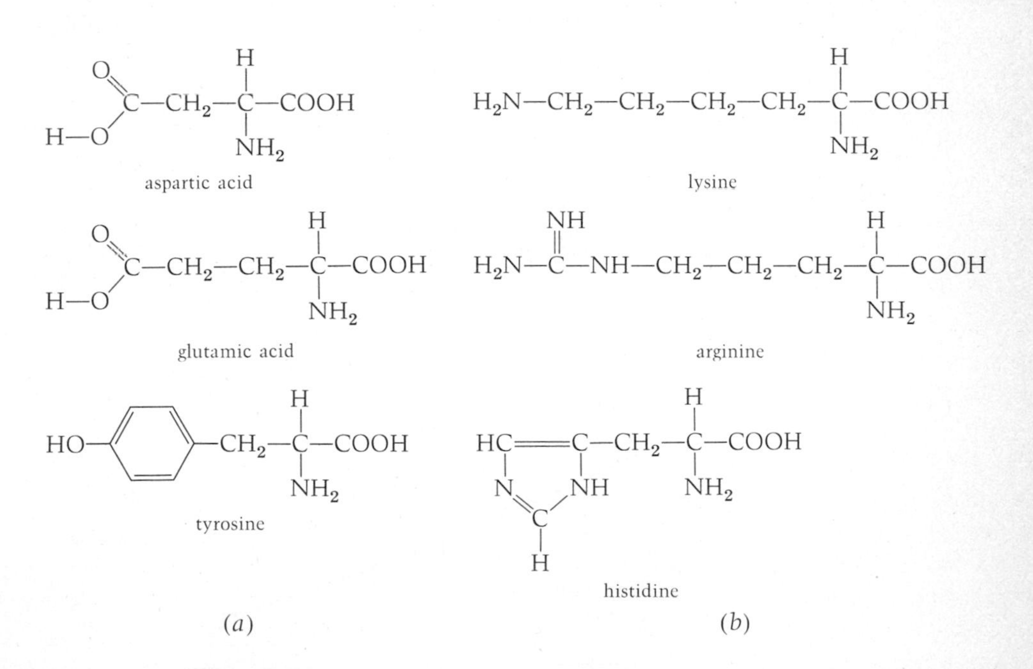

FIGURE 4-15
The amino acids which contain (*a*) acidic and (*b*) basic side chains and produce acidic and basic sites when incorporated in a protein.

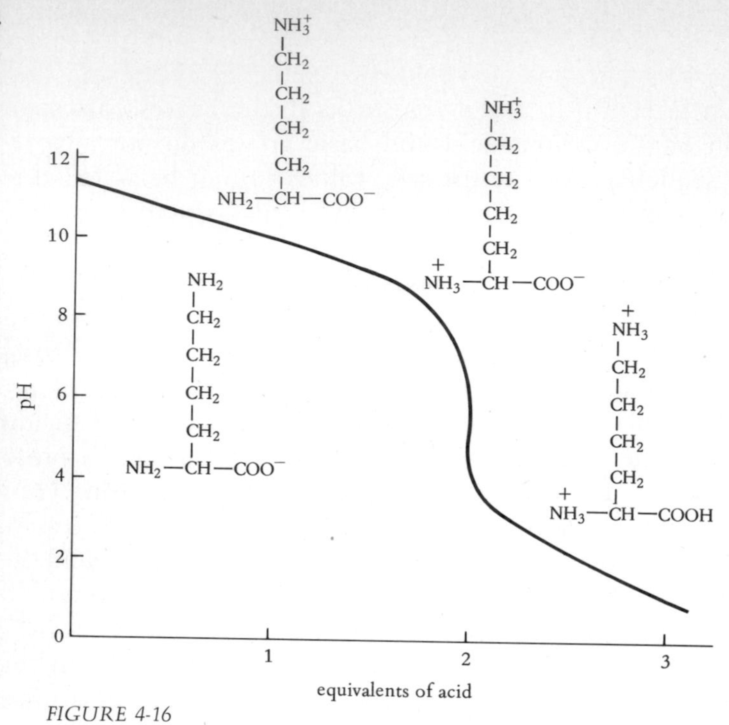

FIGURE 4-16
A titration curve for the sodium salt of lysine. The sequence of species that are formed are shown.

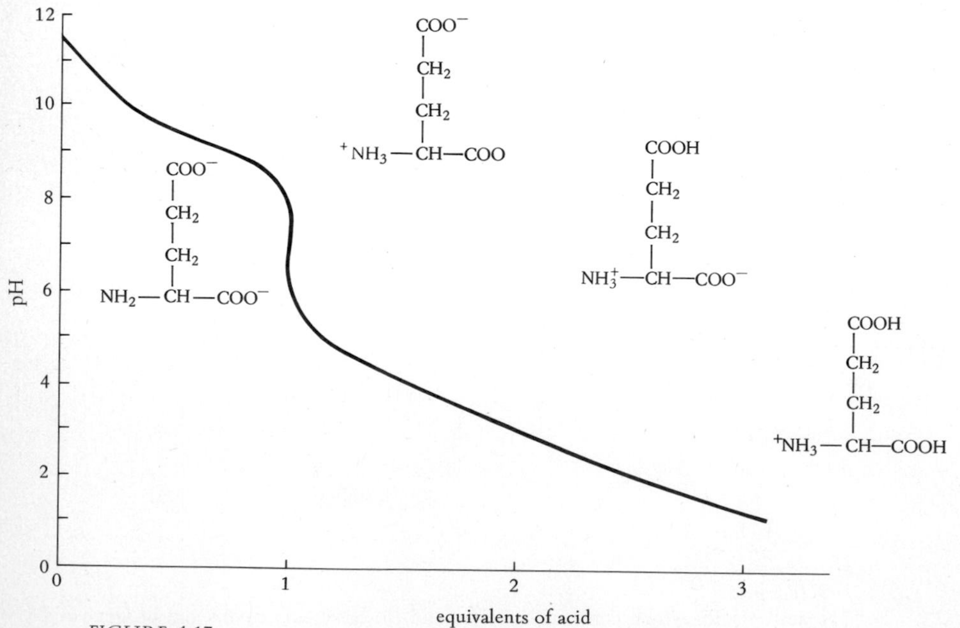

FIGURE 4-17
A titration curve for the sodium salt of glutamic acid. The sequence of species that are formed are shown.

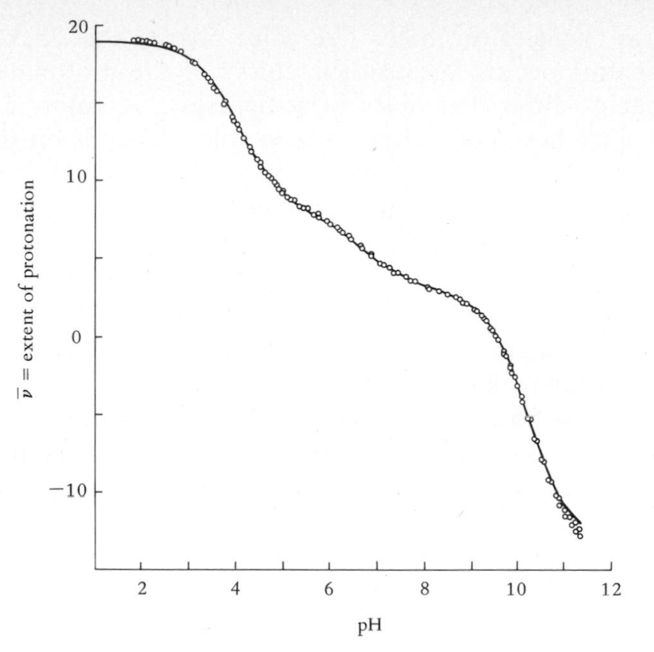

FIGURE 4-18
The degree of protonation of ribonuclease in a 6 *M* guanidine hydrochloride solution at 25°C as a function of pH. The isoionic point of ribonuclease under these conditions is at pH 9.54. [*From Y. Nozaki and C. Tanford, J. Am. Chem. Soc.,* **89:**742 (1967).]

The second procedure uses additional reagents or measurement techniques to provide more detailed information than pH measurements. A particularly powerful method depends on spectral measurements made during the course of the titration. In Sec. 4-7 spectrophotometric titrations will illustrate the use of visible-ultraviolet spectra in studies of molecular species in solution.

4-6 INDICATORS AND VISIBLE-ULTRAVIOLET ABSORPTIONS

In acid-base equilibrium studies we often need a means other than that provided by pH-meter measurements for determining what molecular species are present. That this can be done by using visible spectra, which consist of refinements of the observation of color, is illustrated by the action of acid-base *indicators*. Indicators are weak acids or bases. In solution the equilibrium they establish can be represented by

$$H \cdot In \rightleftharpoons H^+ + In^-$$

or $$In' + H^+ \rightleftharpoons In' \cdot H^+$$

where H · In and In′ represent acid and base indicator molecules. The ratio of the two forms of the indicator depends on the pH of the solution and varies from the value of unity as the pH varies from the value of pK_a of the acid or the conjugate acid of the base. The color change that makes these compounds indicators occurs because the colors of the protonated and nonprotonated species differ. Examples of structures and colors are shown in Fig. 4-19. Thus we have compounds whose color depends on the extent of protonation.

Before exploring practical applications, let us consider the molecular basis of the absorption of visible or ultraviolet radiation. The energy of quanta in these regions, as shown in Fig. 4-20, is much greater than the energy of the microwave and infrared regions studied in Chaps. 1 and 3. We expect, therefore, that absorbed visible-ultraviolet radiation will do more to the absorbing molecule than make it rotate or vibrate with greater energy. Visible-ultraviolet radiation is found to disrupt the electronic structure of the absorbing molecule. In some cases, particularly with small molecules, the disruption causes the molecule to fall apart, in a *dissociation* process.

Phenolphthalein:

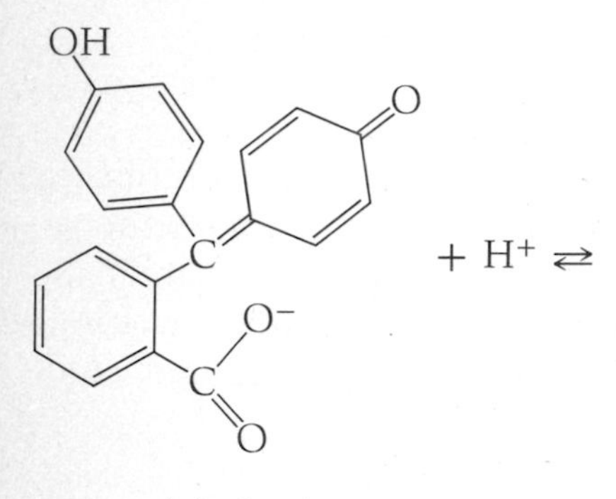

$+ H^+ \rightleftarrows$

OH OH C O C O

red (in base)

colorless (in acid)

Methyl orange:

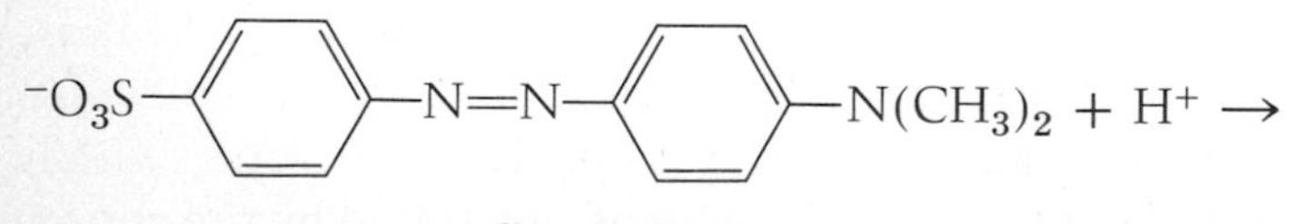

yellow (in base)

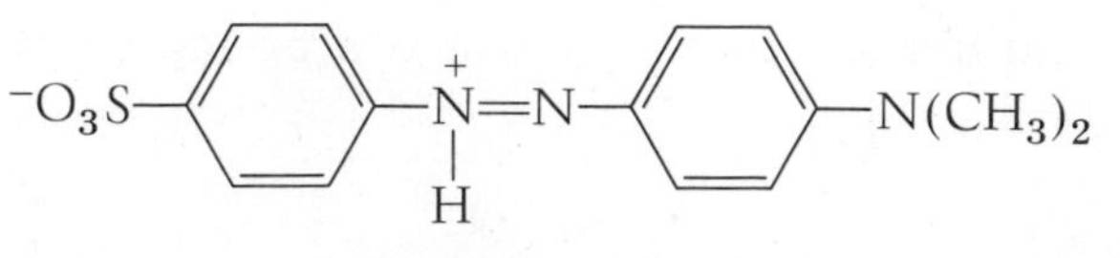

red (in acid)

FIGURE 4-19
The structures and colors of two common indicators.

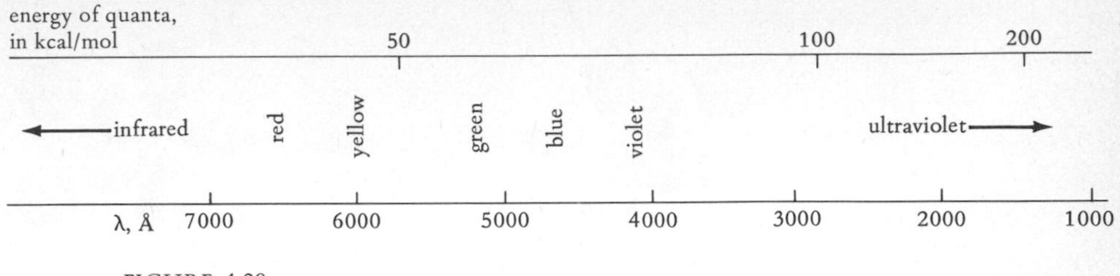

FIGURE 4-20
Wavelength and quantum energies in kilocalories per mole in the visible and ultraviolet spectral regions.

In other cases, an electron of the molecule is hurled out, in an *ionization* process. More often, particularly with large molecules in solution, the absorption of visible or ultraviolet radiation merely changes the electronic structure of the molecule; the molecule is then said to be in an *excited electronic state* rather than in its initial *ground state*.

The initial ground state is usually recovered by the degradation of the excess electronic energy into thermal energy of the molecule and its neighbors. Some molecules, however, reemit some of the absorbed energy in *fluorescence* or *phosphorescence*.

In the ordinarily accessible visible-ultraviolet region, i.e., in the wavelength region out to about 2000 Å, or 200 nanometers (nm), single-bonded compounds generally show no absorptions. The bonding in such compounds apparently requires quanta of even greater energy than provided by ultraviolet radiation to convert the electronic structure into that of an excited state. Two important types of compounds do exhibit visible-ultraviolet absorption: one class consists of unsaturated and aromatic compounds, and the other consists of compounds containing a metal atom. Attention will be restricted here to the first class.

In treating the electronic basis of absorption by unsaturated and aromatic compounds, it is convenient to describe the bonding in terms of single, or σ-bond, components and π-bond components. This approach leads, for example, to the description of the double bond shown in Fig. 4-21. Of particular importance in spectral studies are compounds that consist of *conjugated* double bonds, a designation that can include the special case of aromatic compounds. For some purposes it is adequate to treat bonding in such compounds in terms of *resonance* structures, as shown in Fig. 4-22. For spectral studies the *molecular-orbital* approach is more rewarding. In qualitative, schematic treatments or in detailed quantum-mechanical analyses it leads to an energy pattern for the electronic states that can be formed by various arrangements of the electrons. The π molecular orbitals of

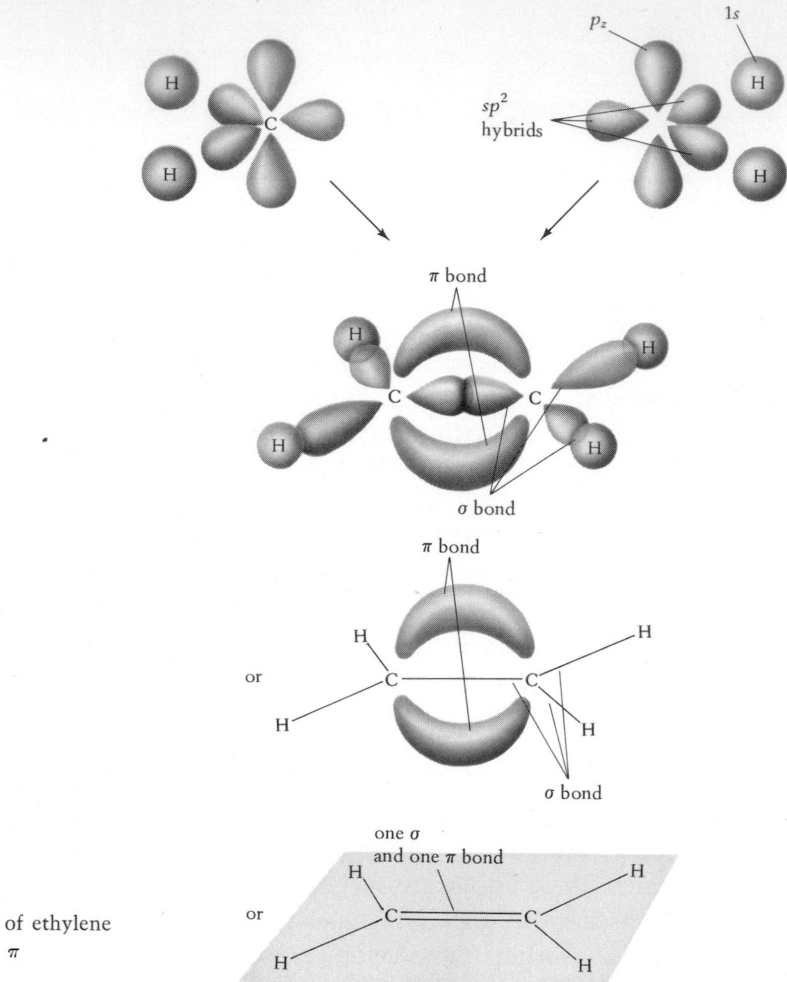

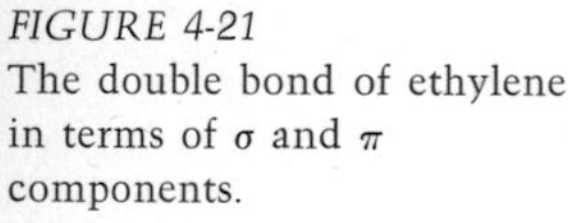
FIGURE 4-21
The double bond of ethylene in terms of σ and π components.

benzene are shown in Fig. 4-23. Absorption bands that are treated on the basis of such orbitals are π-π^* bands, where π^* indicates an excited π-electron state.

A guide to some of the qualitative features of electronic energy patterns is provided by the application of the particle-in-a-box model of Sec. 3-3 to long conjugated molecules. In such molecules the π electrons are said to be *delocalized* and relatively free to move throughout the length of the molecule. They see the molecule to which they are bound as a region of roughly uniform potential energy. The π electrons of the molecule can thus be treated (if we ignore the interaction between them) as independent

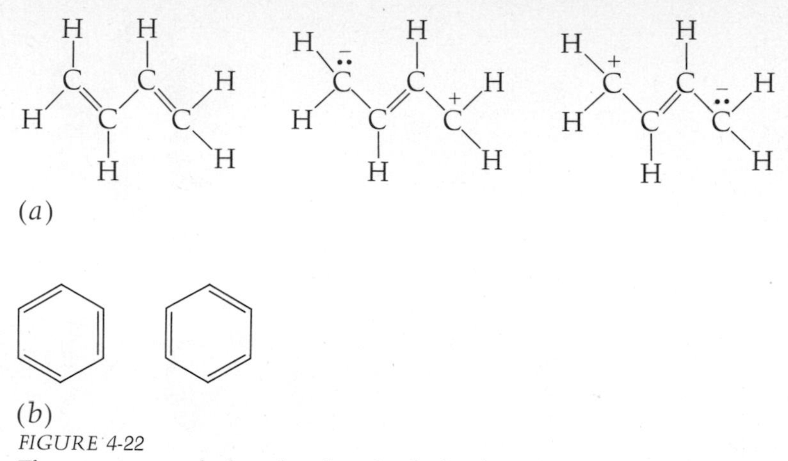

FIGURE 4-22
The resonance method used to describe the bonding in conjugated and aromatic molecules: (*a*) butadiene; (*b*) benzene.

particles confined to a one-dimensional box, the length of the box being approximately the length of the molecule. The allowed energies for the electrons are then given by Eq. **9** or **16** of Sec. 3-3:

$$\epsilon = \frac{n^2h^2}{8ml^2} \qquad n = 1, 2, 3, \ldots$$

It remains to require, according to the Pauli exclusion principle, that in any one molecule not more than two electrons can have the same value of the quantum numbers that determine the energy and probability function of the electron. Here, only two electrons can have a given value of n. With this rule, the calculated states for a conjugated molecule can be filled, as in the example of Fig. 4-24. Furthermore, the energy of the transition of the highest-energy electron to the next higher available state can be calculated and compared with the energy of the quanta that are observed to be absorbed. Generally, as in Fig. 4-24, there is the rough agreement that is all that can be expected on the basis of such a simple model.

More important than the numerical results is the recognition, from the form of $\epsilon = n^2h^2/8ml^2$, that the larger the region in which the electrons are free to move, the more closely spaced the allowed electronic states will be. Thus, the more conjugated a system is the more likely it becomes that absorption will occur in the accessible ultraviolet or even the visible region. This idea is confirmed by the highly extended and conjugated structures that make up the indicator molecules of Fig. 4-19.

The presence of heteroatoms, as in molecules containing oxygen or nitrogen atoms as well as carbon atoms, complicates this simple idea

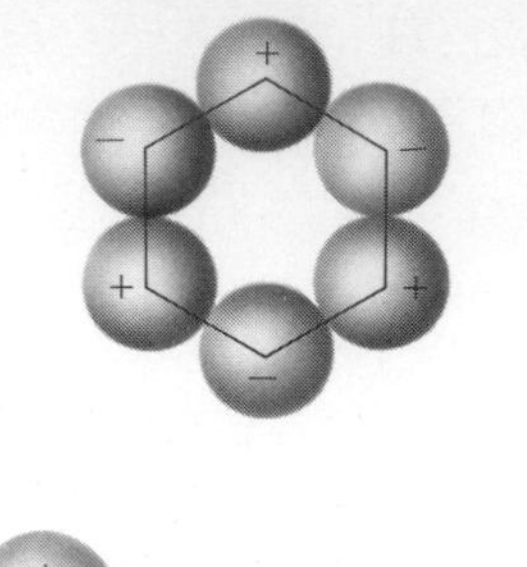

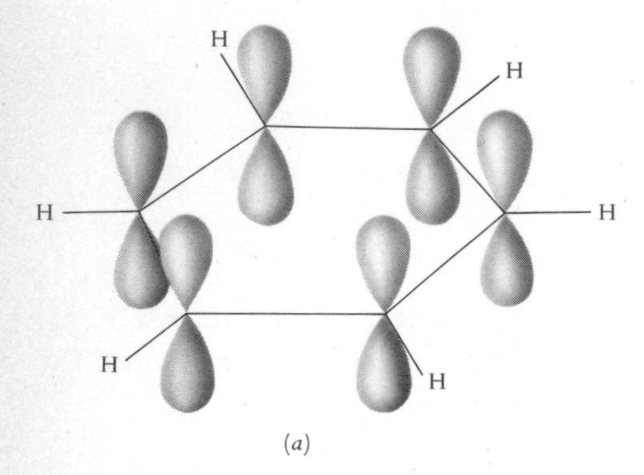

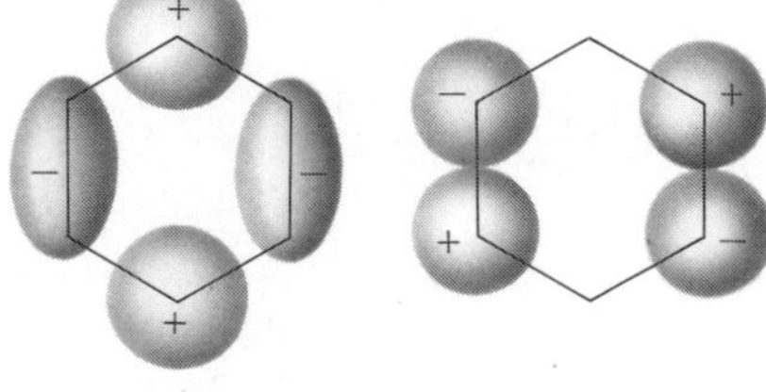

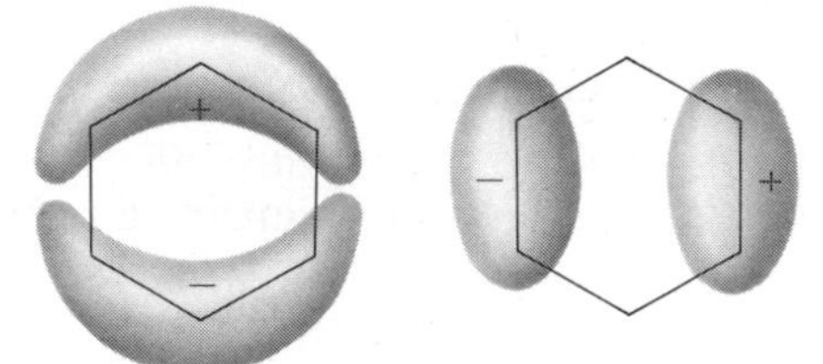

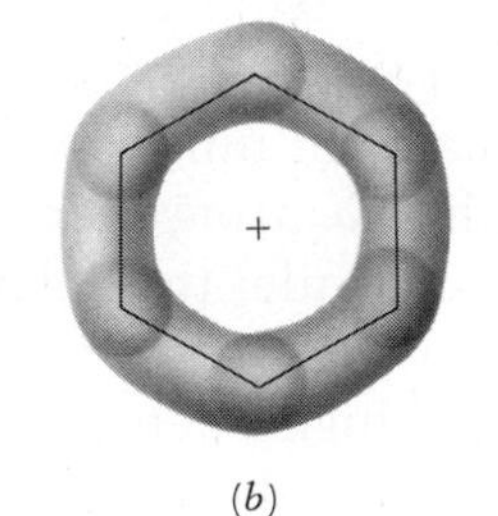

FIGURE 4-23
Molecular orbitals: (*a*) the π orbitals on each carbon atom that are not involved in the three σ bonds of each orbital; (*b*) the molecular orbitals that can be constructed, arranged in order of energy. As with a particle in a box, the more nodes in a wave function, the higher the energy.

of π-π^* transitions in uniform potential systems. First, even the crude approximation that the π electrons experience a uniform potential throughout the molecule must be relinquished. Second, the presence of N and O atoms raises the possibility that the nonbonding electrons of these atoms might be involved in the spectral transition. Absorption bands that

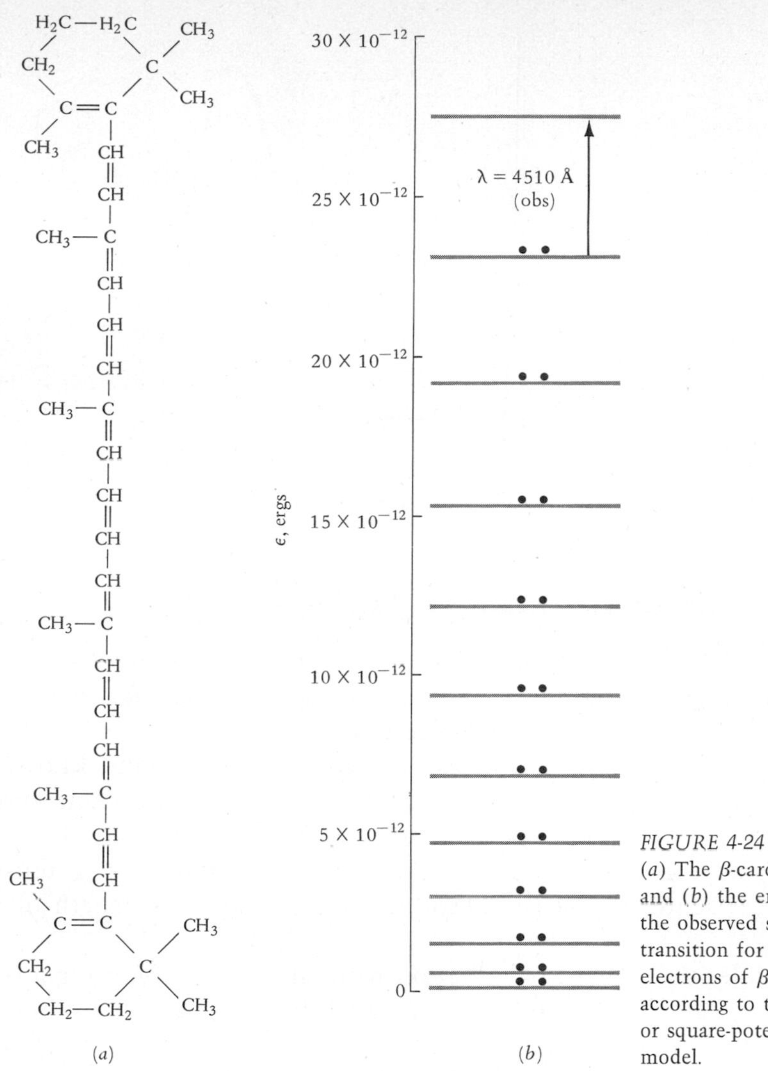

FIGURE 4-24
(*a*) The β-carotene molecule and (*b*) the energy levels and the observed spectroscopic transition for the 22 π electrons of β-carotene according to the free-electron or square-potential-well model.

are attributed to such involvement are called n-π^* bands, the n implying nonbonding electrons localized on the heteroatoms.

Even more disrupting is the presence of a charge, as from the addition or loss of a proton in acid-base reactions. For the addition of a proton and the introduction of a positive charge, the added influence that can be expected is a tightening of the electron binding, a closing up of the energy spacing of the electronic states, and a shift of absorption bands to shorter wavelengths. For the loss of a proton the opposite effect is likely. These effects are illustrated in Fig. 4-25.

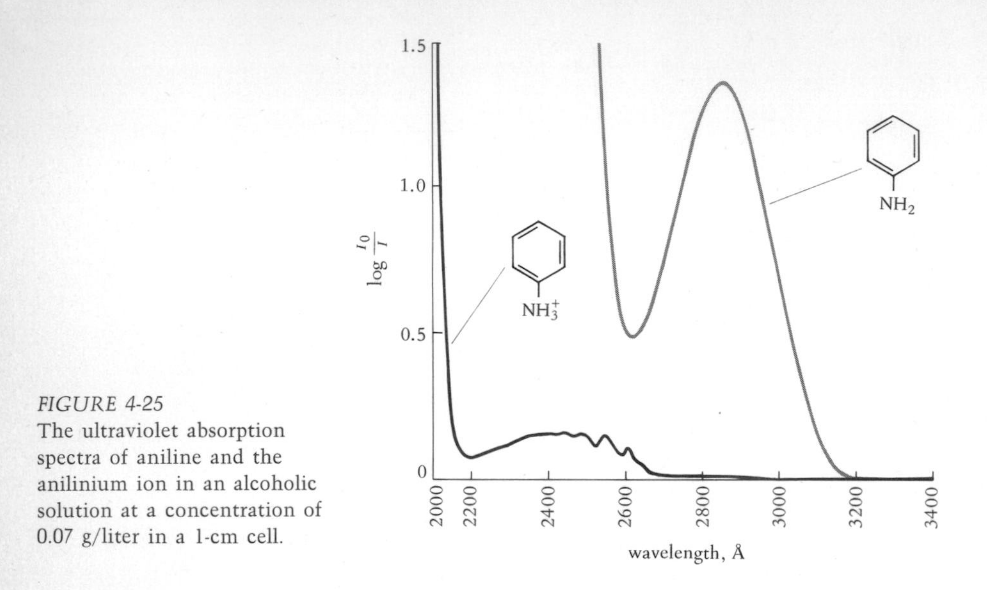

FIGURE 4-25
The ultraviolet absorption spectra of aniline and the anilinium ion in an alcoholic solution at a concentration of 0.07 g/liter in a 1-cm cell.

4-7 SPECTROPHOTOMETRIC TITRATIONS

Figure 4-25 illustrates that both the extent of absorption of radiation and the wavelength at which the maximum absorption occurs are affected by protonation. The titration of a molecular species that is so affected can be followed by making spectral measurements. Such titrations, known as *spectrophotometric titrations,* can be valuable supplements to pH titrations. To see how spectra can be used to determine the amount of an absorbing species we must investigate the relation between radiation intensity, amount or concentration of the absorbing species, and the length of the sample cell.

Consider the passage of radiation through a cell containing a solution of concentration c mol/liter that absorbs the radiation. The intensity of the radiation entering the cell is I_0, and after passing through a length of the solution the intensity has been reduced to I. These variables often are related by the assumption of an exponential falloff in intensity given by the *Beer-Lambert law*

$$I = I_0 e^{-\epsilon' lc} \qquad \text{or} \qquad \ln \frac{I_0}{I} = \epsilon' lc$$

$$\text{and} \qquad I = I_0 10^{-\epsilon lc} \qquad \text{or} \qquad \log \frac{I_0}{I} = \epsilon lc$$

The proportionality factors ϵ' and ϵ, often referred to as *extinction coeffi-*

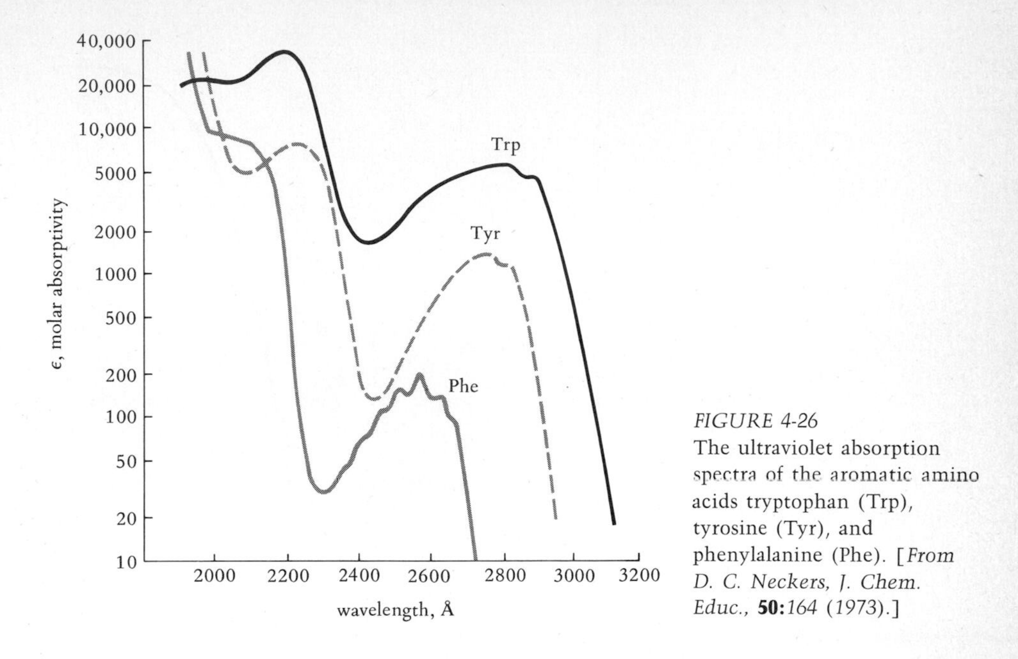

FIGURE 4-26
The ultraviolet absorption spectra of the aromatic amino acids tryptophan (Trp), tyrosine (Tyr), and phenylalanine (Phe). [*From D. C. Neckers, J. Chem. Educ.*, **50:***164* (*1973*).]

cients or *absorptivity*, are measures of the extent to which the solute absorbs radiation at the specified wavelength. Spectra are often presented with an ordinate scale of percent transmission ($I/I_0 \times 100$ percent), or absorbance [$A = \log (I_0/I)$]. From such data, if values of l and c for the spectrum are also given, ϵ' or ϵ values can be deduced.

The amino acid side chains that have ultraviolet absorptions with suitable wavelengths and extinction coefficients to permit their protonation to be followed by spectrophotometric titrations are shown in Fig. 4-26. The absorption at 295 nm versus pH for ribonuclease (Fig. 4-27) illustrates the additional information provided by such titrations. The absorption at this wavelength is a measure of the nonprotonated tyrosine side chains. In more acidic solutions, i.e., with pH values lower than pH 9, all of the six tyrosine side chains known to be in ribonuclease are apparently protonated. As the pH is raised, absorption at 295 nm increases, corresponding to the formation of unprotonated tyrosine side chains.

It follows that the loss of protons with increasing pH in the region of pH 9 to 11 can be attributed to the tyrosine groups. In this way the contributors to the various pH segments of a curve of degree of protonation versus pH, like that of Fig. 4-18, can be deduced. Furthermore, these experimental results allow effective pK_a values to be assigned to the acid-base groups in a protein molecule.

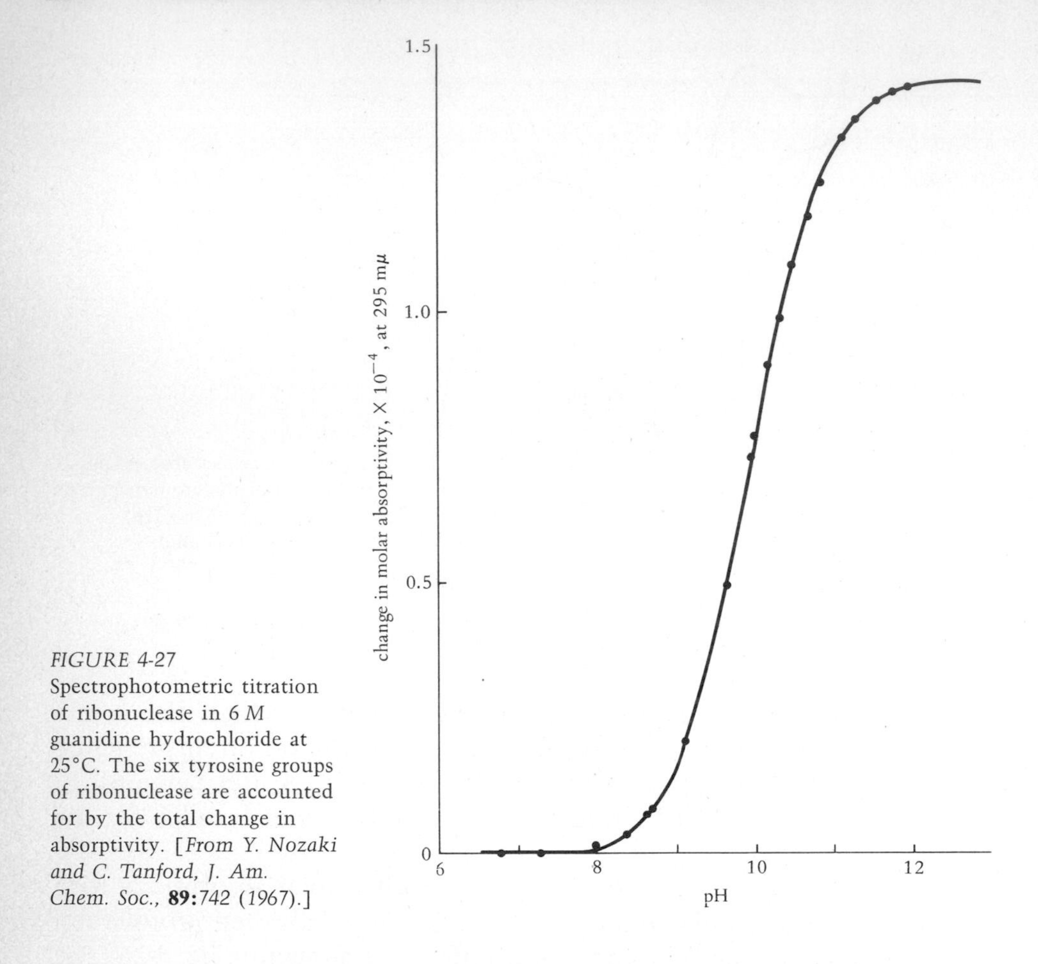

FIGURE 4-27
Spectrophotometric titration of ribonuclease in 6 *M* guanidine hydrochloride at 25°C. The six tyrosine groups of ribonuclease are accounted for by the total change in absorptivity. [*From Y. Nozaki and C. Tanford, J. Am. Chem. Soc.,* **89:**742 (1967).]

4-8 INTERPRETATIONS OF MULTIPLE BINDING IN TERMS OF INDIVIDUAL BINDING SITES

The preceding sections have discussed experiments revealing the nature of the species involved in titrations and the extent of their involvement. Additional progress can be made by considering model compounds that can bind species, such as protons, on some number of sites, n, per molecule. Initially this number and the effect of occupancy of some of these sites on the binding power of the remaining sites are unknown. Equations can be derived, on the basis of certain simplifying assumptions, that allow these features to be deduced from observables.

Let us assume we are dealing with the binding of some species A to a protein molecule P. As illustrated for the binding of protons in Sec. 4-5, the average number of species A bound to a protein molecule at any concentration of A can be deduced. This observable is denoted by $\bar{\nu}$; that is,

$\bar{\nu}$ = average number of A molecules bound to a P molecule

If there are n binding sites on the protein molecule, and if $\bar{\nu}$ is the average number of sites occupied, we can write

$$\frac{\text{Number of sites occupied}}{\text{Number of sites vacant}} = \frac{\bar{\nu}}{n - \bar{\nu}} \qquad \mathbf{49}$$

Now let us suppose that the binding power of any one site is completely independent of whether other sites are vacant or occupied. Then each site would appear to an A molecule to be a potential reactant, just as if each site were on an individual reactant molecule. The equilibrium that would be set up would be described by the equation

Vacant site + A $\rightleftarrows$ occupied site

The equilibrium expression for this reaction is

$$K = \frac{[\text{occupied sites}]}{[\text{vacant sites}][\text{A}]} \qquad \mathbf{50}$$

With Eq. **49** this can be converted to

$$K = \frac{\bar{\nu}}{(n - \bar{\nu})[\text{A}]}$$

or

$$\frac{\bar{\nu}}{n - \bar{\nu}} = K[\text{A}] \qquad \mathbf{51}$$

Measurements of $\bar{\nu}$ and [A] can be used to deduce n and K if the independent action of the sites is valid. Graphical treatments that are usually applied are suggested by rearranging Eq. **51** to

$$\bar{\nu} = n - \frac{1}{K}\frac{\bar{\nu}}{[\text{A}]} \qquad \mathbf{52}$$

or

$$\frac{1}{\bar{\nu}} = \frac{1}{n} + \frac{1}{nk}\frac{1}{[\text{A}]} \qquad \mathbf{53}$$

A system that appears to be satisfactorily treated by this noncooperative-site approach is shown in Fig. 4-28.

Now let us go to the other extreme of the assumptions that can be made about the effect of the occupancy of the sites of a molecule on their binding power.

In *cooperative binding* the binding power of all the sites is enhanced when one or more sites on the molecule are occupied. Such a situation,

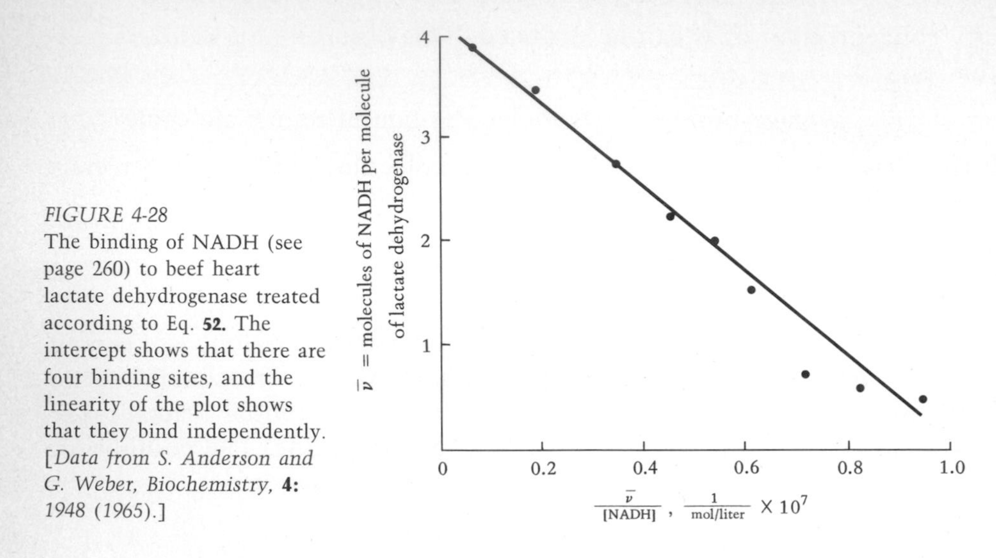

FIGURE 4-28
The binding of NADH (see page 260) to beef heart lactate dehydrogenase treated according to Eq. **52.** The intercept shows that there are four binding sites, and the linearity of the plot shows that they bind independently. [*Data from S. Anderson and G. Weber, Biochemistry,* **4:** *1948 (1965).*]

often found in biologically important molecules, is just the opposite of that found in simple multiple-site compounds. Imagine, for example, the sequential binding of protons to a phosphate ion. The first is bound very tightly, to form the HPO_4^{2-} species. Subsequent binding to form $H_2PO_4^-$ and then H_3PO_4 is progressively weaker. We say that PO_4^{3-} is a better base than HPO_4^{2-}, which, in turn, is a better base than $H_2PO_4^-$. For many complex biological molecules the opposite occurs, and binding is enhanced by the prior occupation of some of the binding sites.

In the limit of highly cooperative binding on a molecule P with n sites, the only species of consequence are those with no sites occupied and those with all sites occupied. The equilibrium is described by

$$P + nA \rightleftarrows PA_n \qquad \textbf{54}$$

The corresponding equilibrium expression is then

$$K = \frac{[PA_n]}{[A]^n\,[P]} \qquad \textbf{55}$$

The ratio $[PA_n]/[P]$ can still be identified with the ratio [occupied sites]/[vacant sites] and thus, according to Eq. **49,** with $\bar{\nu}/(n - \bar{\nu})$. The assumption of highly cooperative sites thus leads us to

$$K = \frac{\bar{\nu}}{[A]^n\,(n - \bar{\nu})} \qquad \textbf{56}$$

or

$$\frac{\bar{\nu}}{n - \bar{\nu}} = K[A]^n \qquad \textbf{57}$$

This result is to be compared with the noncooperative-site result of Eq. **51.**

If the value of n is known or can be estimated, the cooperative and noncooperative nature of the n sites can be recognized by making plots guided by the logarithmic forms of Eqs. **51** and **57:**

Noncooperative:

$$\log \frac{\bar{\nu}}{n - \bar{\nu}} = \log K + \log [A] \qquad \textbf{58}$$

Cooperative:

$$\log \frac{\bar{\nu}}{n - \bar{\nu}} = \log K + n \log [A] \qquad \textbf{59}$$

On this basis, plots of $\log [\bar{\nu}/(n - \bar{\nu})]$ versus $\log [A]$ would be expected to yield straight lines in either case, but the slope would be unity for the noncooperative case and equal to n, the sites per molecule, in the limit of completely cooperative action. Such graphs are called *Hill plots*.

It should be recognized, however, that Eqs. **57** and **59** represent the limit of perfect cooperative action. In actual cases this limit cannot be expected, and thus the slope of the plot of $\log \bar{\nu}/(n - \bar{\nu})$ versus A will be somewhere between 1 and the value of n. Furthermore, at very low concentrations of the binding reagent cooperative action will be specially ineffective, and a slope near the noncooperative value of unity must be expected. Likewise when almost all sites are occupied, the cooperative aspects are again obscured.

The best known of the cooperative binding systems is that of oxygen and hemoglobin. A clear contrast is shown in Fig. 4-29*a*, between the oxygen uptake of hemoglobin, with four binding sites, and the normal uptake by myoglobin with one such site. The assumption of four oxygen sites on a hemoglobin molecule and one on a myoglobin molecule allows the $\bar{\nu}$-versus-oxygen-pressure curves of Fig. 4-29*b* to be drawn. Then the Hill plots of Fig. 4-29*c* show the cooperative nature of the hemoglobin binding. The binding power of the fourth oxygen site is enhanced by a factor estimated at 200 or 300 when the first three sites are occupied. It is known that the overall structure of the hemoglobin molecule changes as oxygen is added. This gives us a basis for understanding the enhanced binding power, but the mechanism responsible is not known.

PROBLEMS

1 What is the concentration of undissociated butyric acid in an 0.300 *M* aqueous butyric acid solution at 25°C? What is the hydrogen-ion concentration? (Use the K_a data of Table 4-1.)

2 An 0.0100 *M* solution of benzoic acid in water at 25°C is found to have a hydro-

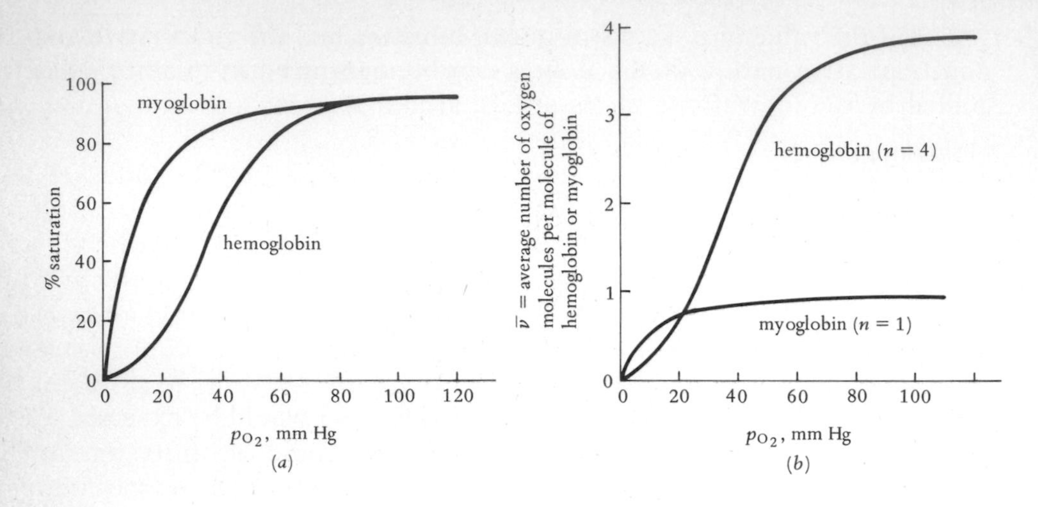

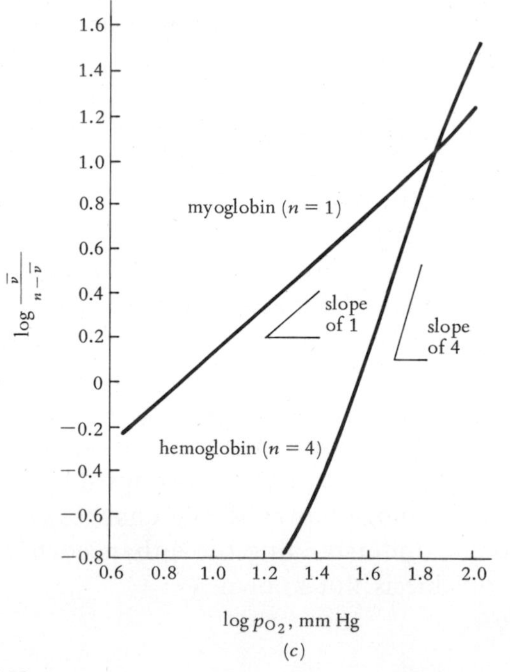

FIGURE 4.29
The oxygen uptake of hemoglobin and myoglobin and the interpretation of the data in terms of the number of sites and cooperative effects. (*a*) Experimental pressure versus saturation curves; (*b*) interpretation in terms of number of sites; (*c*) data treated according to the plots suggested by Eqs. **58** and **59**.

gen-ion concentration of 7.6×10^{-4} *M*. What is the value of the acid dissociation constant K_a of benzoic acid?

3 What is the concentration of H^+ and OH^- in an 0.50 *M* 25° aqueous solution of aniline hydrochloride?

4 The bicarbonate ion has a pK_a value of 10.3. What is the conjugate base of the bicarbonate, and what is the value of K_b of this base?

5 What are the pH values of the following solutions of strong acids and bases? (*a*) 0.01 *M* HCl, (*b*) 0.025 *M* HNO_3, (*c*) 1 *M* HCl, (*d*) 2.2×10^{-3} *M* NaOH.

6 What are the hydrogen-ion and hydroxide-ion concentrations in solutions with the following pH values? (*a*) pH = 2, (*b*) pH = 9, (*c*) pH = 3.76, (*d*) pH = 12.1.

7 What is the degree of dissociation of lactic acid in an 0.20 *M* solution? Check your answer by referring to Fig. 4-1.

8 What fraction of ammonia molecules are protonated, i.e., in the form of the ammonium ion, in a solution that is labeled "0.1 *M* ammonium hydroxide"? What would be a better label?

9 What is the pH of an 0.20 *M* sodium carbonate solution? (K_a of HCO_3^- is 4.7 NO^{-11}.)

10 The alkaloid morphine is a nitrogen base with a pK_b value of 6.1. What is the value pK_a of its conjugate acid? What would be the pH of a saturated solution, 10^{-3} molar, of morphine in water?

11 Calculate three points for a titration curve for the titration of 100 ml of a solution containing 1.01 g of triethylamine (pK_a of conjugate acid = 10.8). The titration is carried out using an 0.1 *M* HCl solution. By comparison with Fig. 4-6, sketch the titration curve.

12 On the same graph plot ν, the extent of protonation, against pH for bases with pK_b values of 4 and 8.

13 How would you prepare a liter of a buffer solution at pH 7.0 from the reagents $Na_2HPO_4 \cdot 2H_2O$ and $NaH_2PO_4 \cdot H_2O$? (The pK_a values of phosphate species are given in Sec. 4-5.)

14 Verify that the maximum buffer index value occurs at the pH equal to the pK_a of the acid in a buffer made from a weak acid and a salt of the weak acid. Use calculus or assume a value of K_a and calculate several buffer-index values at hydrogen-ion concentration near this K_a value.

15 Prepare diagrams like that for the phosphoric acid species given in Fig. 4-11 that show the relative amounts of carbonate, bicarbonate, $pK_a = 10.2$, and carbonic acid (or carbon dioxide), $pK_a = 6.3$, as a function of pH.

16 Prepare a graph like that for phosphate given in Fig. 4-12 to show the extent of protonation ν of carbonate as a function of pH.

17 Use the titration curve of Fig. 4-16 to deduce the extent of protonation of lysine as a function of pH.

18 If the π electrons of ethylene are treated by the particle-in-one-dimension model, what size box would have to be assumed to account for the observed absorption

at about 1700 Å? Is this length reasonable compared with the size of the ethylene molecule?

19 A solution containing 1.5 millimol of phenylalanine per liter of water in a 1-cm-long cell absorbs 50 percent of the radiation of wavelength 2570 Å. What is the value of the molar absorptivity or extinction coefficient ϵ at this wavelength?

20 From the data of Fig. 4-25 calculate the absorptivity of aniline at 2850 Å.

REFERENCES

Acid-Base Equilibria

WILLIAMS, V. R., and H. B. WILLIAMS: "Basic Physical Chemistry for the Life Sciences," W. H. Freeman and Company, San Francisco, 1967.

BUTLER, J. N.: "Solubility and pH Calculations," Addison-Wesley Publishing Company, Inc., Reading, Mass., 1964. This is a short version of "Ionic Equilibrium," by the same author and publisher. The approach is mathematical, but it is especially well organized and clear.

SEGEL, I. H.: "Biochemical Calculations," John Wiley & Sons, Inc., New York, 1968. Chapters 1 to 3 provide instruction and practice problems in the many variations of acid-base equilibria calculations that occur.

Multiple Equilibria

VAN HOLDE, K. E.: "Physical Biochemistry," Prentice-Hall, Inc., Englewood Cliffs, N.J., 1971.

TANFORD, CHARLES: "Physical Chemistry of Macromolecules," chap. 8, John Wiley & Sons, Inc., New York, 1961.

NOZAKI, Y., and C. TANFORD: Titration of Ribonuclease, *J. Am. Chem. Soc.*, **89:**742 (1967) and other papers by Tanford and coauthors referred to there.

GURD, F. R. N.: Binding of Protons and Other Ions, chap. 15 in S. J. Leach (ed.), "Physical Principles and the Techniques of Protein Chemistry," pt. B, Academic Press, Inc., New York, 1970. A rather complete bibliography of articles on the subject is included.

STEINHARDT, J., and S. BEYCHOK: Interaction of Proteins with Hydrogen Ions and Other Small Ions and Molecules, chap. 8 in Hans Neurath (ed.), "The Proteins," vol. 11, Academic Press, Inc., New York, 1964.

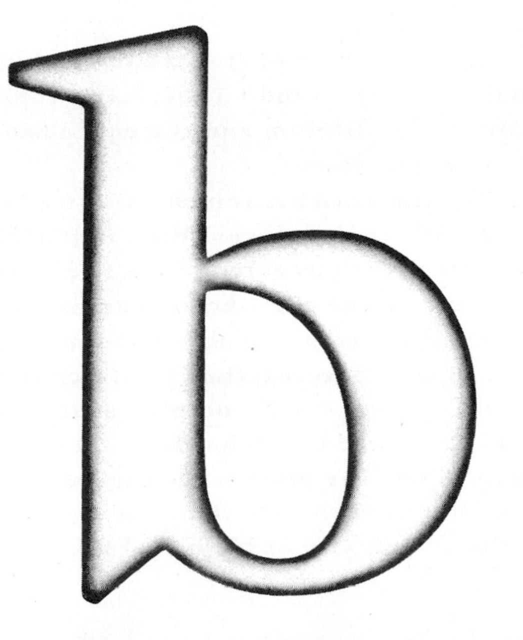

Energy and Equilibria

In Part a, we accepted substances and their molecules to be what they are, i.e., not static, but apparently immutable. But life, and much of chemistry, involves, or even consists of, processes that create new substances and new molecules out of old.

Here, in Part b, we treat the new and the old, and the relation between them. We reserve for Part c the studies of the processes themselves that transform the old into the new.

The products and reactants of chemical reactions are related in a number of ways that interest us. One is the relative amounts or concentrations reached when the reaction has run its course. We deal with this in terms of the *equilibrium* the reaction reaches. Many interests lead us to investigate the difference in the *energy* of the products and the reactants. Less familiar than the energy but equally important is the *entropy*. A study of the difference in energy and entropy of products and reactants turns out to be very fruitful for the treatment of chemical equilibria and chemical reactions.

All these points of view are brought together in the subject of *thermodynamics*, on which we now embark.

As in Part a, the first chapters (Chaps. 5 to 7) serve to establish the base for the more applied topics of Chaps. 8 and 9. These latter topics center on the chemical drive toward equilibrium and the equilibrium state itself in chemical and physical processes.

Chemists often seem unduly concerned with chemical equilibria and the many numerical problems this topic spawns. More important for the life sciences is an appreciation of the measures we use to express the tendency of systems to move toward the equilibrium condition or equilibrium concentrations. The importance remains even though, in biological systems, the drive of any one reaction to its equilibrium conditions is often thwarted by the overall complex scheme of reactions. Life, in this regard, can be viewed as a relentless war on the tendency of each of the reactions of the life process to reach equilibrium. Recognition of the driving force of these reactions then gives us considerable understanding of the complexity of biological systems.

5 Energy Changes in Chemical Reactions

In some chemical transformations the energy release or the energy demand is one of the principal reasons for our interest in the process. Fuel combustion, food metabolism, and photosynthesis are major examples. The study of the energetics of such processes is clearly merited.

But an understanding of the energetics of chemical reactions has a much broader value, as the following two chapters will show. The direction in which reactions tend to proceed and the equilibrium state which they tend to reach depend, in part, on the energy change associated with the reaction. This reason and the importance of the war on equilibrium that is waged in living systems underlie our study of the energetics of chemical reactions.

5-1 CALORIMETRY AND THE FIRST LAW OF THERMODYNAMICS

The molecules and the molecular-level processes of both animate and inanimate systems are often exceedingly complex. The energetics of processes can be studied in a way that avoids these complications. Thermodynamics has a disdain for molecular-level complexities—and even for the molecular level! This is well illustrated by the experimental procedures used to obtain the basic data on the heat changes, or more generally energy changes, that accompany reactions.

The basis for the measurement of energy changes produced by chemical reactions is illustrated by the block diagram of Fig. 5-1. The implication is that the energy change accompanying any process in the *system* can be completely determined by making measurements on two parts of the surroundings of the system, the thermal surroundings, or *thermal reservoir,* and the mechanical surroundings, or *mechanical reservoir.* When an energy change in the system results from some process occurring in the system, these components of the surroundings are convenient because we can simply and directly measure the energy changes that occur in them. (This is not true of the system itself, where the change may involve a complicated chemical or biological process.) For example, the energy change in the thermal surroundings might be calculated from a measured temperature change and the heat capacity of the thermal reservoir. The energy change in the mechanical surroundings might be calculated from a measured change in the height of a weight of a weight-and-pulley system and the mass of the weight.

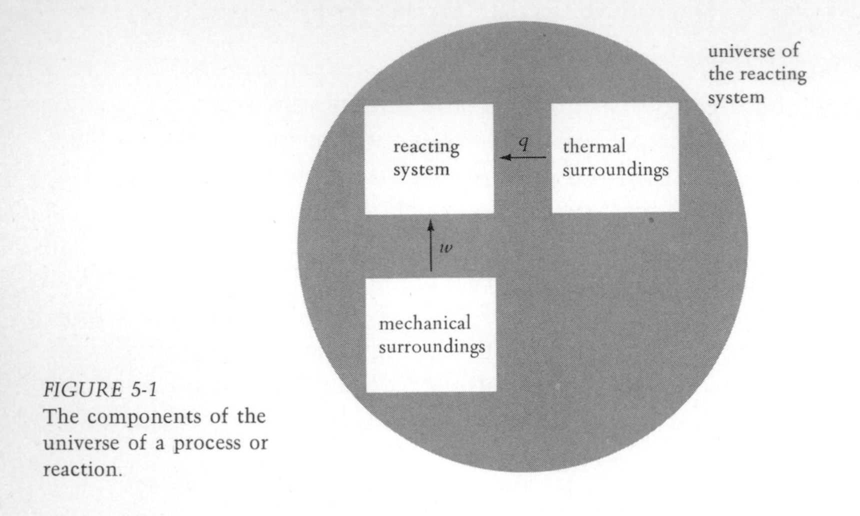

FIGURE 5-1
The components of the universe of a process or reaction.

The three blocks of Fig. 5-1 constitute the *universe* of the system, the term being appropriate in that everything that is affected by the process occurring in the system is encompassed by the thermal and the mechanical surroundings.

Measurements of energy changes in the two types of surroundings can be used to deduce the energy change in the process occurring in the system even if this process involves complicated chemical or biological reactions. The basis for this is the *conservation-of-energy principle,* which holds that the total energy of a universe, such as that of Fig. 5-1, is constant.

In many calorimetric studies the mechanical surroundings play no explicit role, and we can visualize the process of interest which occurs in the system as having an effect only on the thermal reservoir. Then if the amount of heat *gained* by the system is denoted by q, the energy change in the system ΔE would be calculated as

$$\Delta E = q \qquad \mathbf{1}$$

The value of q, the heat that flows into the system from the thermal reservoir, is obtained from measurements on that reservoir. From this, ΔE, the energy change in the system, with its possible complex transformation of substances, is deduced.

Measurements of the heat released, or absorbed, by a chemical reaction are made with a variety of devices known as *calorimeters.* One of the most widely used is the *bomb calorimeter* (Fig. 5-2), in which combustion reactions are carried out. Such reactions are particularly suited to calorimetric studies because they occur rapidly, with a large heat production, and often go completely to well-defined products. Furthermore,

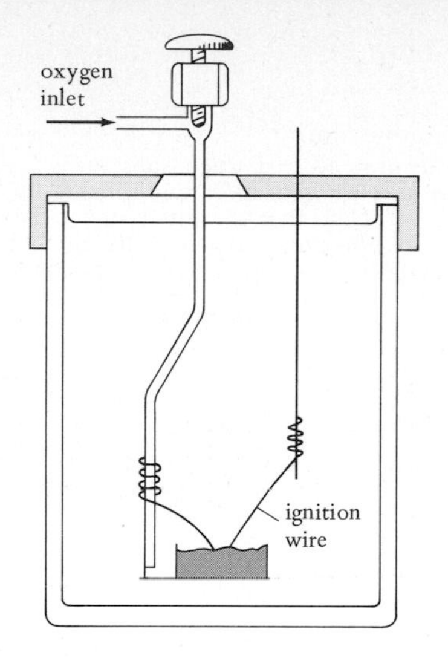

FIGURE 5-2
A bomb calorimeter.

the heat released in complete oxidation reactions is often an important reference quantity. Some values for fuels and foods are given in Table 5-1.

In a later type of calorimeter, also widely applicable, the heat produced or absorbed by the reaction is deduced from measurements of the rate of flow of heat away from or into the reagents and of the time over which this flow occurs. The principal components of this *heat-conduction,* or *heat-burst,* calorimeter are shown in Fig. 5-3. The thermocouple assembly produces a signal that depends on the temperature difference between the reaction liquid and the constant-temperature block. This temperature difference is proportional to the rate of heat flow. (We are more accustomed to this statement in the reverse direction, i.e., the heat flow is proportional to the temperature difference.) A plot of the temperature difference, or rate of heat flow, versus time can be subject to a graphical integration to produce the total heat given out or absorbed by the reaction. The calorimeter can be designed with a many-junction thermocouple so that very small heat effects can be measured with considerable accuracy. Many biologically important reactions for which only small amounts of materials are available can be studied by this technique.

In some calorimetric studies and in many of the reactions that occur under conditions of general interest, we cannot simply ignore energy exchanges with the mechanical surroundings. Such is the case for reactions occurring in systems whose volume is not fixed. Then as the process proceeds, the system generally expands or contracts, and as it does so against the external pressure, work is done by or on the system. Then a work term enters into the accounting-of-energy calculation.

TABLE 5-1

Heats of Combustion

Values for the heat evolved when substances burn at constant pressure to yield gaseous carbon dioxide, liquid water, and nitrogen gas. Note that the listed values correspond to $-\Delta H$ for the combustion reactions. The abbreviations (*g*), (*l*), and (*c*) designate gas, liquid, and crystalline states.

compound	formula	kcal/mol
Benzene (*l*)	C_6H_6	782.3
Citric acid, anhydric (*c*)	$C_6H_8O_7$	474.5
Cyclohexane (*l*)	C_6H_{12}	937.8
Diethyl ether (*l*)	$C_4H_{10}O$	651.7
Ethane (*g*)	C_2H_6	368.4
Ethyl acetate (*l*)	$CH_3COOC_2H_5$	536.9
Ethyl alcohol (*l*)	C_2H_5OH	327.6
Ethylamine (*l*)	$C_2H_5NH_2$	408.5
Ethylene (*g*)	C_2H_4	331.6
Formaldehyde (*g*)	CH_2O	134.1
l-Fructose (*c*)	$C_6H_{12}O_6$	675.6
d-Glucose (*c*)	$C_6H_{12}O_6$	673.0
Glycerol (*l*)	$(CH_2OH)_2CHOH$	397.0
Glycine (*c*)	H_2NCH_2COOH	234.5
Glycogen (*c*)	$(C_6H_{10}O_5)_x$	4186.8†
Glycylglycine (*c*)	$C_4H_8O_3N_2$	470.7
n-Heptane (*l*)	C_7H_{16}	1149.9
n-Hexane (*l*)	C_6H_{14}	989.8
Lactic acid (*l*)	$CH_3CHOHCOOH$	326.0
Lactose (*c*)	$C_{12}H_{22}O_{11}$	1350.8
Methane (*g*)	CH_4	210.8
Methyl alcohol (*l*)	CH_3OH	170.9
Methylamine (*l*)	CH_3NH_2	256.1
Palmitic acid (*c*)	$C_{16}H_{32}O_2$	2398.4
Pyridine (*l*)	C_5H_5N	658.5
Starch (*c*)	$(C_6H_{10}O_5)_x$	4178.8†
Stearic acid (*c*)	$C_{18}H_{36}O_2$	2711.8
Sucrose (*c*)	$C_{12}H_{22}O_{11}$	1349.6
Tyrosine (*c*)	$C_9H_{11}O_3N$	1070.2
Urea (*c*)	$(NH_2)_2CO$	151.6

† In kcal/kg.

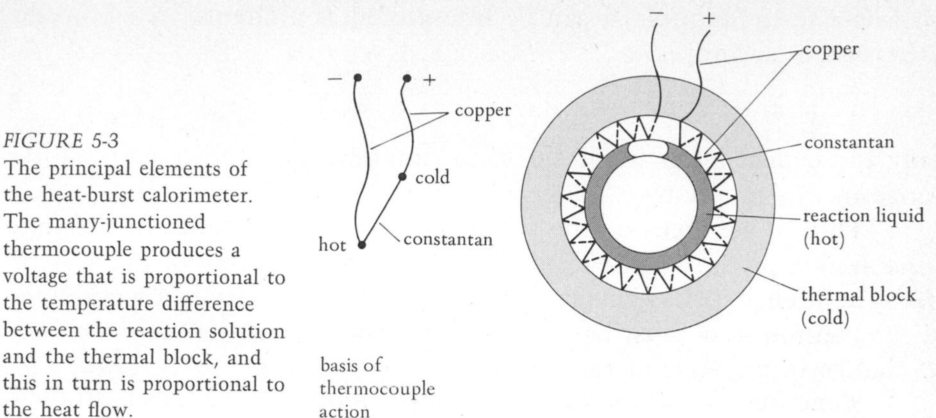

FIGURE 5-3
The principal elements of the heat-burst calorimeter. The many-junctioned thermocouple produces a voltage that is proportional to the temperature difference between the reaction solution and the thermal block, and this in turn is proportional to the heat flow.

We introduce w to denote the work done *on* the system. (Note that work done *by* the system, as in an expansion, would give a negative sign to the work. But be warned that there is an arbitrariness in the definition of w that has led also to the definition for w as work done by the system. For some reason q has been spared this ambiguity.) Now we can express the basis for the deduction of the energy change ΔE of the system more generally as

$$\Delta E = q + w \tag{2}$$

That is, the energy change ΔE for the process occurring in the system is given by the heat that flows from the thermal reservoir and the work that is done on the system by the mechanical surroundings.

Why do we write the simple lowercase symbols q and w but use the compound symbol ΔE? Consider an initial state a and a final state b of the system. Suppose, furthermore, that there are a variety of ways in which the system can be changed from state a to state b. (By "state" we mean a completely specified condition such as one described by indicating the amount of a specified substance, or substances, the concentration of these substances if the system is a solution, the crystalline form if it is a solid, the temperature, the pressure, and so forth. Often a state can be quite simply specified—as by specifying 100 g of pure water at 25°C and 1 atm. All additional variables are then fixed.) If you could study these processes in a calorimeter, you might find different values of q and w for each process. You would also find, however, that no matter how you varied these processes the sum $q + w$ would always be the same. You could conclude that ΔE for these processes depends only on states a and b and not on any feature of the processes that transforms the system. As a result, you could claim that E is a *state function*. It can be interpreted as a change in a property of the system according to

$$\Delta E = E_b - E_a \qquad \mathbf{3}$$

Using the capital letter implies a *state function*, and the Δ stands for a change in this function.

Thus, if we focus on the system, we see that a change in the state of the system produces a definite change ΔE in the energy of the system. This system change may involve various amounts of heat and work since only the sum $q + w$ must be the same for any process that accomplishes the change in the state of the system.

Sometimes, as when the second law of thermodynamics is studied, we treat each of the three components of the universe of Fig. 5-1 on an equal basis. Then we deal with ΔE, $\Delta E_{\text{th surr}}$, and $\Delta E_{\text{mech surr}}$, where each quantity deserves the state-function indication if we understand that the change in energy of each component of the universe is for some particular change in state of that component. Thus $\Delta E_{\text{th surr}}$ is a state function of the thermal reservoir, the state of the reservoir being indicated perhaps by a thermometer reading. And $\Delta E_{\text{mech surr}}$ is similarly dependent on the state of the mechanical reservoir, indicated, for example, by the height of the weight in a pulley-and-weight system.

Of course, the quantities $\Delta E_{\text{th surr}}$ and $\Delta E_{\text{mech surr}}$ are not functions of the state of the *system*. If we deal with changes of the state of the system, we must use q to encompass the values of $-\Delta E_{\text{th surr}}$ and w to encompass the values of $-\Delta E_{\text{mech surr}}$ that could satisfy all the processes implied in Eq. **2.**

The chemically convenient form of the principle of conservation of energy is known as *the first law of thermodynamics*. In an isolated assembly, like that of Fig. 5-1, the total energy is constant, and a change in energy in the system can be deduced from Eq. **2**

$$\Delta E = q + w$$

Furthermore, the energy change so deduced is determined not by the particular process that occurs but rather by the initial and final states of the system. If these states are indicated by a and b, this implies that Eq. **3** holds, i.e.,

$$\Delta E = E_b - E_a$$

Although the w term in the first-law expression is sometimes of considerable importance, as in batteries, fuel cells, and muscle action, here it is more of a nuisance. When a process occurs in a container open to

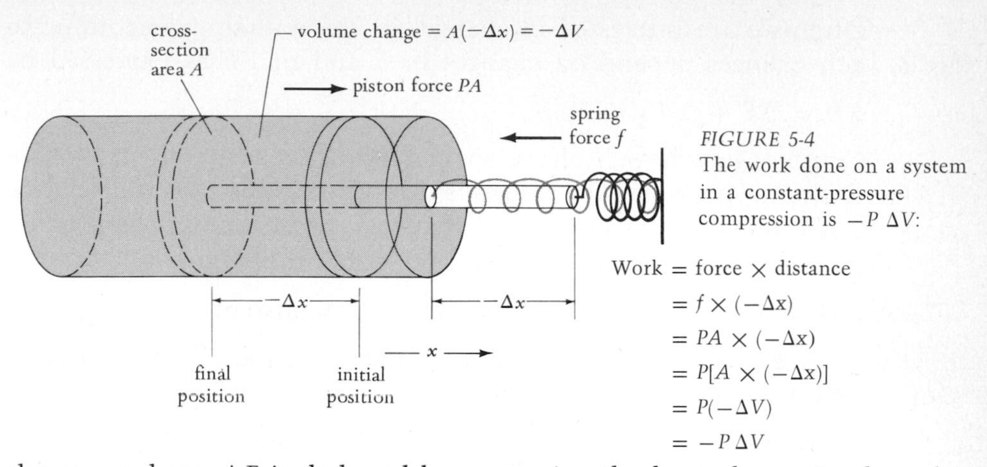

FIGURE 5-4
The work done on a system in a constant-pressure compression is $-P\,\Delta V$:

$$\begin{aligned}\text{Work} &= \text{force} \times \text{distance}\\ &= f \times (-\Delta x)\\ &= PA \times (-\Delta x)\\ &= P[A \times (-\Delta x)]\\ &= P(-\Delta V)\\ &= -P\,\Delta V\end{aligned}$$

the atmosphere, ΔE is deduced by measuring the heat change in the calorimeter, i.e., the thermal reservoir, and adding or subtracting the work of compression or expansion. This second term is often quite small, but it cannot be ignored.

When a system is compressed, work is done on the system by the mechanical surroundings. As Fig. 5-4 shows, a constant-pressure compression that produces a volume change ΔV does an amount of work equal to $-P\,\Delta V$ on the system. (The minus sign enters because in a compression ΔV is negative, and for w to be positive, as it must be when work is done on the system, we must write $w = -P\,\Delta V$.) Thus the work related to a volume change against a constant external pressure is given by

$$w = -P\,\Delta V \qquad \text{[constant pressure]} \qquad \mathbf{4}$$

Now we see that we can cope a little better with the w term, and for constant-pressure processes we can write

$$\Delta E = q - P\,\Delta V \qquad \mathbf{5}$$

For constant-volume processes, i.e., with $\Delta V = 0$, the final work term of Eq. **5** vanishes, and we have the simple relation $\Delta E = q$.

An even more convenient procedure can be developed. Let us arbitrarily define a new property of the system that is related to the E property, which we now call the *internal energy*. The new property H is called the *enthalpy*. This term may well be unfamiliar, and it may help at first to keep in mind that it is an indication, like E, of the energy of the system. Let us arbitrarily define H as

$$H = E + PV \qquad \mathbf{6}$$

Notice that since E is a property of the system, i.e., a state function, and so is V and thus PV, H is another state function.

Often we are interested in changes in the enthalpy. According to Eq. **6,** such changes depend on changes in E and in PV as expressed by

$$\Delta H = \Delta E + \Delta(PV)$$

Consider now how H changes when a process occurs in the system at a fixed external pressure, like processes that occur exposed to atmospheric pressure. If P remains constant but V changes, the change in the product PV will be $P\,\Delta V$. Thus, for the constant-pressure processes

$$\Delta H = \Delta E + P\,\Delta V \qquad \text{[constant pressure]*} \qquad \mathbf{7}$$

In view of Eq. **5,** that is, $\Delta E = q - P\,\Delta V$ or $\Delta E + P\,\Delta V = q$, we see that we can write for a *constant-pressure process*

$$\Delta H = q \qquad \text{[constant pressure]} \qquad \mathbf{8}$$

Thus for a constant-pressure process the calorimeter change, i.e., the effect on the thermal reservoir of Fig. 5-1, gives the change in H for the system directly. (This is the basis for the alternate term *heat content* for H.)

The enthalpy function is quite generally used as a measure of the energy of a system, and it is particularly convenient for constant-pressure processes.

5-2 THERMOCHEMICAL EQUATIONS

Although oxidation reactions have a remarkable importance in supplying thermodynamic data, the enthalpy changes accompanying many other reaction types are also of interest. The description of the reaction is then most succinctly given by the chemical equation for the reaction.

Some additional information must be added to make these descriptions adequate for thermochemical purposes. In particular, since the enthalpy change in a reaction depends on the physical states of the reagents, these states must be included in the description of the reaction. Usually this is done by attaching (g), (l), or (c) to the name or chemical symbol to indicate the *gas, liquid,* or *crystalline* states, for example, $O_2(g)$, $H_2O(l)$, or glycine(c). Also important are reagents dissolved in water to form aqueous solutions. Then (aq) is used, but in some cases it must be replaced by a more specific statement on the concentration.

The use of chemical equations to describe processes of interest implies the use of *moles* (rather than fixed mass amounts such as a gram) for the amount of material consumed or produced. The formula of a material is often taken to imply 1 mol of the material, and this implication

*Terms in square brackets indicate that the relation is only valid under the conditions specified.

often is used when the chemical formulas are part of a balanced chemical equation. Thus we write

$$CH_4(g) + 2O_2(g) \rightarrow CO_2(g) + 2H_2O(l)$$

The amounts implied are 1 mol of methane, 2 mol of oxygen, 1 mol of carbon dioxide, and 2 mol of water. Now if we report a ΔH value along with this reaction, as

$$CH_4(g) + 2O_2(g) \rightarrow CO_2(g) + 2H_2O(l) \qquad \Delta H = -212.79 \text{ kcal}$$

we mean that this is the enthalpy change in a system containing these molal amounts.

Example 5-1

A sample of crystalline urea weighing 1.372 g is burned in a bomb calorimeter for which the net heat capacity is 1176 cal/deg. A temperature rise from 21.34 to 24.29°C is observed. The products of the reaction are carbon dioxide gas, liquid water, and nitrogen gas.

a What is the change in E of the reacting system for the combustion process? What is the value of ΔE in calories per mole?

b What volume change would have occurred if the combustion of 1 mol of urea had been carried out at a constant pressure of 1 atm?

c What would the value of w have been, i.e., how much work would have been done by the atmosphere on the system, if the process had been carried out at a constant 1 atm pressure?

d How much heat would have been given to the thermal surroundings in a constant-pressure process?

e What is the value of ΔH for the combustion of urea in kilocalories per mole?

Solution

a Since in a bomb-calorimeter experiment w is zero, $\Delta E = q$. If heat flows from the system, q must be given a negative sign. For 1.372 g of urea,

$$\Delta E = -2.95 \text{ deg} \times 1176 \text{ cal/deg}$$
$$= -3470 \text{ cal}$$

For 1 mol of urea,

$$\Delta E = \frac{3470 \text{ cal}}{1.372 \text{ g}} \times 60.05 \text{ g/mol}$$
$$= -151{,}900 \text{ cal/mol}$$

b The reaction is

$$H_2N\overset{\overset{\displaystyle O}{\|}}{C}NH_2(s) + \tfrac{3}{2}O_2(g) \rightarrow CO_2(g) + 2H_2O(l) + N_2(g)$$

The only appreciable contributors to a volume change are gases. The net effect of the process is to produce $\frac{1}{2}$ mol of gas. Thus, at 1 atm and about 22°C,

$$\Delta V = 0.5(22.4)\left(\tfrac{295}{273}\right) = 12.1 \text{ liter}$$

c Because of the increase in volume calculated in part (*b*), work would have been done *by* the system on the atmosphere. We express this with a negative value of w. Thus

$$\begin{aligned} w &= -12.1 \text{ liter} \times 1 \text{ atm} \\ &= -12.1 \text{ liter atm} \\ &= -12.1 \frac{1.987 \text{ cal}}{0.08206 \text{ liter atm}} = -300 \text{ cal} \end{aligned}$$

d For a constant-pressure process the energy produced by the reaction must be accounted for by considering both the thermal and mechanical surroundings. According to $\Delta E = q + w$, where we now know that $\Delta E = -151{,}900$ and $w = -300$ cal, we deduce

$$\begin{aligned} q &= \Delta E - w \\ &= -151{,}900 + 300 \\ &= -151{,}600 \text{ cal} \end{aligned}$$

Or, 151,600 cal per mole of urea would go from the system to the surroundings.

e Since ΔH is the heat given out in a constant-pressure process,

$$\Delta H = -151.6 \text{ kcal/mol}$$

5-3 STANDARD HEATS OF FORMATION

Many reactions can be carried out in calorimeters of one design or another so that a value of ΔH can be placed alongside the equation for the reaction. Such data can then be made available to those whose interests are in the reactions rather than in the calorimetric measurements themselves. But there are so many reactions in which we might be interested that compiling data for all reactions of interest would be a hopeless task. Far better would be the presentation of data for the individual *substances* that might be involved in these reactions. The number is large but not impossibly so. But what numbers are to be tabulated? It is, after all, only the enthalpy *changes* accompanying reactions that can be measured.

We proceed by considering the enthalpy change that would occur when a compound is formed from its elements, and we use such data as if they gave the enthalpy content of the compound. This enthalpy content of compounds compared to their elements is known as the *standard enthalpy* (or *heat*) *of formation* of the compound. Ambiguity in the state of the compound is avoided by writing (*g*), (*l*), (*c*), or (*aq*) along with the description of the compound. Ambiguity about the physical state of the elements from which the compounds are formed is avoided by requiring that the elements be in the chemical form and physical state in which they normally occur under the standard conditions of 1 atm pressure and 25°C. (This implies, in particular, that carbon will be considered to be graphite.)

Standard heats of formation are indicated by the symbol ΔH_f°, the Δ reminding us that the values are for the formation of the compound from the elements, the superscript ° that the substance is in its *standard state*, and the small capital H that we are dealing with 1 mol of the substance. For pure substances this implies that they are under a pressure of 1 atm. For materials in solution, in particular those labeled (*aq*), for now we take the standard state to mean a 1 molal (m) concentration. (A refinement of this solution standard state will have to be made in the following section.)

The standard heats of formation ΔH_f° of all *elements* in their normal form and state are necessarily zero; i.e., no enthalpy is involved in the formation of an element from itself.

Standard heats of formation of compounds are the enthalpy changes, with all compounds in their standard states, for reactions such as

$$C(\text{graphite}) + O_2(g) \rightarrow CO_2(g) \qquad \Delta H = -94.05 \text{ kcal}$$

A few such enthalpies can be deduced directly from calorimetric studies, but more often an indirect route must be taken to the required data.

Let us see where the value listed for ΔH_f° for glucose comes from. Clearly the formation of this substance from its elements is not a reaction that is feasible for calorimetric study. As in many other cases, we can turn to heat-of-combustion data. The value of 3.736 cal/g for the heat of combustion of glucose and the glucose molal mass of 180.16 g lets us display the combustion reaction and corresponding enthalpy change as

$$C_6H_{12}O_6(c) + 6O_2(g) \rightarrow 6CO_2(g) + 6H_2O(l) \qquad \Delta H = -673.0 \text{ kcal}$$

Now this reaction can be turned around *and combined* with the formation reactions for CO_2 and H_2O which are the combustion reactions for C (graphite) and $H_2(g)$. In turning a reaction around you must change the

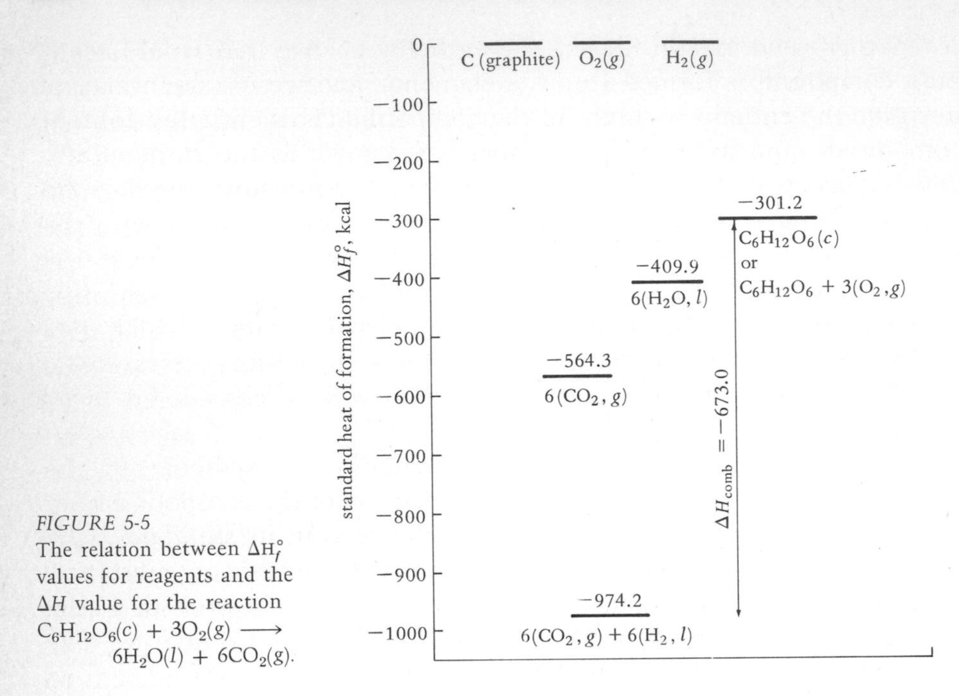

FIGURE 5-5
The relation between ΔH_f° values for reagents and the ΔH value for the reaction $C_6H_{12}O_6(c) + 3O_2(g) \longrightarrow 6H_2O(l) + 6CO_2(g)$.

sign of ΔH. The following measurable heats of reaction can be used to obtain the desired standard heat of formation value:

$$6CO_2(g) + 6H_2O(l) \rightarrow C_6H_{12}O_6(c) + 6O_2(g) \qquad \Delta H = +673.0 \text{ kcal}$$

$$6C(\text{graphite}) + 6O_2(g) \rightarrow 6CO_2(g)$$

$$\Delta H = 6(-94.05) = -564.3 \text{ kcal}$$

$$6H_2(g) + 3O_2(g) \rightarrow 6H_2O(l) \qquad \Delta H = 6(-68.32) = -409.9 \text{ kcal}$$

$$6C(\text{graphite}) + 6H_2(g) + 3O_2(g) \rightarrow C_6H_{12}O_6(c)$$

$$\Delta H = -301.2 \text{ kcal}$$

In similar ways, reactions for which data are available can be combined so that heats of formation of many compounds from their elements can be obtained. Some values are given in Appendixes I and II.

Can the standard heats of formation be used as if they gave the enthalpy content of the compound? The answer, as suggested by the graphical display of Fig. 5-5, is clearly yes. Since the ΔH_f° data for all compounds are referred to the same base, i.e., the elements, the data for different compounds can be used as a measure of the enthalpy of the compounds. Thus such data can be used to deduce the enthalpy change in any reaction involving compounds for which these data are available. Such calculations, furthermore, can be made even for imagined reactions that could not possibly be carried out.

For example, an interest in the bacterial production of protein for synthetic foods might lead one to ask about the heat demanded or produced in a reaction by which glycine would be formed from the simple compounds ammonia, methane, and oxygen. One then could write

$$\underbrace{\underset{-11.02}{NH_3(g)} + \underset{2(-17.89)}{2CH_4(g)} + \underset{\frac{5}{2}(0)}{\tfrac{5}{2}O_2(g)}}_{-46.80} \rightarrow \underbrace{\underset{-128.4}{H_2NCH_2\overset{\overset{O}{\|}}{C}OH(c)} + \underset{3(-68.32)}{3H_2O(l)}}_{-333.4}$$

ΔH_f°, kcal: -11.02, $2(-17.89)$, $\frac{5}{2}(0)$; -128.4, $3(-68.32)$

Then

$$\Delta H = -333.4 - (-46.80) = -286.6 \text{ kcal}$$

Thus an appreciable amount of heat would be given off. (This does not imply that the reaction has a tendency to proceed in the direction written. Further considerations, dealt with in Chaps. 6 and 7, are needed to deduce the direction in which reactions tent to proceed.)

Example 5-2

Show that the tabulated value for the heat of combustion of urea (Table 5-1) is consistent with the heat-of-formation value given in Appendixes I and II.

Solution

An organized way to proceed in many such problems consists of writing the reaction and inserting ΔH_f° values beneath each reagent. Thus

$$\underbrace{\underset{-79.58}{H_2N\overset{\overset{O}{\|}}{C}NH_2(c)} + \underset{0}{\tfrac{3}{2}O_2(g)}}_{-79.58} \rightarrow \underbrace{\underset{-94.05}{CO_2(g)} + \underset{2(-68.32) = -136.64}{2H_2O(l)} + \underset{0}{N_2(g)}}_{-230.69}$$

ΔH_f°, kcal: -79.58, 0; -94.05, $2(-68.32) = -136.64$, 0

$$\Delta H = -230.69 - (-79.58) = -151.11 \text{ kcal}$$

The result is in satisfactory agreement with the 151.6 value of Table 5-1.

5-4 STANDARD HEATS OF FORMATION FOR SOLUTES IN AQUEOUS SOLUTIONS

In most reactions of biological interest and, in fact, in many of the reactions of chemical interest, one or more of the reagents exist as solutes in water, i.e., as an *aqueous* solution. Data that bear on the energetics and the equilibrium states must therefore encompass such dissolved materials.

We can begin without complication. If a ΔH_f° value is available for a pure substance, a corresponding value for the substance in solution is obtained from the measurable heat of solution. Thus, when urea is dissolved in water to form a dilute solution, 3.66 kcal of heat must be added per mole of dissolved urea to maintain a constant temperature. We can thus write

$$NH_2\overset{\overset{\displaystyle O}{\|}}{C}NH_2(c) \xrightarrow{H_2O} NH_2\overset{\overset{\displaystyle O}{\|}}{C}NH_2(aq) \qquad \Delta H = +3.66 \text{ kcal}$$

If, further, we recognize that ΔH_f° for the crystalline form is tabulated as -79.58 kcal/mol, we can calculate

$$\begin{aligned} \Delta H_f^\circ(\text{urea},aq) &= \Delta H_f^\circ(\text{urea},c) + 3.66 \\ &= -79.58 + 3.66 = -75.92 \text{ kcal/mol} \end{aligned}$$

There are, however, some loose ends in the above deduction.

First, we must recognize that the heat of solution generally depends on the concentration of the solution being formed. The value of 3.66 kcal for the heat of solution is, in fact, the limiting value obtained by extrapolating measured heats of solution to the limit of infinite dilution. Then we see that the ΔH_f° value of urea(aq) is to be associated with an infinitely dilute solution.

The troublesome dependence of heat of solution on solution concentration arises because the heat of dilution is not zero. Generally there is a heat effect when solutions are diluted. The simple situation of a zero heat of dilution is never exactly followed and therefore is hypothetical. It characterizes, in part, an *ideal solution*. We make frequent use of the concept of ideal solutions so that we can put data on a common, organized basis and leave the individual characteristics of solutions to be dealt with separately.

With the concept of ideal solutions, characterized here by an independence of ΔH_f or solution concentration, we can also interpret ΔH_f° as the heat of formation that would be assigned to the solute if its solution behaved ideally as it is brought from infinite dilution to a concentration of 1 mol per 1000 g of water, that is, 1 m. Thus we say that the standard state for solutes is the 1 m concentration of a hypothetical ideal solution. *Real* use can be made of such data, in spite of the *hypothetical* state because the data are those deduced for the limit of infinite dilution. (The idea of a 1 m ideal solution state will have some merits when other thermodynamic functions are introduced.)

One additional step is worth taking when we think of the solute as being ionized in solution. For example, the heats of solution of HCl(g) could be measured and could be extrapolated to infinite dilution. This value could be combined with that for $\Delta H_f^\circ(\mathrm{HCl},g)$ to give the value -40.02 kcal/mol for $\Delta H_f^\circ(\mathrm{HCl},aq)$. Greater use can be made of separate values for $H^+(aq)$ and $Cl^-(aq)$, but there is no experimental way of separating any of the measured enthalpy changes so that components of the total can be assigned to the two ions. This very fact allows us to make an arbitrary assignment for one ionic species and then deduce values for other species on this basis. It is agreed to set

$$\Delta H_f^\circ(\mathrm{H}^+,aq) = 0 \qquad (9)$$

Then the result given above for HCl can be used with

$$\Delta H_f^\circ(\mathrm{HCl},aq) = \Delta H_f^\circ(\mathrm{H}^+,aq) + \Delta H_f^\circ(\mathrm{Cl}^-,aq)$$

to obtain

$$-40.02 = 0 + \Delta H_f^\circ(\mathrm{Cl}^-,aq)$$

and $\quad \Delta H_f^\circ(\mathrm{Cl}^-,aq) = -40.02$ kcal/mol

Once a start is made, values for other ionic species, as given in Appendixes I and II, are readily obtained. For example, the $\Delta H_f^\circ(\mathrm{NaCl},aq)$ valuc of -97.30 kcal/mol yields, with the above Cl^- value,

$$\begin{aligned}\Delta H_f^\circ(\mathrm{Na}^+,aq) &= \Delta H_f^\circ(\mathrm{NaCl},aq) - \Delta H_f^\circ(\mathrm{Cl}^-,aq)\\ &= -97.30 - (-40.02)\\ &= -57.28 \text{ kcal/mol}\end{aligned}$$

Then, the ΔH_f° value for NaOH(*aq*) of -112.24 kcal/mol gives

$$\begin{aligned}\Delta H_f^\circ(\mathrm{OH}^-,aq) &= \Delta H_f^\circ(\mathrm{NaOH},aq) - \Delta H_f^\circ(\mathrm{Na}^+,aq)\\ &= -112.24 - (-57.28)\\ &= -54.96 \text{ kcal/mol}\end{aligned}$$

The consistency of such data can often be checked, as we demonstrate here by calculating the heat of neutralization expected for a strong acid and strong base. From ΔH_f° data we write

$$\mathrm{H}^+(aq) + \mathrm{OH}^-(aq) \rightarrow \mathrm{H_2O}(l)$$

$$\Delta H_f^\circ\text{, kcal:} \quad 0 \qquad -54.96 \qquad -68.32$$

Thus $\Delta H = -68.32 - (-54.96) = -13.36$ kcal

Direct measurement of acid-base reactions confirms that $-13{,}360$ cal is the enthalpy of neutralization in the limit of infinite dilution.

It is important to note, as in this example, that although many water molecules are implied by the abbreviation (*aq*) and many of these are affected by the reaction, they do not enter into the calculation. Their contribution is included in the value assigned to the species indicated as (aq).

A feature of the tabulated ΔH_f° data for weak acids and bases, and their ions, can be illustrated by showing the origin of the data for a particular example. Take, as chemists always do, acetic acid. In moderately dilute solutions, say 0.01 m, very little of the acid is dissociated, and heat-of-solution data can be used to obtain for the undissociated aqueous species

$$\Delta H_f^\circ(CH_3COOH,aq) = -115.98 \text{ kcal/mol}$$

If we make the assumption that acetic acid solutions behave ideally, this value corresponds to the heat of formation of undissociated acetic acid either in an infinitely dilute solution or in an ideal solution of any concentration, 1 m, for example. (Notice that another hypothetical but in no way troublesome feature has entered. Acetic acid could not be undissociated in an infinitely dilute solution.)

Measurement of the heat of neutralization of acetic acid gives the value -13.44 kcal/mol, that is,

$$CH_3COOH(aq) + OH^-(aq) \rightarrow CH_3COO^-(aq) + H_2O(l) \qquad \Delta H = -13.44 \text{ kcal}$$

The divergence of this value from the -13.360 value for strong acids can be associated with the heat of ionization of acetic acid or with the difference in the heats of formation of the parent acid and the acetate ion.

If to the above equation we add

$$H_2O(l) \rightarrow H^+(aq) + OH^-(aq) \qquad \Delta H = +13.36 \text{ kcal}$$

we obtain

$$CH_3COOH(aq) \rightarrow CH_3COO^-(aq) + H^+(aq) \qquad \Delta H = -0.08 \text{ kcal}$$

This result can also be used to obtain $\Delta H_f^\circ(CH_3COO^-,aq)$. We write

$$\Delta H_f^\circ(CH_3COO^-,aq) + \Delta H_f^\circ(H^+,aq) - \Delta H_f^\circ(CH_3COOH,aq) = -0.08$$

or $$\Delta H_f^\circ(CH_3COO^-,aq) = 0 - 115.98 - 0.08 = -116.06 \text{ kcal/mol}$$

Notice that the standard heats of formation of acetic acid and the acetate ion are a little different, in contrast to the situation for strong acids, hydrochloric for example, for which the tabulated values of the heats of formation of the parent acid and the anion of the acid are identical.

Similar features can be recognized in the tabulated values for the parent and anionic forms of weak acids included in Appendixes I and II.

5-5 TEMPERATURE DEPENDENCE OF THE HEAT OF REACTIONS: HEAT CAPACITIES

Enthalpy data, from which heats of reaction can be deduced, are almost always given at 25°C. To extend such data to other temperatures of interest, the temperature dependence of heats of reactions must be deduced. We begin by investigating the temperature dependence of the enthalpy of substances.

Measurements can be made of the *heat capacity* of a compound, the heat absorbed by the compound per unit temperature rise. From such studies carried out at constant pressure, we can obtain the *constant-pressure heat capacity* C_p as

$$C_p = \lim_{T\to 0} \frac{q}{\Delta T} \qquad \text{[constant pressure]} \qquad \mathbf{10}$$

or with the relation of Eq. **8** as

$$C_p = \lim_{T\to 0} \frac{\Delta H}{\Delta T} = \frac{dH}{dT} \qquad \mathbf{11}$$

Notice that Eq. **11** does not need the qualification that the pressure be held constant. A change in enthalpy is, according to Eq. **8,** equal to the heat that would be absorbed by the system in a constant-pressure process. Thus, without qualification, we can write $C_p = dH/dT$, or, for 1 mol, $\mathrm{C}_p = d\mathrm{H}/dT$.

Although C_p values for compounds depend on temperature, use can often be made of values listed for standard states, indicated by C_p°, and a temperature of 25°C. These heat-capacity values, where available, are given for the substances of Appendixes I and II. (Notice that $\Delta\mathrm{H}_f^\circ$ values are given in units of kilocalories per mole but the C_p° values are in calories per mole degree.)

If greater ranges of temperature must be dealt with, the dependence of heat capacity on temperature must be recognized. This is usually done with an empirical equation of a form like

$$\mathrm{C}_p^\circ = a + bT + cT^2 \qquad \mathbf{12}$$

where a, b, and c are constants whose values must be given for each compound.

The dependence of the heat of a reaction on temperature can be illustrated by relating the heat of reaction to standard heats of formation and standard heat capacities. We write

$$\Delta H = (\Delta H_f^\circ)_{\mathrm{prod}} - (\Delta H_f^\circ)_{\mathrm{react}}$$

and $$\frac{d}{dT}(\Delta H) = \frac{d}{dT}(\Delta H_f^\circ)_{\text{prod}} - \frac{d}{dT}(\Delta H_f^\circ)_{\text{react}}$$
$$= (C_p^\circ)_{\text{prod}} - (C_p^\circ)_{\text{react}}$$
$$= \Delta C_p^\circ \qquad \mathbf{13}$$

Again, the Δ sign implies the difference in the value of the indicated quantity for the molal amounts expressed in the equation written for the reaction.

This derivative equation, Eq. **13,** can be integrated. The temperature range of interest is often from 25°C, or 298 K, to some general value of T. Then we have

$$\int_{T=298}^{T} d(\Delta H^\circ) = \int_{T=298}^{T} \Delta C_p^\circ \, dT$$

or $$\Delta H_T^\circ = \Delta H_{25^\circ\text{C}}^\circ + \int_{T=298}^{T} \Delta C_p^\circ \, dT \qquad \mathbf{14}$$

If a constant value of ΔC_p° can be assumed in this temperature region, we have simply

$$\Delta H_T^\circ = \Delta H_{25^\circ\text{C}}^\circ + \Delta C_p^\circ (T - 298) \qquad \mathbf{15}$$

Typical heat-capacity data would show that heats of reaction are usually changed only slightly by small temperature changes, e.g., that from 25°C to body temperature 37°C. As a result, tabulated ΔH_f° values can often be used without change in the study of processes or experiments at this higher temperature. But other reaction characteristics can be changed appreciably by this small temperature difference, the rate at which reactions proceed being most important of these.

Example 5-3

The standard heats of formation of glucose and lactic acid are −304.60 and −165.88 kcal/mol, respectively. The molal heat capacities of the two compounds are 52.31 and 30.5 cal/deg mol. What is the heat of reaction at 25°C for the formation of lactic acid from glucose? What value would be estimated for this heat if the reaction proceeded at the physiological temperature of 37°C?

Solution

The reaction is described by

$$C_6H_{12}O_6 \rightarrow 2\left(\begin{array}{c} \text{H} \\ | \\ \text{CH}_3\text{—C—COOH} \\ | \\ \text{OH} \end{array}\right)$$

Thus we can write

$$\begin{array}{lll} & \text{Glucose} \rightarrow & 2 \text{ lactic acid} \\ \Delta H_f^\circ, \text{kcal:} & -304.60 & 2(-165.88) = -331.76 \end{array}$$

Therefore, at 25°C,

$$\Delta H^\circ = -331.76 - (-304.60) = -27.16 \text{ kcal}$$

Thus heat is evolved, in a constant-pressure conversion, to the extent of 27.16 kcal per mole of glucose consumed.

The corresponding value at 37°C is estimated by first calculating ΔC_p° for the reaction as

$$\begin{array}{lll} & \text{Glucose} \rightarrow & 2 \text{ lactic acid} \\ \Delta C_p^\circ, \text{cal/mol, deg:} & 52.31 & 2(30.5) = 61.0 \end{array}$$

Thus $\Delta C_p^\circ = 61.0 - 52.3 = +8.7$ cal/deg. Equation **15** now lets us write

$$\begin{aligned} \Delta H_{310}^\circ &= \Delta H_{298}^\circ + \Delta C_p^\circ(310 - 298) \\ &= -27{,}160 + 8.7(12) \\ &= -27{,}160 + 104 \\ &= -27{,}060 \text{ cal} \end{aligned}$$

The result, rounded off to the 10-cal place in keeping with the given ΔH_f° data, shows that the heat-of-reaction difference at the two temperatures is small but not negligible.

PROBLEMS

1 The combustion of diborane, B_2H_6, proceeds according to the equation

$$B_2H_6(g) + 3O_2(g) \rightarrow B_2O_3(s) + 3H_2O(g)$$

For the reaction at constant pressure, 464 kcal of heat is liberated per mole of diborane. Combustion of elemental boron also proceeds to the product, B_2O_3, and 283 kcal is released per gram atom of boron. What is the standard heat of formation of diborane?

2 Comparison of the fuel value of fuels and foods is done on a weight rather than mole basis. Compare the fuel value, calculated as calories per gram, for a hydrocarbon, a carbohydrate, and a fatty acid. Use data given in Table 5-1. Include in your comparison the fact that a representative fuel value for protein is 5.2 kcal/g.

3 It has been estimated that a man expends about 0.5 kcal of energy per pound of body weight in walking 1 mi. On the basis of pounds per mile moved, compare the energy needed by a walking man and by a car. Assume any particular car you like or take as representative a 1-ton car and a 15 mi/gal fuel consumption.

4 Compare the energy released per mole of foodstuff with the energy released per mole of oxygen in the metabolic oxidation of carbohydrates, fats, and proteins. Use Table 5-1 data for sugars and for fatty acids and the value 5.2 kcal/g for protein. A representative protein composition is given by the formula $C_{4.3}H_{6.6}NO$.

5 Verify that the value listed for ΔH_f° for sucrose in Appendix II is consistent with the heat-of-combustion value of Table 5-1.

6 Measurements can be made of the heat absorbed when various amounts of NaCl are added to 1000 g of water. From such data the relation ΔH for this solution reaction and m, the number of moles of NaCl added to this fixed amount of water, can be deduced. At 25°C the results for ΔH in calories can be expressed by the equation

$$\Delta H = 923m + 476.1m^{3/2} - 726.1m^2 + 243.5m^{5/2}$$

a How much heat would be absorbed, i.e., what is the value of ΔH, for the addition of 1 mol of NaCl to 1000 g of water?

b A quantity often used is the heat absorbed or evolved for the addition of 1 mol of solute to an infinite amount of solvent so that an infinitely dilute solution is formed. The heat absorbed per mole of NaCl added to any molality can be obtained by differentiating the ΔH-versus-m equation to obtain $d(\Delta H)/dm$. Obtain this derivative expression. What is the value of $d(\Delta H)/dm$ at $m = 0$? (This is the desired heat evolved per mole of NaCl added in the limit of zero salt concentration.)

c What is the value of $d(\Delta H)/dm$ at $m = 1$? Compare the values of $d(\Delta H)/dm$ at $m = 0$ and at $m = 1$ with the value found in part (*a*) for the heat evolved in the addition of 1 mol of NaCl to 1000 g of water. Explain qualitatively the relation between the "differential" heats and the "integral" heats you are comparing here.

7 From the heat of formation data of Appendix I, deduce the heat of solution of NaCl. To what value obtained in the preceding problem should this be compared?

8 The addition of CO_2 to a solution containing calcium ions can lead to the formation of insoluble $CaCO_3$ according to the equation

$$Ca^{2+}(aq) + CO_2(g) + H_2O(l) \rightarrow CaCO_3(s) + 2H^+(aq)$$

How much heat would be absorbed or given out if this reaction produced 1 g of $CaCO_3$?

9 Use the values listed for ΔH_f° in Appendix II to calculate the heat of the various ionization steps of succinic acid.

10 Discuss the basis of the difference in the listed values of ΔH_f° for $NH_3(aq)$ and $NH_4^+(aq)$ as given in Appendix I.

11 If a value of ΔH_f° were included in Appendix I for NaOH(*aq*), what would that value be? Confirm that this is correct by using your value to deduce the heat of neutralization of a strong acid with this base. (A value of $-13{,}360$ cal for ΔH for such reactions should be deduced.)

12 Using ΔH_f° and C_p° data of Appendixes I and II, calculate for any reaction, other than that used in Example 5-3, the heat of the reaction at 100°C. (Assume that the C_p° values, or at least the calculated ΔC_p° value, can be taken as constant over the temperature range 25 to 100°C.)

13 *a* Deduce the value of the standard heat of formation for water vapor at 100°C. Use the values given in Appendix I for the heat of formation of water vapor at 25°C and for the heat capacity of water vapor.

b Another route to the quantity asked for in part (*a*) begins with liquid water at 25°C and considers the heating of the liquid to 100°C and then the vaporization of the liquid at this temperature. What must the value of the heat of vaporization of water at 100°C be for this route to give the same value for ΔH_f° of water vapor at 100°C as obtained previously?

REFERENCES

Thermodynamics Books on thermodynamics treat the material of this and the following two or three chapters. These three texts are particularly recommended.

KLOTZ, I. M., and R. M. ROSENBERG: "Introduction to Chemical Thermodynamics," 2d ed., W. A. Benjamin, Inc., Menlo Park, Calif., 1972.

WALL, F. T.: "Chemical Thermodynamics," 2d ed., W. H. Freeman and Company, San Francisco, 1965.

NASH, L. K.: "Elements of Chemical Thermodynamics," Addison-Wesley Publishing Company, Inc., Reading, Mass., 1962.

Thermodynamics of Biochemical Systems

KLOTZ, I. M.: "Energetics in Biochemical Reactions," Academic Press, Inc., New York, 1957.

LEHNINGER, A. L.: "Bioenergetics," W. A. Benjamin, Inc., Menlo Park, Calif., 1965.

BROWN, H. D. (ed.): "Biochemical Microcalorimetry," Academic Press, Inc., New York, 1969. A collection of 17 authoritative and informative accounts of the techniques and biochemical and biological applications of calorimetry. Included is an extensive tabulation of thermodynamic values for compounds important in biochemical systems.

BENZINGER, T. H.: Ultrasensitive Reaction Calorimetry, chap. 3 in P. Alexander and H. P. Lundgren (eds.), "Analytical Methods of Protein Chemistry," Pergamon Press, New York, 1969. An excellent detailed account of the technique and results of heat-burst calorimetry.

6 Entropy and the Direction of Chemical Change

A very large part of chemistry is concerned, in one way or another, with the tendency of a system to move in the direction of an equilibrium state. The thermodynamic treatment developed in response to this interest is beautifully illustrated by studying some of the processes crucial to the living state. Life cannot upset the overall drive of nature toward equilibrium, but the intricacies of this drive are so manipulated in living systems that the chemically "unnatural" state of these systems is maintained. Added appreciation of how the apparently nonequilibrium living state is maintained can be gained with a knowledge of an important tool of thermodynamics: the entropy function.

6-1 THE SECOND LAW OF THERMODYNAMICS: GENERAL STATEMENTS

In Chap. 5 the conservation-of-energy principle was stated and developed into a useful form, embodied in the equation $\Delta E = q + w$, the first law of thermodynamics. The same strategy will be used here, and we begin with general statements that express the idea but not the form of the chemically useful second law of thermodynamics.

Although the second law can be stated in a number of different ways, all statements can be shown to generalize our knowledge that natural processes tend to go to a state of equilibrium. The second law sums up our experiences with equilibria, just as the first law summed up our experience with energy. The drive toward equilibrium, it turns out, can be expressed in terms of the heat and work terms used in the first-law studies.

The classic general statements that lead to the second law seem to deal with heat-work transformations. Thus, Lord Kelvin's statement is that "it is impossible by a cyclic process to take heat from a reservoir and convert it into work without at the same time transferring heat from a hot to a cold reservoir." A moment's thought about all types of engines will show that there is always a hot and a cold source. A steam engine, for example, could not be made to produce work if it were not for the high pressure and high temperature of the steam *compared* with the surroundings.

This statement is seen to be related to equilibria when it is realized that work can be obtained from a system only when the system is not already at equilibrium. If a system is at equilibrium, no process tends to occur spontaneously and there is nothing to harness to produce work.

Kelvin's statement recognizes that the spontaneous process is the flow of heat from a higher to a lower temperature and that only from such spontaneous processes can work be obtained.

Another statement, given by Clausius, is that "it is impossible to transfer heat from a cold to a hot reservoir without at the same time converting a certain amount of work into heat." This statement is readily illustrated by the operation of a refrigerator. Again, we recognize that the spontaneous flow of heat is from a high to a low temperature and that the reverse is possible only when work is expended.

Our interest in such statements is aroused not by these heat-engine examples but by the possibility of a law which says something about the properties associated with the position of equilibrium toward which natural processes tend to proceed.

6-2 ENTROPY AND ANOTHER STATEMENT OF THE SECOND LAW OF THERMODYNAMICS

Is there a thermodynamic property that lets us deduce whether a spontaneous process will change a system from one state to another? If there is, how can it be evaluated?

Chemical terminology implies that in spontaneous processes there is a *driving force* that tends to make the reaction or process occur. Of obvious interest is the study of properties associated with the tendency of processes or chemical reactions to proceed, and our object is to relate the driving force of chemical reactions to other measurable quantities.

A process or reaction with no tendency to proceed in either direction is called *balanced* or *reversible*. By contrast, a spontaneous process is unbalanced and *irreversible:* no small change in the conditions can overcome the natural driving force.

In the search for a driving force it is profitable to think again of the system and its thermal and mechanical surroundings. Now, however, $\Delta E_{\text{th surr}}/T$ turns out to be a more useful quantity than does $\Delta E_{\text{th surr}}$ itself. This quantity is sufficiently important to merit a symbol $\Delta S_{\text{th surr}}$ and a name, the change in *entropy* of the thermal surroundings. To deal with a system at various temperatures, we must imagine thermal surroundings at a number of different temperatures or the reservoir must operate over a range of temperatures. If a range of temperatures is involved, the entropy change of the thermal surroundings is calculated by a suitable integration of

$$dS_{\text{th surr}} = \frac{dE_{\text{th surr}}}{T} \quad \mathbf{1}$$

We are at liberty to ascribe any features to this new entropy function that we like, so long as the function is self-consistent and allows us to form an expression that is useful. In this vein we further specify that for all processes

$$dS_{\text{mech surr}} = 0 \tag{2}$$

Now we are at a stage comparable to that in the development of the first law where we knew how to determine the energy changes in the thermal and mechanical surroundings. The next step is to use this information to learn about the change in the property (now entropy) of the system and to see how it can be used.

We proceed by making two statements which together express the law of nature known as the *second law of thermodynamics* (a law of nature is a generalization; you accept it not as a result of a derivation but only if you find that it generalizes the individual observations you can make):

1 *When a process is carried out reversibly, the entropy change in the universe of the process is zero.* (Since we know how to calculate $\Delta S_{\text{th surr}}$ and we have required $\Delta S_{\text{mech surr}}$ to be zero, this statement provides a means for deducing the entropy change ΔS of the system.)
2 *For processes that proceed irreversibly, i.e., out of balance and therefore spontaneously, the entropy of the universe of the process increases.* (Thus if ΔS and $\Delta S_{\text{th surr}}$ are known for some change, the possibility that this process will occur spontaneously can be deduced by inspecting the sign of $\Delta S + \Delta S_{\text{th surr}}$.)

The first statement can lead to a value for the entropy change of the system, as we illustrate by an example.

Example 6-1

What is the entropy change suffered by 1 mol of liquid water at 1 atm and 100°C when it is converted to vapor at the same temperature and pressure?

Solution

The problem can be displayed by the diagram of Fig. 6-1, and the numerical entries result from using the heat of vaporization, 9720 cal, to obtain

$$\Delta S_{\text{th surr}} = \frac{-9720}{373} = -26 \text{ cal/deg} \tag{3}$$

Then, since the vaporization can be imagined to occur reversibly with only an infinitesimal temperature difference between the reservoir and the system, the requirement that $\Delta S_{\text{univ}} = 0$ leads to

$$\Delta S_{\text{th surr}} + \Delta S = 0$$

$$\text{or} \quad \Delta S = -\Delta S_{\text{th surr}} = +26 \text{ cal/deg} \tag{4}$$

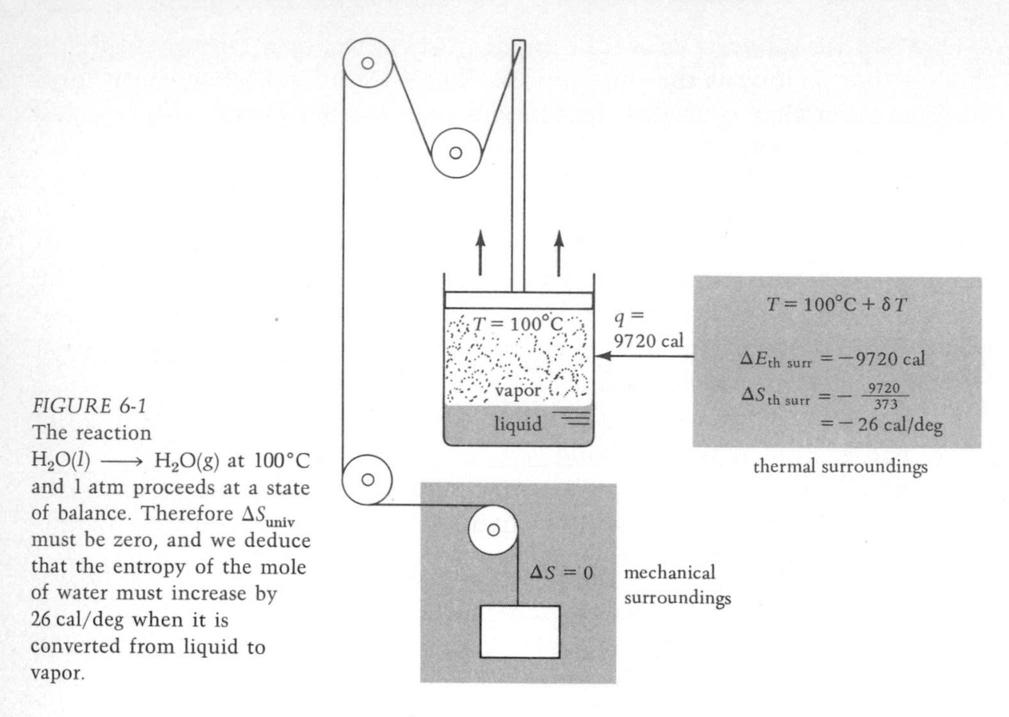

FIGURE 6-1
The reaction $H_2O(l) \longrightarrow H_2O(g)$ at 100°C and 1 atm proceeds at a state of balance. Therefore ΔS_{univ} must be zero, and we deduce that the entropy of the mole of water must increase by 26 cal/deg when it is converted from liquid to vapor.

Turning to a process we know to be spontaneous, let us see whether (as the second statement claims) the entropy of the universe of the process increases.

Example 6-2

Consider a vessel containing liquid water in equilibrium with steam at 1 atm and, therefore, 100°C to be brought in contact with a vessel containing liquid water and ice at 1 atm and, therefore, 0°C. What is the entropy change for the flow of some amount of heat Q from the hot to the cold vessel? (The water-and-steam and water-and-ice containers are used so that their temperature will remain constant even though heat, if not too much, is transferred. If temperatures were allowed to change, the calculations would be similar in principle, but more involved.)

Solution

First a reversible way of performing this energy transfer must be devised so that the constancy of the entropy of all parts in a reversible process can be used to calculate the entropy change of the system, which consists of the two containers. (Heat flowing directly from a high to a low temperature corresponds to an irreversible process, and for such processes we have no direct way of calculating this entropy change.) A reversible process results if we connect each of the two parts of the system to separate heat reservoirs, as shown in Fig. 6-2*b*, one at a temperature infinitesimally lower than 100°C, the other infinitesimally higher than 0°C. The heat flows indicated in the figure give the net result, as far as the system is concerned, of transferring energy from the hot to the cold body. The addition of the thermal reservoirs, however, makes the process reversible.

The entropy change suffered by the thermal reservoirs is now calculated as

$$\begin{aligned}\Delta S_{\text{th surr}} &= \Delta S_{\text{hot surr}} + \Delta S_{\text{cold surr}} = \frac{Q}{373} - \frac{Q}{273} \\ &= Q(0.0027 - 0.0037) = -0.0010Q \\ &= \text{a negative quantity}\end{aligned} \tag{5}$$

Since the process of Fig. 6-2*b* is reversible, $\Delta S_{\text{univ}} = 0$, and thus

$$\begin{aligned}\Delta S = -\Delta S_{\text{th surr}} &= -(-0.0010Q) = +0.0010Q \\ &= \text{a positive quantity}\end{aligned}$$

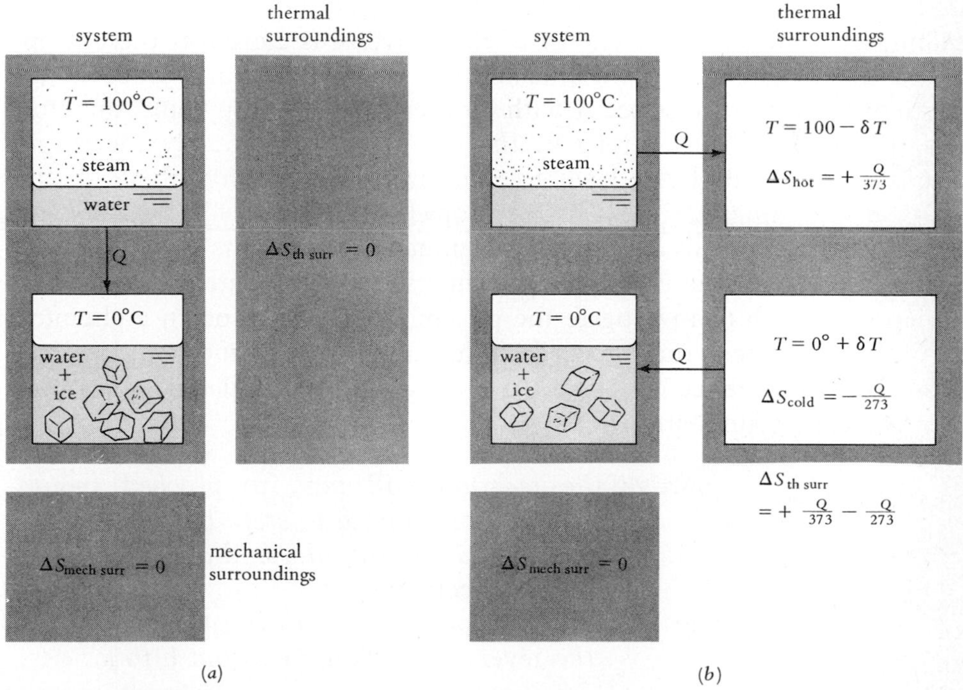

FIGURE 6-2
The entropy changes for the transfer of an amount of heat Q from 100 to 0°C (*a*) by the direct irreversible flow of the heat and (*b*) by a balanced reversible process.

The entropy change of the system has been calculated by means of the arrangement of Fig. 6-2*b*, but, as can be shown, the entropy change of the *system* is independent of the way in which the process is performed and depends only on the initial and final state; i.e., the entropy of the system is a function of the state of the system. Therefore, the result gives the entropy change ΔS of the system when heat Q is transferred from the hot end to the cold end by *any* process, including that of Fig. 6-2*a*.

Now we can consider the entropy change of the universe for the direct heat transfer, as in the arrangement of Fig. 6-2*a*. Since no thermal surroundings are involved, an entropy change results only from that which occurs in the system. This we have deduced to be the positive quantity $+0.0010Q$. Thus, for the direct heat transfer

$$\Delta S_{univ} = \Delta S = \text{a positive quantity} \qquad \mathbf{7}$$

Thus, if we did not know that heat will flow spontaneously from hot to cold, this result and the statements of the second law would tell us it would happen.

Some of the importance attached to entropy as a result of such deductions can be seen by its position alongside energy in the famous maxim of Clausius: "The energy of the universe is constant; the entropy of the universe tends toward a maximum." Since all natural processes are spontaneous, they must occur with an increase of entropy, and therefore the sum total of the entropy in *the* universe is continually increasing. Recognition of this trend leads to some interesting philosophical discussions; e.g., Eddington's idea that "entropy is time's arrow."

We are now in a position to summarize the results of this and the preceding section, thereby indicating the use to which entropy can be put. Suppose we wish to investigate the possibility of a reaction in a chemical or biological system that proceeds from one state *a* to another state *b*. If the entropy difference ΔS_{univ} for the process can be calculated, we can use the following statements:

If ΔS_{univ} *is positive, the reaction will tend to proceed spontaneously from state a to state b.*

If ΔS_{univ} *is zero, the system is at equilibrium and no spontaneous process will occur.*

If ΔS_{univ} *is negative, the reaction will tend to go spontaneously in the reverse direction, i.e., from b to a.*

That these properties of entropy sum up our experience with naturally occurring phenomena has been illustrated by the simple heat of vaporization and heat flow examples given above.

For the type of process used in these examples, it is certainly cumbersome and unnecessary to introduce the entropy function. On the other hand, when dealing with a chemical reaction

Reactants → products

one would be greatly aided by a thermodynamic property which could be determined for the reactants, the products, and the surroundings involved and which would tell whether the reaction would tend to proceed spontaneously.

These considerations all depend on using results obtained from measurements on the thermal surroundings to deduce, for a reversible process, the entropy change occurring in the system. (The procedure is parallel to that used in applications of the first law, where measurements of the energy changes in the thermal and mechanical surroundings made it possible to deduce the change in energy in the system.) But again, there generally are a number of reversible ways of going from some initial state of the system to some final state. Have we any assurance that the values calculated for ΔS for all these ways will be the same; i.e., are we entitled to interpret a value for ΔS deduced for some process that takes the system from state a to state b as $S_b - S_a$?

The proof that $\Delta S = S_b - S_a$ and therefore that we can speak of the entropy of a system just as we can speak of its volume, or mass, or enthalpy can be shown by working out the entropy changes in the steps of a cyclic process. It is found that if the entropy is not a state function, the general statements of the second law can be violated. This deduction will not be worked out here, and we shall proceed with the idea that a system in a specified state has a definite amount of entropy.

6-3 ENTROPY AND THE THIRD LAW OF THERMODYNAMICS

The third law of thermodynamics is based on what has been found when attempts are made to achieve very low temperatures. These experiences are so much less common than those leading to the first and second laws and the statements of the third law are so qualified that it scarcely merits comparable billing with the first two. But it does lead to a procedure for determining the entropy of chemical substances.

The general experience is that the absolute zero of temperature is unattainable. From this general statement, in a way that will not be shown here, a chemically useful statement is deduced. (The same step, you recall, was used in the treatments of the first two laws, where descriptions of our experiences with energy and with spontaneous processes were used to construct useful functions and relations.)

In this way, we come to the chemically useful statement of the third law of thermodynamics, quoted from a classic thermodynamics text:

> If the entropy of each element in some crystalline state be taken as zero at the absolute zero of temperature, every substance has a finite positive entropy; but at the absolute zero of temperature the entropy may become zero, and does so become in the case of perfect crystalline substances.†

The assignment of zero to the entropy of all properly structured crystalline materials at absolute zero is allowed by the third law of thermodynamics. This makes it possible to deduce the entropy values of compounds at other temperatures. All that is necessary are calorimetric data on heat capacities and heats of phase transitions.

For example, the difference in the entropy of a sample between 0 K and a temperature T can be deduced from the defining equation for entropy by considering nearly reversible additions of heat from a variable-temperature heat reservoir. Then, since $\Delta S_{\text{univ}} = \Delta S + \Delta S_{\text{th surr}} = 0$ and $\Delta S = -\Delta S_{\text{th surr}}$, we can write

$$\Delta S = S_T - S_0 = -\int_0^T \frac{dE_{\text{th surr}}}{T}$$

or

$$S_T = S_{T=0} - \int_{T=0}^{T=T} \frac{dE_{\text{th surr}}}{T}$$

$$= S_{T=0} + \int_{T=0}^{T=T} \frac{dq}{T} \qquad \mathbf{8}$$

where dq is the heat transferred from the thermal reservoir to the sample in the system.

The third law lets us assign the value zero to $S_{T=0}$ and thus for the evaluation of the entropy of the sample at some temperature T leaves us the task set by the equation

$$S_T = \int_{T=0}^{T=T} \frac{dq}{T} \qquad \mathbf{9}$$

†G. N. Lewis and Merle Randall, "Thermodynamics," 2d ed., revised by K. S. Pitzer and Leo Brewer, McGraw-Hill Book Company, New York, 1961.

If we imagine that the sample is heated from near $T = 0$ to the temperature of interest by being exposed to thermal surroundings at temperatures kept just infinitesimally above that of the sample, we can perform the heating in the desired balanced way. Furthermore, if this process is carried out with one mole of the sample at constant total pressure, we have, from Sec. 5-1,

$$dq = d\mathrm{H} = \mathrm{C}_p \, dT \tag{10}$$

Then we have

$$\mathrm{S}_T = \int_{T=0}^{T=T} \frac{\mathrm{C}_p \, dT}{T} \tag{11}$$

The integration can be carried out if the necessary values for C_p are available. This integration can be performed graphically from a plot of C_p/T versus T. The method is illustrated with the data for glycine (Table 6-1) in Fig. 6-3. Since heat-capacity measurements are usually not taken down below about 15 K, attainable with liquid hydrogen, an extrapolation to absolute zero is necessary.

Taking a compound from near absolute zero to some temperature such as 25°C usually involves a number of phase transitions. At each of these transitions heat is absorbed. The entropy change corresponding to such transitions can be calculated (as illustrated in Example 6-1) from the

TABLE 6-1
Data for Determining the Entropy of Crystalline Glycine at 25°C†

T, **K**	C_p, **cal/mol deg**	T, **K**	C_p, **cal/mol deg**
10	0.061	160	14.74
20	0.572	180	16.02
30	1.682	200	17.28
40	3.108	220	18.52
60	6.012	240	19.80
80	8.427	260	21.14
100	10.34	280	22.48
120	11.95	300	23.84
140	13.40		

†From J. O. Hutchens, A. G. Cole, and J. W. Stout, *J. Am. Chem. Soc.*, **82:**4813 (1960).

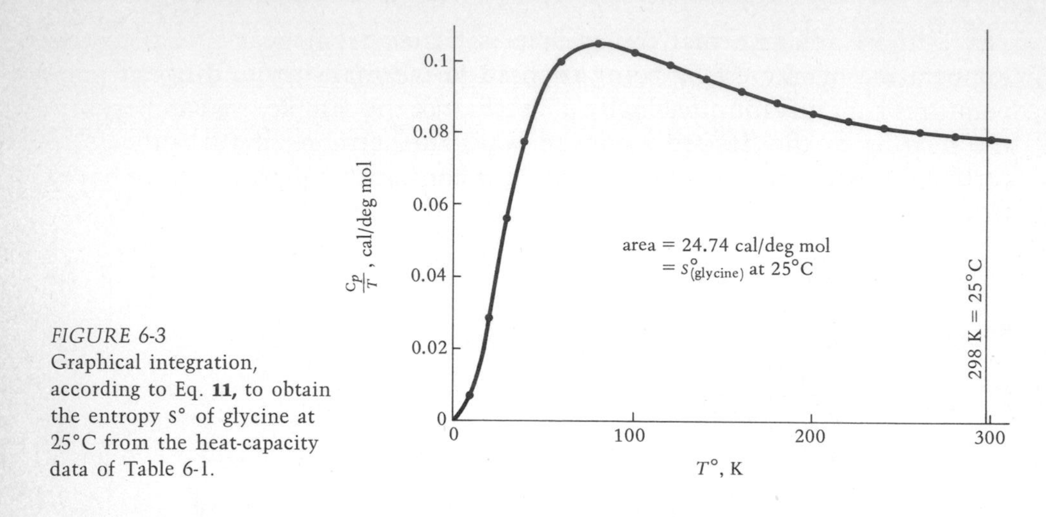

FIGURE 6-3
Graphical integration, according to Eq. **11,** to obtain the entropy S° of glycine at 25°C from the heat-capacity data of Table 6-1.

measured enthalpy change for the transition by the expression

$$\Delta S_{trans} = \frac{\Delta H_{trans}}{T_{trans}} \tag{12}$$

The entropy obtained by adding up all the contributions from absolute zero is usually reported at 25°C. Appendixes I and II give the results that have been obtained for the listed compounds.

Such third-law entropy values are of value because they constitute part of the thermodynamic basis for treating chemical equilibria.

6-4 ENTROPY AS THE INDICATOR OF THE DIRECTION OF CHEMICAL REACTIONS

The data for the entropies of compounds in Appendixes I and II can be used along with heat-of-formation data to give the total entropy change of the system and the thermal surroundings. Inspection of the sign of this ΔS_{univ} quantity reveals the direction in which the reaction would proceed.

Here we are restricted to dealing with pure materials at standard conditions. In later sections, this limitation will be overcome so that reactions of reagents in solution at various concentrations and temperatures can also be investigated.

Consider, for example, the formation of glucose from carbon dioxide and water, i.e., photosynthesis. If you set aside your knowledge of the photosynthetic process, you might ask whether this reaction could be expected to proceed spontaneously, the reagents being CO_2 at 1 atm pressure and liquid water and the products being solid glucose and O_2 at 1 atm. The answer is given by the sign of ΔS_{univ} for the reaction

$$6CO_2(g) + 6H_2O(l) \rightarrow C_6H_{12}O_6(s) + 6O_2(g)$$

The heat of the reaction needed for the calculation of $\Delta S_{\text{th surr}}$ is first obtained:

$$\begin{array}{lcccc} & 6CO_2(g) \quad + & 6H_2O(l) & \rightarrow C_6H_{12}O_6(s) \quad + & 6O_2(g) \\ \Delta H_f^\circ\text{, kcal:} & 6(-94.05) & 6(-68.32) & -304.60 & 6(0) = 0 \\ & = -564.30 & = -409.92 & & \\ & \multicolumn{2}{c}{-974.22} & \multicolumn{2}{c}{-304.60} \end{array}$$

Thus

$$\Delta H = -304.60 - (-974.22) = +669.62 \text{ kcal}$$
$$= +669{,}620 \text{ cal}$$

and $\Delta E_{\text{th surr}} = -\Delta H = -669{,}620$ cal

Then $$\Delta S_{\text{th surr}} = \frac{\Delta E_{\text{th surr}}}{T} = -\frac{669{,}620}{298.16} = -2246 \text{ cal/deg} \qquad \textbf{13}$$

The entropy change in the system itself can be calculated from the entropy values of Appendixes I and II.

$$\begin{array}{lcccc} & 6CO_2(g) \quad + & 6H_2O(l) & \rightarrow C_6H_{12}O_6(s) \quad + & 6O_2(g) \\ s^\circ\text{, cal/deg:} & 6(51.06) & 6(16.72) & 50.7 & 6(49.00) \\ & = 306.36 & = 100.32 & & = 294.00 \\ & \multicolumn{2}{c}{406.68} & \multicolumn{2}{c}{344.7} \end{array}$$

Thus $\Delta S = 334.7 - 406.7 = -62$ cal/deg **14**

Now we can calculate the total entropy change accompanying the reaction as

$$\Delta S_{\text{univ}} = \Delta S + \Delta S_{\text{th surr}}$$
$$= -62 + (-2246)$$
$$= -2308 \text{ cal/deg} \qquad \textbf{15}$$

The negative sign shows that the reaction cannot proceed spontaneously. The photosynthesis process brings an added driving force to bear so that this "unnatural" synthesis can in fact occur.

6-5 INTRODUCTION TO THE MOLECULAR BASIS OF ENTROPY

In the preceding developments it was quite unnecessary to attempt to reach any understanding of entropy. The definition of entropy shows how changes in this function can be calculated, and the discussions of the preceding sections have shown how entropy changes are related to the spontaneity of reactions. With this information one understands entropy as well as one understands a familiar quantity like work, for instance. One's appreciation of work is pretty much limited to its definition as force times distance and to a familiarity with some of its qualitative features. Entropy differs only in that it is a less familiar function.

It is very natural (and for an investigation of the molecular world very profitable) to attempt to find a molecular interpretation of entropy. What is it about the molecules of one substance that gives it a greater or lesser entropy than another substance? What is it about the molecules of a vapor and a liquid that gives the vapor a greater entropy than the liquid with which it is in equilibrium? In this section a qualitative approach to the answers to such questions will be explored. A more detailed and quantitative answer will be developed in the following section.

It is not immediately obvious what molecular phenomenon is responsible for the entropy of a system. Some idea of what should be calculated can be obtained by trying to discover a quantity that would tend to increase when a system moves spontaneously toward the equilibrium position. A nonchemical example will reveal such a quantity.

Consider a box containing a large number of pennies. Suppose that the pennies are initially arranged with all the heads showing. If the box is shaken, the chances are very good that some arrangement of higher probability, i.e., with a more nearly equal number of heads and tails, will result. This system of pennies has a natural, or spontaneous, tendency to go from a state of low probability to one of high probability. The driving force that operates in this system is the probability. The system tends to change toward its equilibrium position, and this change is accompanied by an increase in the probability. Such an example suggests that the entropy might be identified with some function like the probability. The next section will show in a more rigorous manner that entropy is, in fact, closely related to the probability.

An example showing the probabilities of various arrangements of four pennies in a box will be sufficient to illustrate that a quantitative interpretation can be given to the term *probability*. When a box containing four pennies is shaken, each penny has an equal chance of showing a head or a tail. The number of ways, and therefore the probability, of getting a total of one, two, three, or four heads can be calculated as shown in Table 6-2. The probability of getting two heads is seen, for example, to be 6 times as great as that of getting no heads. If a very large number of boxes are shaken, the number of boxes showing two heads would be nearly 6 times as great as the number showing no heads. The larger number of boxes showing two heads is due to the larger number of arrangements, or the greater probability of the two-head result, and does not depend in any way on an energy factor.

The molecular equivalent of the numbers of heads showing is the number of quantum states available which, when occupied, correspond to a given description of the system. The equilibrium of A and B in which B has the higher entropy, for example, can be understood in terms of the fact that for some reason more quantum states corresponding to B are available. Therefore, there are more ways of distributing the atoms in these states so that a molecule of type B is formed than there are ways of arranging the atoms in the quantum states so that a molecule of type A is formed. The tendency for A to change over to B, even if no energy driving force exists, is understood, therefore, to be due to the driving force that takes the system from a state of lower probability, i.e., of fewer quantum states and fewer possible arrangements, to one of higher probability, i.e., one of more available quantum states and more possible arrangements. A substance for which the molecules have more available quantum states has the higher probability and therefore the higher entropy.

The molecular explanation of the entropy change in a process is basically quite simple. In practice, it is often easy to see which of two thermodynamic states has the more available quantum states, or energy levels, but difficult to proceed to a quantitative measure of the corresponding increase in probability. Consider the liquid-to-vapor transition, for example.

That many more quantum states are available to the gas molecules than to the liquid molecules can be seen by thinking of the molecules in each case as being particles in a box. The treatment of Sec. 3-3 led to the one-dimensional allowed-energy expression $\epsilon = n^2h^2/8ml^2$. This shows that the allowed-energy separation factor $h^2/8ml^2$ decreases as l, the size of the

TABLE 6-2
Number of Arrangements of Four Pennies
An H means a head showing, T a tail showing.

Description	**Arrangements**	**No. of arrangements**
4 heads, 0 tails	HHHH	1
3 heads, 1 tail	HHHT, HHTH, HTHH, THHH	4
2 heads, 2 tails	HHTT, HTHT, HTTH, THHT, THTH, TTHH	6
1 head, 3 tails	TTTH, TTHT, THTT, HTTT	4
0 heads, 4 tails	TTTT	1

region in which the molecules are free to move, increases. Since gas-phase molecules are free to move throughout the volume of their container, their allowed translational states are closely spaced on an energy scale. The corresponding motion for the molecules of a liquid is restricted to a volume of the order of their own size, and on this basis we expect the probability and the entropy of a gas to be greater than that of the liquid with which it is in equilibrium. To proceed to a quantitative deduction of this entropy difference we would have to know more about the energies of the states of the molecules of the liquid. The particle-in-a-box model is too crude to be refined, and no other satisfactory route to the liquid-state molecular energies is presently available.

We have reached the conclusion that entropy stems from the distribution of molecules throughout the states available to them—and the greater the number of states over which this distribution occurs, the greater the entropy. This idea is often expressed by relating entropy to *disorder*. The term disorder is very descriptive and valid as long as molecular-level disorder is meant. Then it properly leads one to expect a high entropy when the molecules of the sample can assume a large number of different positions or energies, i.e., can occupy a large number of different states. This disorder and the corresponding entropy are often greater for systems that are clearly more mixed up molecularly, e.g., gases compared with liquids and liquids compared with solids. Likewise, we expect a large, structurally organized molecule like a protein to be less capable of disorder than its component small molecules. Often, however, the molecular-level disorder is hard to anticipate without a more detailed look into the basis for molecular disorder or probability.

6-6 ENTROPY AND PROBABILITY

Let us advance the idea that entropy is related to probability. This extension could lead to expressions that would enable us to calculate the entropy of some gaseous and solid substances from their molecular properties. The development will be carried only to the stage where the molecular properties that influence the entropy can be recognized.

The entropy of 1 mol of any substance, i.e., an amount of the substance containing an Avogadro's number of molecules, is given by the simple-looking expression

$$\mathrm{S} = k \ln \mathrm{W} \qquad \mathbf{16}$$

where k is the Boltzmann constant and W is the probability associated with the system of $\mathfrak{N}$ molecules. To proceed we must investigate the meaning of the word probability in this context. Further, we must see how to express W in terms of the properties of the molecules of the system.

We can begin by extending the idea of Sec. 6-5 that probability depends on the number of states accessible to the molecules of the substance. For any system of molecules there are very many states the molecules might occupy. Thus, for gas molecules there are many allowed translational states, each with energies that can be calculated from an expression like that of Eq. **16** of Sec. 3-3. Further, there are a variety of possible rotational states, as revealed by the studies of Sec. 1-8. For molecules of liquids and solids, we lack convenient expressions for the energies of the states corresponding to the free translation and rotation of gas-phase molecules, but presumably there are allowed states for these motions, as there are also as a result of the vibrational motion of the atoms of the molecules.

The states that exist and can be occupied by molecules are at various energies. Those at extremely high energies compared to the average energy of the molecules are out of reach and not really available even though they exist. What states are occupied in the equilibrium or most probable distribution that molecules adopt in a system at some given temperature?

The answer depends on the *Boltzmann distribution* expression, which allows us to calculate the number of molecules per state for states at various energies. Consider, as shown in Fig. 6-4, two energy levels i and j consisting of g_i and g_j states, respectively. The number of molecules N_i and N_j at these two energies is given in terms of the number of molecules *per state* N_i/g_i and N_j/g_j. The energy difference between these two sets of states is $\epsilon_i - \epsilon_j$. The Boltzmann distribution expression gives the ratio of the molecules per state at the two energies in terms of the ratio of this energy difference to the reference thermal-energy term kT. The dependence is exponential, and the Boltzmann equation is

$$\frac{N_i/g_i}{N_j/g_j} = e^{-(\epsilon_i - \epsilon_j)/kT} \qquad \mathbf{17}$$

Frequently we compare the populations of the available states to that of the lowest state. If we take the energy of the lowest state ϵ_0 to be zero, we have $\epsilon_i - \epsilon_0 = \epsilon_i - 0 = \epsilon_i$ and

$$\frac{N_i/g_i}{N_0/g_0} = e^{-\epsilon_i/kT}$$

or

$$\frac{N_i}{g_i} = \frac{N_0}{g_0} e^{-\epsilon_i/kT} \qquad \mathbf{18}$$

Typically we deal with a system with some known total number of molecules, Avogadro's number $\mathfrak{N}$, for example. We then can deal with the number of molecules per state rather than ratio of numbers in states of various energies. We first write

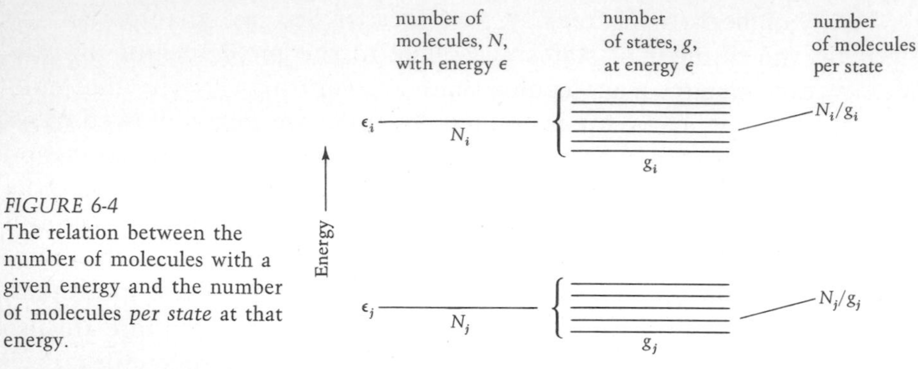

FIGURE 6-4
The relation between the number of molecules with a given energy and the number of molecules *per state* at that energy.

$$\mathfrak{N} = N_0 + N_1 + N_2 + \cdots$$

$$= g_0 \frac{N_0}{g_0} + g_1 \frac{N_1}{g_1} + g_2 \frac{N_2}{g_2} + \cdots \qquad \mathbf{19}$$

Replacement of all the N_i/g_i terms by means of Eq. **18** then leads to

$$\mathfrak{N} = g_0 \frac{N_0}{g_0} + g_1 \left(\frac{N_0}{g_0} e^{-\epsilon_1/kT} \right) + g_2 \left(\frac{N_0}{g_0} e^{-\epsilon_2/kT} \right) + \cdots$$

$$= \frac{N_0}{g_0} (g_0 + g_1 e^{-\epsilon_1/kT} + g_2 e^{-\epsilon_2/kT} + \cdots)$$

$$= \frac{N_0}{g_0} \sum_i g_1 e^{-\epsilon_i/kT} \qquad \mathbf{20}$$

The summation is important enough to warrant a name, the *partition function*, and a symbol q. Then we have

$$\mathfrak{N} = \frac{N_0}{g_0} q$$

or

$$\frac{\mathfrak{N}}{q} = \frac{N_0}{g_0} \qquad \mathbf{21}$$

Now we can return to Eq. **18** and write

$$\frac{N_i}{g_i} = \frac{\mathfrak{N}}{q} e^{-\epsilon_i/kT} \qquad \mathbf{22}$$

Equation **22** gives the number of molecules that will occupy the ith state, which is at energy ϵ_i when $\mathfrak{N}$ molecules distribute themselves throughout the available states in the most probable way.

With the information given by Eq. **22** we can ask how probable is this most probable distribution? We must answer this question in order to be able to insert a value for W in the relation $S = k \ln W$.

A problem identical with that of the probability of a distribution of molecules is that of the probabilities of various distributions of marbles thrown randomly into a box with various-sized compartments. Figure 6-5 suggests a specific marble-compartment analog to the molecular-energy-level problem. If one constructed such a compartmentalized box and had a suitably random throwing device, one could verify that the expression for the probabilities of the different arrangements contains two factors.†

The first factor involves the relative sizes of the compartments. The probability of each marble's landing in a given compartment is proportional to the size of the compartment. This means that in the expression for the probability of a particular distribution of the four marbles of Fig. 6-5 there is a term that is the product of the sizes of the compartments occupied by each marble or, more conveniently, the product of the sizes of the compartments each raised to the power corresponding to the number of marbles occupying that compartment in that particular distribution. (The corresponding factor for the molecules distributed throughout energy levels will be written $g_1^{N_1}\, g_2^{N_2}\, g_3^{N_3} \ldots$.) This first factor implies that the most probable distribution will be the one in which all the marbles are in the largest compartment, and this is clearly not the case.

†This expression is worked out in introductory chapters of texts on probability theory.

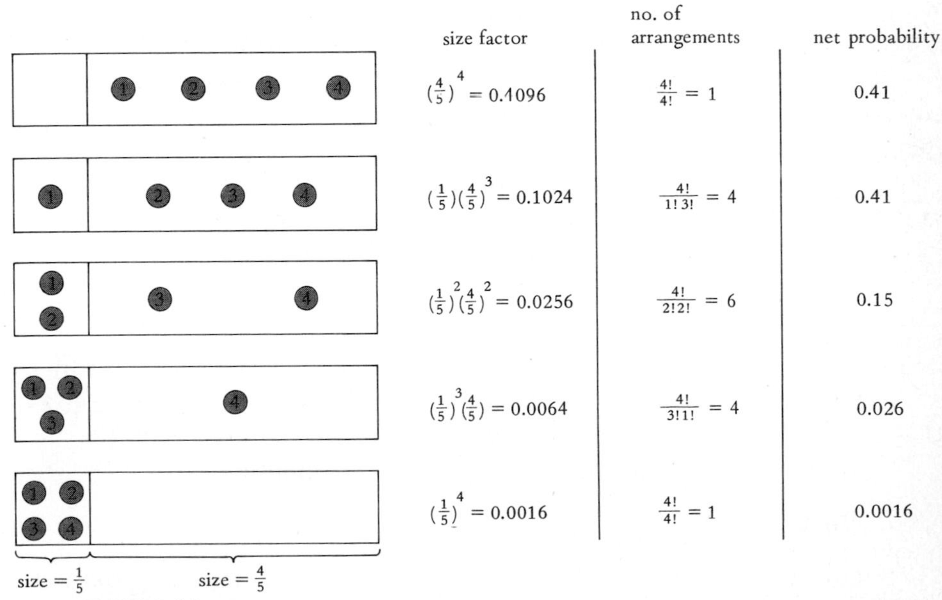

FIGURE 6-5
The relative probabilities of the possible arrangements of numbered marbles in boxes of size $\frac{1}{5}$ and $\frac{4}{5}$.

The second factor expresses the tendency of the marbles to distribute themselves. It can be formed by seeing the total number of ways the individual marbles can be rearranged without altering the total numbers in each box.

Like molecules, marbles of a given type are indistinguishable, and we need an expression for the number of arrangements that can be recognized for such particles. In proceeding to this result we temporarily attach numbers to the marbles so that we can keep track of them.

If you arrange a set of N numbered marbles as 1, 2, 3, . . . , you will convince yourself that you can make $N!$ arrangements. For example, for 3 numbered marbles you can form 1, 2, 3; 2, 3, 1; 3, 1, 2; 2, 1, 3; 1, 3, 2; and 3, 2, 1. The total is 6, or 3!, arrangements. In the marble example of Fig. 6-5 there are four numbered marbles and thus $N! = 4!$ different arrangements. The arrangements that correspond to rearrangements within a compartment are not significant. To arrive at the number of significantly different ways the numbered marbles can be arranged in the compartments of Fig. 6-5 we divide 4! by the meaningless interchanges within each compartment. This factor is given in the third column of Fig. 6-5. If the marbles are numberless, i.e., indistinguishable from each other, the factor for the arrangements is in all cases reduced by the factor $N!$ (here 4!). The molecular counterpart with $N_1, N_2, N_3, \ldots$ molecules in the compartments, which are now energy levels, is a term $1/N_1!N_2!N_3! \cdots$.

The total probabilities for the marble distributions are shown in Fig. 6-5. If you are not familiar with such probability expressions, you should convince yourself of the validity of the results by an appropriate experiment.

The corresponding result for the probability of the distribution of molecules throughout energy levels that correspond to $g_1, g_2, \ldots$ states is

$$\mathrm{W} = (g_1^{N^1} g_2^{N^2} g_3^{N^3} \cdots) \frac{1}{N_1!N_2!N_3! \cdots} \qquad \mathbf{23}$$

It is the value of this probability for the most probable distribution that we need for insertion in $\mathrm{S} = k \ln \mathrm{W}$. To proceed we must recast Eq. **23** so that the awkward $\mathrm{N}_i!$ terms are removed.

First, we take the logarithm of Eq. **23,** to obtain

$$\ln \mathrm{W} = N_1 \ln g_1 + N_2 \ln g_2 + \cdots - \ln N_1! - \ln N_2! + \cdots \qquad \mathbf{24}$$

The factorial terms can then be eliminated by using Stirling's approximation for large numbers, namely,

$$\ln \mathrm{x}! = \mathrm{x} \ln \mathrm{x} - \mathrm{x} \qquad \mathbf{25}$$

Then Eq. **24** becomes

$$\ln W = N_1 \ln g_1 + N_2 \ln g_2 + \cdots - N_1 \ln N_1 - N_2 \ln N_2 \cdots + N_1 + N_2 + \cdots$$

$$= N_1\left(1 + \ln \frac{g_1}{N_1}\right) + N_2\left(1 + \ln \frac{g_2}{N_2}\right) + \cdots$$

$$= \sum_i N_i\left(1 + \ln \frac{g_i}{N_i}\right)$$

$$= \mathfrak{N} + \sum_i N_i \ln \frac{g_i}{N_i} \qquad \textbf{26}$$

Now the Boltzmann distribution expression of Eq. **22** can be inserted so that we calculate the probability of the most probable distribution. The reciprocal of Eq. **22** gives us

$$\frac{g_i}{N_i} = \frac{q}{\mathfrak{N}} e^{\epsilon_i/kT}$$

and $$\ln \frac{g_i}{N_i} = \ln \frac{q}{\mathfrak{N}} + \frac{\epsilon_i}{kT} \qquad \textbf{27}$$

Insertion in Eq. **26** now yields

$$\ln W = \mathfrak{N} + \left(\ln \frac{q}{\mathfrak{N}}\right) \Sigma N_i + \Sigma \frac{N_i \epsilon_i}{kT}$$

$$= \mathfrak{N} + \mathfrak{N} \ln \frac{q}{\mathfrak{N}} + \frac{1}{kT} \Sigma N_i \epsilon_i \qquad \textbf{28}$$

This most probable probability expression can now be put into $S = k \ln W$ to yield (remembering that $R = \mathfrak{N}k$)

$$S = k \ln W$$

$$= R + R \ln \frac{q}{\mathfrak{N}} + \frac{\Sigma N_i \epsilon_i}{T} \qquad \textbf{29}$$

The final summation terms can be recognized as the energy the system has in excess of that it would have if all molecules were in the lowest energy states. We call this the *thermal energy*. For example, in Sec. 1-4 the contribution from the translational motion of the molecules of a gas was found to be $\frac{3}{2}RT$. A similar $\frac{1}{2}RT$ per degree of freedom applies to the rotational motion of most gas molecules. (Exceptions are gases at very low temperatures and gases composed of molecules with very low moments of inertia, H_2

and the hydrogen halides, for example.) For other motions, such as vibrations, and for all motions of molecules in liquids and solids, the allowed energies are more widely spaced, and the thermal energy will be some smaller quantity that will be more difficult to calculate. But in all cases we can write

$$S = R + R \ln \frac{q}{\mathfrak{N}} + \frac{\text{thermal energy}}{T} \qquad \mathbf{30}$$

The principal entropy-determining term is the second one, and thus an understanding of the partition function q provides the key to a molecular-level understanding of entropy.

Recall that q is defined, Eq. **21,** as the series

$$q = \sum_i g_i e^{-\epsilon_i/kT}$$

$$= g_0 + g_i e^{-\epsilon_1/kT} + g_2 e^{-\epsilon_2/kT} + \cdots \qquad \mathbf{31}$$

Moreover, g_i is the number of states at energy ϵ_i, and ϵ_i is measured from the energy of the lowest energy states.

How can q, and thus S, be large? This can happen if the number of states at each energy level, i.e., the g_i's, is large. It can also happen if the $e^{-\epsilon_i/kT}$ terms are not too small. The latter is the case when energy levels are closely spaced, giving ϵ_i values that are small so that ϵ_i/kT will be small and $e^{-\epsilon_i/kT}$ will not fall off far below the value of unity, which corresponds to e^0. The net result is the conclusion (stated in Sec. 6-5) that a substance has a high entropy if many states are available to the molecules of the substance, i.e., if there are many states with ϵ_i values that are less than kT.

Note that the $e^{-\epsilon_i/kT}$ terms become small if the energy term ϵ_i is large or if the temperature is low. Either effect leads to states that are relatively inaccessible because their energy is too high compared with the average thermal energy kT to be reached by many of the molecules of the sample. The limit is reached when T approaches absolute zero; for then all $e^{-\epsilon_i/kT}$ terms go to zero, and only the lowest allowed energy level is occupied. Then the system has the least disorder and the least possible entropy.

PROBLEMS

1 If a process proceeds spontaneously and in doing so absorbs heat from the surroundings, what can you say about the entropy change in the system? Illustrate your answer by an example.

2 If a process proceeds spontaneously and gives out heat, what can you say about the entropy change in the system? Give an example to illustrate your answer.

3 If 100 g of water is converted to ice by exposure to a heat sink that is at a temperature infinitesimally below 0°C, what is the entropy change of the heat reservoir, of the water-ice system, and of the universe of the process? The heat of fusion of water is 79.7 cal/g.

4 Suppose that 100 g of water is converted to ice by exposing the water to a large heat sink at the temperature of liquid nitrogen, 77 K.

a Imagine a reversible way in which the freezing of the water and the transfer of heat to the cold reservoir could be carried out. When the process occurs in this manner, what are the entropy changes of the system, the thermal surroundings, and the universe of the process?

b Now consider the direct, irreversible transfer of heat from the water-ice system to the 77 K reservoir. What are the entropy changes in the system, the thermal surroundings, and the universe of the process?

5 Listed entropies of vaporization usually imply the increase in the entropy of 1 mol of the substance when it is converted from liquid to vapor at its normal boiling point. Calculate the entropies of vaporization of the following substances from the given data.

	normal boiling point, °C	ΔH_{vap}, **cal/mol**
Methane	−161.4	2216
Ethyl ether	+34.6	6210
Carbon tetrachloride	+76.7	7170
Benzene	+80.1	7353
Mercury	+356.6	14,166

Your results should lead you to a generalization about the entropy of vaporization that is known as *Trouton's rule*. [Some substances conform less well to the rule than the compounds listed above. Nonconformers are very-low-boiling substances like H_2 (bp, −252.7°C, ΔH_{vap}, 216 cal/mol) and hydrogen-bonded compounds like acetic acid (bp, 118.2°C, ΔH_{vap}, 5830 cal/mol) and water (bp, 100°C, ΔH_{vap}, 9720 cal/mol).]

6 Compare the entropy change for the fusion of ice (which can be calculated from the melting point of 0°C and the heat of fusion of 1436 cal/mol) with the entropy of vaporization of water.

7 The following heat-capacity data have been reported for *l*-alanine [J. O. Hutchens, A. G. Cole, and J. W. Stout, *J. Am. Chem. Soc.*, **82:**4813 (1960)]: By a graphical treatment and the assumption of applicability of the third law, deduce the molar entropy of *l*-alanine at 25°C. Compare with the value listed in Appendix II.

T, K	C_p, cal/mol deg	T, K	C_p, cal/mol deg
10	0.118	140	16.49
20	0.920	180	19.89
40	4.151	220	23.00
60	7.414	260	26.22
80	10.19	300	29.36
100	12.56		

8 From the data of Prob. 7 and the entropy of *l*-alanine at 25°C listed in Appendix II, estimate the entropy of *l*-alanine at the physiological temperature of 37°C.

9 Use the standard heat of formation and entropy data of Appendix II to decide the spontaneous direction of the following reaction, all reagents being at standard conditions at 25°C:

$$2\text{Glycine}(c) \rightarrow \text{glycylglycine}(c) + H_2O(l)$$

What bearing does this result have on the synthesis of proteins from amino acids?

10 The expression for the allowed vibrational energy levels of a vibrating particle was given in Sec. 3-4 as

$$\Delta\epsilon = \frac{h}{2\pi}\sqrt{\frac{k}{m}}$$

where k measures the rigidity with which the particle is held in its equilibrium position and m is the mass of the particle. The atoms of a simple crystal vibrate about their lattice sites. Although the vibrations of the particles couple with one another, this formula is a guide to the energy-level spacing.

a What force-constant and mass features would give atoms of a crystal few accessible vibrational states and thus a low entropy?

b In graphite the carbon atoms are held in a tight structure in two dimensions, whereas in diamond they are firmly held in three dimensions. Explain why the molar entropy of one is greater than that of the other.

c The molar entropies at 25°C of tungsten, gold, and lead are 8.0, 11.4, and 15.5 cal/deg, respectively. Their atomic weights are not very different. On what basis can these very different entropies be understood?

d KCl forms an ionic crystal, which, since the atomic masses of K and Cl are not very different, can be treated as if the crystal contained a single type of atom with an atomic mass of 37. The molar entropy of KCl at 25°C is 19.8 cal/deg, and $\frac{1}{2}$ mol of KCl, which contains a total of an Avogadro's number of ions, has an entropy of $\frac{1}{2} \times 19.8 = 9.9$ cal/deg at 25°C. Compare this value with that of metals with comparable atomic masses and comment on the rigidity imposed by ionic bonds compared with that imposed by metallic bonds.

11 Make qualitative statements and illustrate them with specific compounds, about (*a*) the comparison of the entropy for a substance, or similar substances, in the gas, liquid, and solid states and (*b*) the comparison of the entropy, for a given state, liquid or gas, for substances that contain molecules with few atoms and those with molecules with many atoms.

REFERENCES

The general references listed for Chap. 5 and:

BENT, H. A.: "The Second Law," Oxford University Press, New York, 1965. An informative and even entertaining discussion of entropy and the second law of thermodynamics.

7 Free Energy

As Chap. 6 illustrated, the thermodynamic quantities ΔH and ΔS can show the direction in which a process will proceed spontaneously. Now a single and more convenient function that indicates this direction will be introduced. Moreover, this function, the *free energy*, lends itself to the treatment of the equilibrium state toward which the process moves. The interrelation of thermodynamic properties and this measure of the driving force toward the equilibrium states is the most important chemical accomplishment of thermodynamics.

7-1 THE CONVENIENT FUNCTION: FREE ENERGY

The entropy change that must be considered if the direction of chemical change is to be deduced is that of the universe of the change. We have already seen that this entropy change is the sum of the change occurring in the system and the change occurring in the thermal surroundings. All the information used to calculate these two quantities comes from tables of entropies and enthalpies of the reagents. These allow the entropy change of the system to be calculated and the heat of reaction (and thus the entropy change of the thermal surroundings) to be deduced. These calculations make use of properties of the system and one can, therefore, proceed without thinking explicitly about the surroundings.

Consider, first, a reaction occurring at constant temperature for which a balanced equation can be written and for which the necessary heat of formation and entropy data are available. We can calculate

$$\Delta S_{\text{univ}} = \Delta S + \Delta S_{\text{th surr}}$$
$$= \Delta S - \frac{\Delta H}{T} \qquad \mathbf{1}$$

It is customary to express this quantity in a different way. To convert the expression to one with the more familiar units of energy, we multiply it by T, the constant temperature of the reaction, to change the right side to $T\,\Delta S - \Delta H$. Then we introduce a symbol and a name for this collection of terms by writing

$$-\Delta G = T\,\Delta S - \Delta H$$
or
$$\Delta G = \Delta H - T\,\Delta S \qquad [T = \text{const}] \qquad \mathbf{2}$$

where ΔG is called the change in *free energy* or, more completely, the *Gibbs free energy*. The function ΔG offers certain conveniences.

Notice, however, that the sign change has reversed the entropy-of-the-universe implications, so that now we have

$$\Delta G = \begin{cases} - & \text{reaction can proceed as written} \\ + & \text{reverse reaction can proceed} \end{cases} \qquad \mathbf{3}$$

The free-energy function itself, from which these constant-temperature changes in free energy stem, is defined as

$$G = H - TS \qquad \mathbf{4}$$

Since H and TS are properties of the system, so also is G.

The quantity ΔG leads to an additional way of looking at the driving force of a reaction. To be specific, consider a reaction in which the entropy of the system decreases. The final term, $-T\,\Delta S$, in Eq. **2** is then positive. The reaction can proceed spontaneously only if ΔH is negative and large enough to overcome $-T\,\Delta S$, thus giving ΔG a negative sign. Or, if we are thinking of converting the energy from the reaction for mechanical or other purposes, at least enough energy to compensate for the $-T\,\Delta S$ term must be left as ΔH and delivered to the thermal surroundings. Only energy in excess of this amount can be used as a special driving force. Thus, the free energy is a measure of the work that can be drawn from a system after the entropy demands have been met by payment of heat to the thermal reservoir. This is the basis for the adjective free in the expression free energy.

Since the free-energy change measures the useful work that might be obtained from a process at constant temperature and pressure, it is a measure of the spontaneity of the process. We have the very important result that *the decrease in free energy of a constant-temperature, constant-pressure process is the measure of the tendency of the process to proceed spontaneously.* This statement is consistent with the sign implications of Eq. **3.**

If a change in a system is considered and a *decrease* in free energy is calculated for that change, spontaneous processes that carry out that change can be expected; moreover, any such process could be harnessed to deposit energy in the mechanical-energy reservoir. If a different change in the system is considered and an *increase* in free energy is calculated for that change, no spontaneous process can occur to carry out that change. In fact, processes carrying the system in the reverse direction will be those which can occur spontaneously. Finally, if the free energy is calculated to be unaffected for a considered change, the system will show no tendency to proceed from one state to the other.

It should be mentioned that although the free-energy function G, the Gibbs free energy, is suitable for direct application to constant-pressure processes, another free-energy function is more convenient for constant-volume processes. This function, known as the *Helmholtz free energy A* is defined as

$$A = E - TS \qquad 5$$

A development like that performed on G would show that in a constant-volume process the decrease in A corresponds to the driving force of the reaction. Thus H and G are functions that are convenient for constant-pressure processes, whereas E and A are more convenient for constant-volume processes. In our introductory thermodynamics only the applications of H and G will be developed.

7-2 STANDARD FREE ENERGIES FOR PURE SUBSTANCES

The preceding considerations show that free energy can be correlated with the tendency of the system to proceed to a state of equilibrium. In view of this fact, it would be very useful to have a tabulation of free energies of chemical compounds so that the free-energy change of a possible reaction could be easily calculated.

Free energies, like any other energies, must have some reference point. The same procedure is followed as for enthalpies. A zero value is assigned to the free energies of the common form of the elements at 25°C and 1 atm pressure. These, and the free energies of compounds based on these references, are known as *standard free energies of formation*. Some values are listed in Appendixes I and II.

The origin of tabulated standard free energies from standard heats of formation and standard entropies can be illustrated by calculating the value for urea. We consider the formation of urea from its elements and apply the equation $G = H - TS$ or $\Delta G = \Delta H - \Delta(TS)$ to standard conditions and 25°C. Then, with the subscript f to denote the formation from the elements, this relation becomes $\Delta G_f^\circ = \Delta H_f^\circ - T\,\Delta S_f^\circ$.

The value of ΔH_f° of urea(c) is given in Appendix II as $-79{,}580$ cal. The value of ΔS_f° can be obtained from the listed S° values of Appendixes I and II. Thus

$$\text{C(graphite)} + N_2(g) + \tfrac{1}{2}O_2(g) + 2H_2(g) \rightarrow NH_2\overset{\overset{\displaystyle O}{\|}}{C}NH_2(c)$$

	C(graphite)	$N_2(g)$	$\frac{1}{2}O_2(g)$	$2H_2(g)$	$NH_2CONH_2(c)$
S°, cal/deg:	1.36	45.78	$\frac{1}{2}(49.00) = 24.50$	$2(31.21) = 62.42$	25.00
	134.06 (sum of reactants)				

Thus $\Delta S_f^\circ = 25.00 - 134.06 = -109.06$ cal/deg

Now we calculate

$$\Delta G_f^\circ = \Delta H_f^\circ - T\,\Delta S_f^\circ$$
$$= -79.580 - 298.16(-109.06)$$
$$= -47{,}070 \text{ cal}$$

It is this value that is listed as the standard free energy of formation of urea(c) in Appendix II.

Example 7-1

Use the free-energy data of Appendixes I and II to explore the possible spontaneity of the formation of glucose from carbon dioxide and water.

$$6CO_2(g) \quad + \quad 6H_2O(l) \rightarrow C_6H_{12}O_6(c) \quad + \quad 6O_2(g)$$

	$6CO_2(g)$	$6H_2O(l)$	$C_6H_{12}O_6(c)$	$6O_2(g)$
ΔG_f°, kcal:	$6(-94.26)$	$6(-56.69)$	-217.63	0
	$= -565.56$	-340.14		
	-905.70		-217.63	

Solution

$\Delta G = -217.63 - (-905.70) = +688.07$ kcal

The large positive value of ΔG shows that the reaction cannot proceed spontaneously and it is the reverse reaction, the combustion of glucose, for which there is a driving force.

7-3 THE DEPENDENCE OF FREE ENERGY ON PRESSURE AND TEMPERATURE

Standard free energies, as appear in Appendixes I and II, allow predictions to be made of the direction of a reaction for the single conditions of 25°C and 1 atm. For these free-energy data to be of appreciable use, we must develop a way of extending them to other conditions. Here the temperature and pressure dependences will be treated. In later sections pressure dependence will be used to obtain the more interesting concentration dependence.

The free energy is defined by the expression

$$G = H - TS \tag{6}$$

or, since H is defined by $H = E + PV$, by

$$G = E + PV - TS \tag{7}$$

The change in G accompanying any change in these quantities that determine G is given by the differential

$$dG = dE + P\,dV + V\,dP - T\,dS - S\,dT \tag{8}$$

This clumsy equation can be reduced to a more manageable form.

We restrict our attention to states that can be connected by balanced, reversible processes. For these, dS_{univ} must be zero, and the entropy decrease dq/T of the surroundings must equal the entropy increase dS of the system. Thus $dS = dq/T$ or $T\,dS = dq$. If, further, the only work involved is the result of expansion or compression, according to Eq. **4** of Sec. 5-1, we can set $P\,dV = -dw$. Under these conditions of reversibility and only $P\,dV$ work, the first-law relation $dE = dq + dw$ becomes $dE = T\,dS - P\,dV$ or $dE - T\,dS + P\,dV = 0$. These three terms in Eq. **8** can thus be set equal to zero, and we are left with

$$dG = V\,dP - S\,dT \tag{9}$$

If we consider separately constant-pressure and constant-temperature processes, we can recast this result.

At constant pressure:

$$dG = -S\,dT \qquad \text{or} \qquad \frac{dG}{dT} = -S \tag{10}$$

At constant temperature:

$$dG = V\,dP \qquad \text{or} \qquad \frac{dG}{dP} = V \tag{11}$$

It is often useful, both in mathematical developments and in experimental studies, to fix all but one of the variables. This is done in Eqs. **10** and **11** by the constant-pressure and constant-temperature comments. A formal calculus procedure for showing that variables are to be held fixed consists of using ∂ instead of d in the derivative symbol. Then the derivative is said to be a *partial derivative*. The variables that are held constant can be shown by subscripts. With these conventions Eqs. **10** and **11** are written

$$\left(\frac{\partial G}{\partial T}\right)_P = -S \tag{12}$$

and

$$\left(\frac{\partial G}{\partial P}\right)_T = V \tag{13}$$

These two results show how the free energy of a chemical compound depends on the temperature and on the pressure.

Since liquids and solids are quite incompressible, the free-energy change corresponding to an isothermal increase in pressure ΔP for some not too large pressure change can be written (according to Eq. **11** and the assumption of a constant volume) as $V\,\Delta P$. Since the molar volumes of solids and liquids are relatively small, this change in free energy resulting from the application of ordinary pressures to liquids and solids is also relatively small, and for many purposes the pressure dependence of the free energy of liquids and solids can be neglected.

For gases the dependence of free energy on pressure is appreciable and important. For an *ideal gas*, P and V are related by the ideal-gas law $PV = nRT$, or $V = nRT/P$. The integration of Eq. **11** can be performed to give the free-energy change when the pressure is changed, at constant temperature, from P_1 to P_2. Thus

$$G_2 - G_1 = \int V\,dP = nRT \int_{P_1}^{P_2} \frac{dP}{P}$$

$$= nRT \ln \frac{P_2}{P_1} \qquad \textbf{14}$$

Of particular interest is the extent to which the free energy changes from its standard-state value when the pressure changes from 1 atm. If state 1 is the standard state, then

$$P_1 = 1 \text{ atm} \quad \text{and} \quad G_1 = G°$$

$$P_2 = P \quad \text{and} \quad G_2 = G$$

With this notation for states 1 and 2, Eq. **14** can be rewritten for 1 mol as

$$G - G° = RT \ln \frac{P}{1}$$

$$G = G° + RT \ln P \qquad \begin{bmatrix} \text{constant temperature} \\ P \text{ in atmospheres} \end{bmatrix} \qquad \textbf{15}$$

or, in terms of free energies of formation,

$$\Delta G_f = \Delta G_f^\circ + RT \ln P \qquad \begin{bmatrix} \text{constant temperature} \\ P \text{ in atmospheres} \end{bmatrix} \qquad \textbf{16}$$

Thus the free energy of a substance that follows ideal-gas behavior varies from its 1-atm value as a result of the additive $RT \ln P$ term. At higher pressures the gas has a higher free energy—the qualitative result one would expect since the gas then has a greater ability to do work. This shows not only the quantitative dependence on pressure but also that ability to do work enters into the tendency of the substance to react chemically.

Example 7-2

What is the free energy of CO_2 in the atmosphere at its normal sea-level partial pressure of 0.00031 atm?

Solution

The standard free energy of formation of CO_2 is given in Appendix I as −94,260 cal. At 0.00031 atm instead of 1 atm, free energy of formation is

$$\begin{aligned}\Delta G_f &= \Delta G_f^\circ + RT \ln P \\ &= -94{,}260 + 2.303(1.987)(298.16) \log 0.00031 \\ &= -94{,}260 + 1364(-3.509) \\ &= -94{,}260 - 4790 \\ &= -99{,}050 \text{ cal}\end{aligned}$$

Thus reactions that produce CO_2 will have a greater tendency to proceed if this product has its atmospheric partial pressure instead of a 1-atm pressure. Conversely, reactions which must use CO_2 will proceed with less of a driving force.

7-4 FREE ENERGIES OF IDEAL NONELECTROLYTE SOLUTES

To apply free-energy considerations to reactions in solution we need to know the dependence of the free energy on concentration. For some solutes this can be done by extending the pressure-dependence result of the preceding section, and so we start with this approach.

The conversion of one reagent to another that is in equilibrium with the first is a balanced process with no driving force. For such a process $\Delta G = 0$. Thus a reaction or process at equilibrium connects reagents that have equal molar free energies. Thus when a solution is saturated by a gas at some fixed temperature and pressure, the molar free energy of the gas is the same in the gas phase and in the solution.

To be specific, consider oxygen gas dissolved in water at 25.9°C, for which the solubility–gas–pressure data are given in Fig. 7-1. At rather low pressures a typical proportionality appears between gas pressure and concentration of the dissolved material. This behavior, an example of conformity to *Henry's law*, can be expressed as

$$P_{O_2} = k m_{O_2} \qquad \textbf{17}$$

where m_{O_2} is the molality, the moles of oxygen per 1000 g of water.

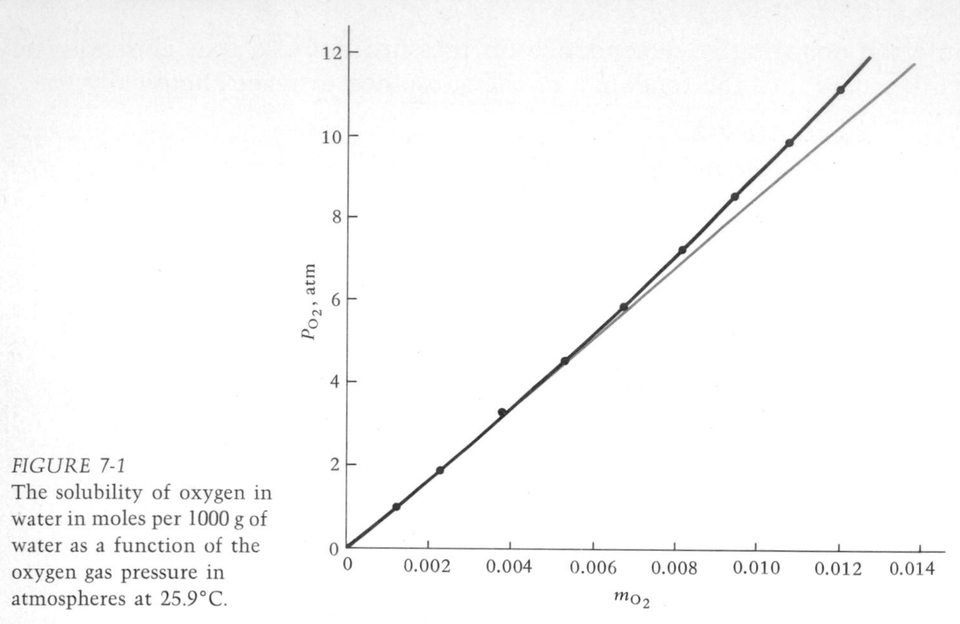

FIGURE 7-1
The solubility of oxygen in water in moles per 1000 g of water as a function of the oxygen gas pressure in atmospheres at 25.9°C.

To deduce the dependence of free energy of the dissolved oxygen on concentration we now consider two solutions with the concentrations m_1 and m_2 that correspond to saturation at pressures P_1 and P_2. From Eq. **14** we write

$$\mathrm{G}_2 - \mathrm{G}_1 = RT \ln \frac{P_2}{P_1}$$

The free-energy difference can be expressed in terms of the concentrations of the dissolved gas by now using the Henry's law relation of Eq. **17** to give

$$\mathrm{G}_2 - \mathrm{G}_1 = RT \ln \frac{km_2}{km_1}$$

$$= RT \ln \frac{m_2}{m_1} \qquad \mathbf{18}$$

Thus, in the region where Henry's law holds, the free energy of the solute depends on an $RT \ln m$ term.

If such dependence were maintained up to 1 m solution (for the O_2 case, Fig. 7-1 shows that it is not), we would choose the 1 m solution as the standard state for the dissolved, or aqueous, material and then we could write

$$\mathrm{G} - \mathrm{G}^\circ = RT \ln \frac{m}{1}$$

or

$$\mathrm{G} = \mathrm{G}^\circ + RT \ln m \qquad \mathbf{19}$$

If we again choose to deal with free energies of formation, we write instead

$$\Delta G_f = \Delta G_f^\circ + RT \ln m \qquad \mathbf{20}$$

we thereby imply that for solutes the standard state to which the ΔG_f° applies is that of the solute in an aqueous solution (unless some other solvent is specified) at a 1 m concentration.

These relations have been deduced for a dissolved gas. But a solute is a solute, and the fact that we are inclined to think of some solutes as dissolved gases and others as dissolved solids or liquids is immaterial. The deduced dependence of free energy on concentration can therefore be applied, at least approximately, to many solutes. This can be done even if the solute is not considered to be a dissolved gas and even if it is so nonvolatile that its vapor pressure is too small to allow us to study its gas-phase form conveniently.

Equations **19** and **20** are taken to represent *ideal*-solute behavior, just as Eqs. **15** and **16** stem from ideal-gas behavior.

Nonideal behavior, as illustrated by solutes that do not conform to Henry's law, spoils the simple dependence of free energy on molality given by Eq. **19** or Eq. **20.** This complicates the selection of standard-state conditions. We then proceed (as will be shown in Sec. 7-6) by modifying the ideal-solute relations developed here.

Example 7-3

The solubility of the amino acid *l*-tryptophan in water at 25°C is reported as 1.14 g per 100 ml of water. Using the ΔG_f° value of $-28{,}500$ cal for the crystalline material and assuming ideal-solution behavior, deduce a ΔG_f° value for the aqueous material.

Solution

A solution in equilibrium with solid *l*-tryptophan contains 1.14 g per 100 ml of water or, since the molecular mass is 204 g, 0.0559 mol per 1000 g of water. Tryptophan in this 0.0599 m solution has the same free energy per mole as solid tryptophan, $-28{,}500$ cal/mol. Equation **20** can now be used to obtain

$$\begin{aligned} -28{,}500 &= \Delta G_f^\circ(aq) + RT \ln 0.0559 \\ &= \Delta G_f^\circ(aq) + 1.987(298)(2.303) \log 0.0559 \\ &= \Delta G_f^\circ(aq) + 1364(-1.253) \end{aligned}$$

$$\text{Thus} \quad \Delta G_f^\circ(aq) = -28{,}500 + 1.710 = -26{,}790 \text{ cal}$$

Notice that the difference between the ΔG_f° values for solid and aqueous depends on the extent to which the solubility varies from 1 *m*.

7-5 FREE ENERGIES OF IDEAL ELECTROLYTE SOLUTES

Some reagents, those we classify as electrolytes, are dealt with in aqueous solutions as the ionic species that stem from the parent substance. Tabulating and using free-energy data for such species requires a few modifications. They can be introduced by again using the equilibrium between a gas and the dissolved material to obtain an insight into the behavior of dissolved materials. The treatment and general features of this behavior will carry over to all solutes.

Hydrogen chloride dissolves in water to give a solution that we usually deal with in terms of the ionic species H^+ and Cl^-. This system provides a suitable example. Measurements of the equilibrium between the gas and the solution have been made for HCl concentrations greater than 4 *m*. Measurements at lower concentrations are difficult because of the very low equilibrium HCl presures that must be worked with, but values can be calculated even for solutions approaching the limit of infinite dilution. Some calculated and observed values are given in Table 7-1.

If we attempt to treat the vapor-phase–solution equilibrium of this

TABLE 7-1
Molal Concentration and Equilibrium Vapor Pressure of HCl†

m	p, **atm**	$\frac{p}{m}$	$\frac{p}{m^2}$
0.001	4.4×10^{-13}	4.4×10^{-10}	4.4×10^{-7}
0.01	4.0×10^{-11}	4.0×10^{-9}	4.0×10^{-7}
0.1	3.1×10^{-9}	3.1×10^{-8}	3.1×10^{-7}
1.0	3.2×10^{-7}	3.2×10^{-7}	3.2×10^{-7}
4.0	2.40×10^{-5}	6.0×10^{-6}	15.0×10^{-7}
5.0	6.974×10^{-5}	1.39×10^{-5}	27.9×10^{-7}
6.0	1.842×10^{-4}	3.07×10^{-5}	51.2×10^{-7}
7.0	4.58×10^{-4}	6.54×10^{-5}	93.4×10^{-7}
8.0	1.11×10^{-3}	1.39×10^{-4}	174×10^{-7}
9.0	2.54×10^{-3}	2.82×10^{-4}	313×10^{-7}
10.0	5.53×10^{-3}	5.53×10^{-4}	553×10^{-7}

† Values at 4 *m* and above are measured [Bates and Kirschman, *J. Am. Chem. Soc.*, **41:**1991 (1919)]; those at lower concentrations have been calculated. (See Pitzer K. S., and L. Brewer: "Thermodynamics," 2d ed., McGraw-Hill Book Company, New York, 1961, pp. 312–319.)

electrolyte in the same way as we treated the oxygen equilibrium, we would investigate the Henry's law proportionality $P = km$. Obedience to this law is shown by constancy of calculated k values. As the third column of Table 7-1 shows, HCl fails to obey this law even as very dilute solutions are approached.

The dilemma this presents is avoided if we recognize that the dissolved material can be treated as $H^+(aq)$ and $Cl^-(aq)$ rather than HCl(aq). Then we see that the gas-phase–solution-phase equilibrium we are studying is described by the equation

$$HCl(g) \rightarrow H^+(aq) + Cl^-(aq) \qquad \textbf{21}$$

Using our earlier nonelectrolyte results as a guide, we can expect, at least in the dilute-solution limit, that we would have the proportionality

$$P \propto m_{H^+(aq)} m_{Cl^-(aq)} \qquad \textbf{22}$$

Further, since $m_{H^+(aq)} = m_{Cl^-(aq)} = m$, where m is the number of moles of HCl dissolved in 1000 g of water, this corresponds to

$$P \propto m^2 \qquad \textbf{23}$$

A test of this proportionality is made, as in Table 7-1, by investigating the constancy of the ratio P/m^2. Approximate constancy of the values at concentrations less than about 1 m suggests that this is indeed a better relation for describing the pressure-concentration data than that based on the simple proportionality of HCl pressure and HCl concentration.

The inconsistency of the P/m^2 ratio, particularly at higher concentrations, points out the nonideality of the system, a common feature of solutions of electrolytes. We proceed here by ignoring these added complexities and deal with ideal solutions, which are approached in the limit of infinite dilution.

We conclude that for the electrolyte HCl, if we assume ideal behavior, we can write

$$\begin{aligned} \mathrm{G}(HCl,aq)_2 - \mathrm{G}(HCl,aq)_1 &= RT \ln \frac{P_2}{P_1} \\ &= RT \ln \frac{m_2^{\,2}}{m_1^{\,2}} \end{aligned} \qquad \textbf{24}$$

where m is the HCl molality. Furthermore, if solutions with various H^+ and Cl^- ion concentrations are studied, the m^2 terms should be interpreted as $m_{H^+} m_{Cl^-}$ and we would write

$$\mathrm{G}(HCl,aq)_2 - \mathrm{G}(HCl,aq) = RT \ln \frac{(m_{H^+} m_{Cl^-})_2}{(m_{H^+} m_{Cl^-})_1} \qquad \textbf{25}$$

If now we deal with free energies of formation and assume that a standard state for the dissolved HCl can be established, we write

$$\Delta G_f(\text{HCl},aq) = \Delta G_f^\circ(\text{HCl},aq) + RT \ln (m_{\text{H}^+} m_{\text{Cl}^-})$$
$$= \Delta G_f^\circ(\text{HCl},aq) + RT \ln m_{\text{H}^+} + RT \ln m_{\text{Cl}^-} \qquad \mathbf{26}$$

The breakup of the concentration-dependent terms and our tendency to deal with ionic species rather than parent electrolytes suggests that the free-energy terms $\Delta G_f(\text{HCl},aq)$ and $\Delta G_f^\circ(\text{HCl},aq)$ can also be broken up into H^+ and Cl^- components. Formally, we can write

$$\Delta G_f^\circ(\text{HCl},aq) = \Delta G_f^\circ(\text{H}^+,aq) + \Delta G_f(\text{Cl}^-,aq) \qquad \mathbf{27}$$

But supposing that we have a value for $\Delta G_f^\circ(\text{HCl},aq)$, how do we break this expression up to obtain values for the ions? No thermodynamic or equilibrium studies can help us here. All such studies lead to combinations of ionic terms that correspond to some parent electrolyte. This fact lets us make an *arbitrary* assignment for the ionic free energy of one ionic species and then deduce all other ionic species values in terms of this reference. We select the H^+ ion and arbitrarily set

$$\Delta G_f^\circ(\text{H}^+,aq) = 0 \qquad \mathbf{28}$$

Then studies of electrolytes will yield other ionic values. For example, Eq. **27** becomes

$$\Delta G_f^\circ(\text{HCl},aq) = 0 + \Delta G_f^\circ(\text{Cl}^-,aq)$$
$$= \Delta G_f^\circ(\text{Cl}^-,aq) \qquad \mathbf{29}$$

The deduced value of -31.35 kcal for ΔG_f° for the electrolyte HCl in aqueous solution then gives

$$\Delta G_f^\circ(\text{Cl}^-,aq) = -31.35 \text{ kcal} \qquad \mathbf{30}$$

With such standard ionic free-energy data, we can return to the expression of Eq. **26** and can show the concentration dependence of ionic free energies as

$$\Delta G_f(\text{H}^+,m) = 0 + RT \ln m_{\text{H}^+}$$
or
$$\Delta G_f(\text{Cl}^-,m) = -31{,}350 + RT \ln m_{\text{Cl}^-} \qquad \mathbf{31}$$

Such ionic terms can be used to recover the free energy of a dissolved electrolyte. Thus, addition of Eqs. **31** gives again

$$\Delta G_f(\text{HCl},m) = \Delta G_f(\text{H}^+,m) + \Delta G_f(\text{Cl}^-,m)$$
$$= -31{,}350 + RT \ln m_{\text{H}^+} m_{\text{Cl}^-}$$

As a result of the interactions of the charged species, solutions of electrolytes show considerable nonideality, which so far has been ignored.

The strategy that yields free-energy values like those of Appendixes I and II for ionic species in aqueous solution has been illustrated, however. Nonideality will be recognized in Secs. 7-6 and 7-7.

Example 7-4

From the ionic free energies of Appendix I, calculate the standard free energy of formation of aqueous calcium carbonate. Use this value and the standard free energy of −269,780 cal for solid calcium carbonate to deduce the solubility of the material.

Solution

The tabulated values for $\Delta G_f^\circ(aq)$ for Ca^{2+} and CO_3^{2-} are −132,180 and −126,170 cal, respectively. In view of the way in which such ionic values are related to values for parent electrolytes, we have

$$\Delta G_f^\circ(CaCO_3,aq) = -132{,}180 - 126{,}170$$
$$= -258{,}350 \text{ cal}$$

This is to be interpreted as the molar free energy that dissolved $CaCO_3$ would have if it behaved ideally and had a concentration of 1 m, that is, if it consisted of ideally behaving Ca^{2+} and CO_3^{2-} ions both at concentration of 1 m.

A saturated solution, with a molality we shall call s, is in equilibrium with solid $CaCO_3$. The molar free energy of dissolved $CaCO_3$ at a concentration of s must be −269,780 cal; the value for solid $CaCO_3$ as calcite.

Now we can relate the free energies of the ions, both present at 1 m concentration, to the free energies when each is present at a concentration of s by

$$\Delta G_f = \Delta G_f^\circ + RT \ln \frac{s^2}{1\,m \times 1\,m}$$

or $\quad -269{,}780 = -258{,}350 + 1364 \log s^2$

This gives

$$\log s^2 = -8.38 \qquad \text{and} \qquad s = 6.5 \times 10^{-5}$$

Thus, in an aqueous solution saturated with $CaCO_3$

$$[Ca^{2+}] = [CO_3^{2-}] = 6.5 \times 10^{-5}$$

The more familiar measure of solubility is the solubility product, which we can also now evaluate as:

$$K_{sp} = [Ca^{2+}][CO_3^{2-}] = 4.2 \times 10^{-9}$$

7-6 FREE ENERGIES OF REAL SOLUTES: NONELECTROLYTES

Now let us recognize that except at very low concentrations, the dependence of free energy of a solute on the concentration is not correctly given by the $RT \ln m$ term of Eq. **19.** If equilibrium gas-phase pressures are used, as in Sec. 7-4, to investigate solute free energies, failure of the system to obey Henry's law would nullify the derivation of this $RT \ln m$ dependence.

Consider again the pressure-concentration data for O_2 in water given in Fig. 7-1. At low pressures, Henry's law $P = km$ is obeyed. Extension of this straight-line segment shows the solubility curve oxygen would have if it behaved in this ideal manner at higher concentrations. Actual behavior leads to the solid curve. This is not described by the equation $P = km$. We could construct a more complex function to describe the curve, but it is preferable to keep this simple form, which we can do by introducing a quantity a, the *activity*, related to the molality. The value of a for any value of m is such that the equation

$$P = ka \qquad \mathbf{32}$$

is satisfied. The activity can be looked on as an *effective concentration*. Values of a can be obtained from the pressure-concentration curve, as shown in Fig. 7-2. Some values are listed in Table 7-2. Thus, by introducing the activity we can maintain the *form* of Henry's law.

Often it is convenient to show the relation of a to m by introducing the *activity coefficient* γ, defined by

$$a = \gamma m \qquad \mathbf{33}$$

The value of γ then shows explicitly, by its deviation from unity, the extent of nonideality of the solute. Values of γ can be deduced from a and m data. Values calculated for dissolved oxygen are included in Table 7-2.

The advantage of introducing the activity a (or of using γ and writing γm instead of a) is that we can express the free-energy difference for the equilibrium gases or the solutes for two solutions by equations that parallel the equations for ideal behavior. Thus, for solutions 1 and 2 we write

$$\begin{aligned} G_2 - G_1 &= RT \ln \frac{P_2}{P_1} \\ &= RT \ln \frac{ka_2}{ka_1} \\ &= RT \ln \frac{a_2}{a_1} = RT \ln \frac{\gamma_2 m_2}{\gamma_1 m_1} \end{aligned} \qquad \mathbf{34}$$

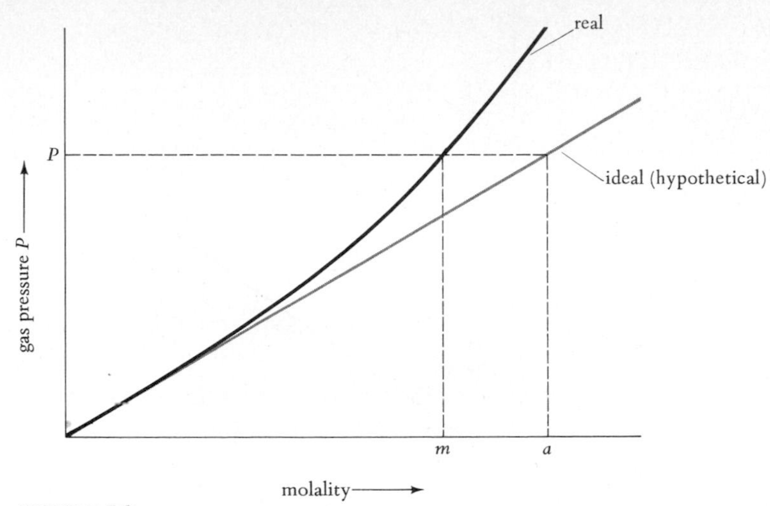

FIGURE 7-2
Activities and molalities related by equilibrium vapor data. A solution of molality m is in equilibrium with a gas of pressure P. Had the solution behaved ideally, it would have required a molality of a for gas at this pressure to be in equilibrium with it. This molality is the activity.

Now we have a relation for the free-energy difference of solutes in solutions of different concentration, even if we are outside the ideal region. (We have, of course, just hidden the complexities in the activities and their relation to molalities.)

Now we must see how we can establish a convenient standard state and how we can deduce the free energy of this state. Let us continue with the O_2 example and start with the fact that the solubility of oxygen at 1 atm

TABLE 7-2
Pressure, Concentration, and Activity-Coefficient Data for O_2 Dissolved in Water at 25.9°C
Values of γ are such that γm is proportional to P.

pressure, atm	m_{O_2}	a_{O_2}	$\gamma_{O_2} = a_{O_2}/m_{O_2}$
1	0.00117	0.00117	1.000
2	0.00233	0.00233	1.001
4	0.00461	0.00467	1.013
6	0.00683	0.00701	1.027
8	0.00894	0.00935	1.046
10	0.01095	0.01169	1.068
12	0.01284	0.01403	1.093

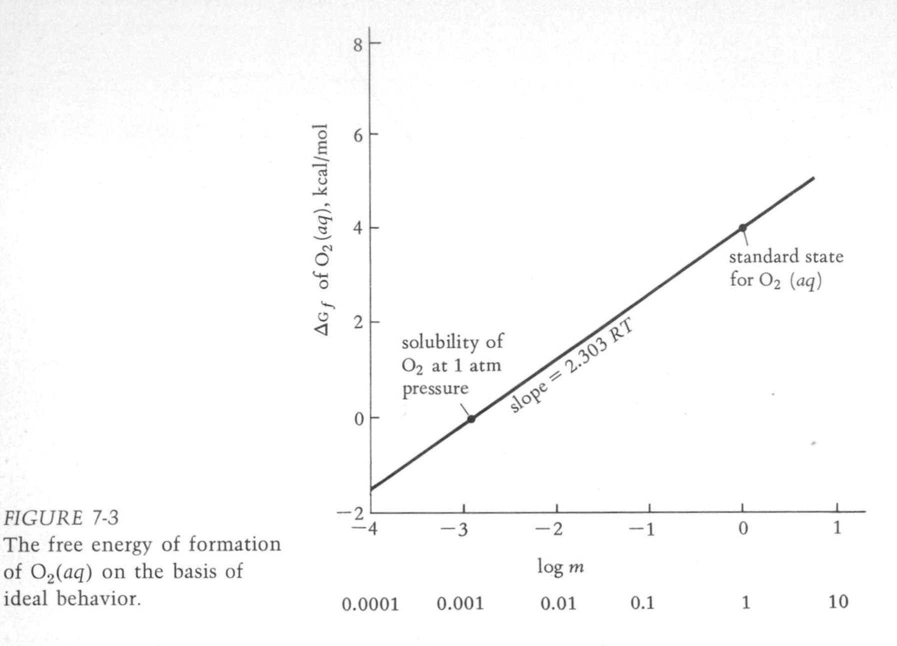

FIGURE 7-3
The free energy of formation of $O_2(aq)$ on the basis of ideal behavior.

pressure and 25°C is 0.00117 mol per 1000 g of water. The free energies of the gas and of the solute at this concentration are equal. If we use standard free energies of formation, they both will have the elemental value of zero.

The development can best be followed by recognizing that ΔG_f will increase, for ideal behavior, according to $RT \ln m$ or to $2.303RT \log m$. Thus if we form a plot of ΔG_f versus $\log m$, we know that the ideal-behavior assumption must correspond to a slope of $2.303RT$. Then the one value of $\Delta G_f = 0$ for $m = 0.00117$ can be used to generate the entire graph.

The standard free energy generally used for solutes is the energy the solute would have if it behaved ideally up to a 1 m concentration. For our oxygen example, which is representative of many solutes that are solids, liquids, or gases, the standard state is thus a hypothetical one because oxygen does not behave ideally at this concentration. We can, however, use the ideal line on Fig. 7-3, to deduce the value of ΔG when $m = 1$, or we can calculate the value of ΔG_f° of aqueous oxygen from Eq. **18** as

$$\begin{aligned}\Delta G_f^\circ(O_2,aq) &= \Delta G_f(O_2,0.00117\ m) + RT \ln \frac{1}{0.00117} \\ &= 0 + 2.303(1.987)(298.16) \log 854.7 \\ &= +4000 \text{ cal} \qquad \textbf{35}\end{aligned}$$

It is values such as this that are tabulated for nonelectrolyte solutes in Appendixes I and II.

Now let us see how such standard free energy data can be used to obtain free energies for the solute in actual solutions. The relation

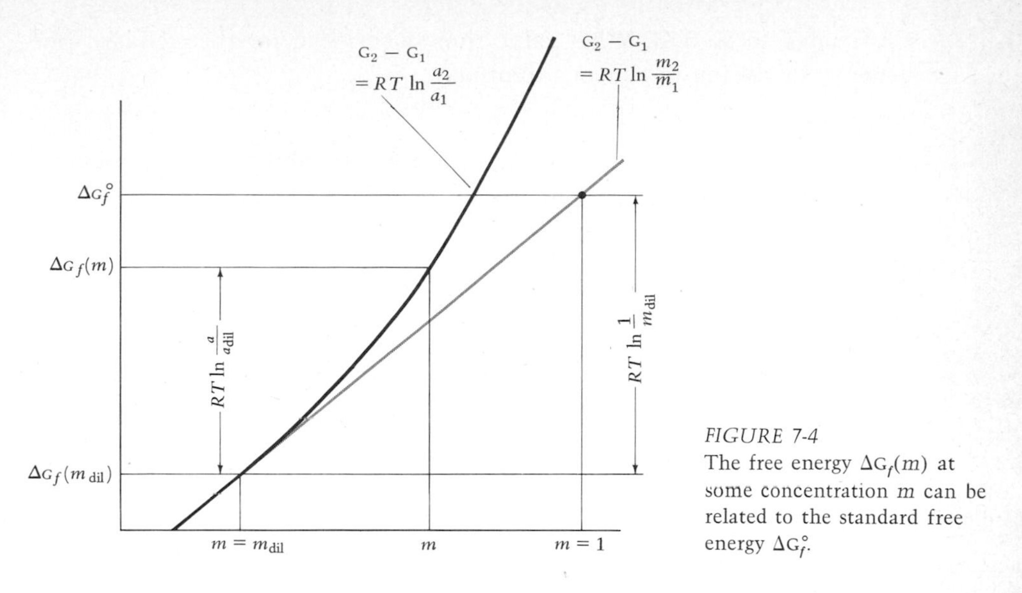

FIGURE 7-4
The free energy $\Delta G_f(m)$ at some concentration m can be related to the standard free energy ΔG_f°.

between the standard state and some real solution can be seen by relating them both to some dilute solution that is in the Henry's law region. Since the dilute solution obeys Henry's law, its activity a_{dil} will be equal to its molality m_{dil}; that is, γ_{dil} equals unity.

Now, by working along the hypothetical ideal-solution line of Fig. 7-4, we write

$$\Delta G_f^\circ - \Delta G_f(m_{dil}) = RT \ln \frac{1}{m_{dil}} \tag{36}$$

Along the real-solution line we have

$$\Delta G_f(m) - \Delta G_f(m_{dil}) = RT \ln \frac{a}{a_{dil}} \tag{37}$$

Subtraction of the equations for these two steps and use of the equality of a_{dil} and m_{dil} give the simple result

$$\begin{aligned} \Delta G_f(m) &= \Delta G_f^\circ + RT \ln a \\ &= \Delta G_f^\circ + RT \ln \gamma m \end{aligned} \tag{38}$$

Thus, if a value of ΔG_f° is available, and if the activity or activity coefficient of the solute in the solution of interest is known, the free energy of the solute of that solution can be calculated.

Example 7-5

The standard free energy of formation of crystalline glycine is $-90{,}270$ cal. The solubility of glycine in water at 25°C is 3.4 mol per 1000 g of water, and the activity coefficient for glycine at this concentration has been estimated to be 0.73. What value should be listed for the standard free energy of glycine in aqueous solution?

Solution

Since the solid is in equilibrium with a 3.4 m solution, the free energy per mole of glycine in the solution must be equal to that of the solid. Thus

$$\Delta G_f(\text{glycine}, 3.4\ m) = -90{,}270 \text{ cal}$$

Then Eq. **38** can be used, with $a = \gamma m$, as

$$\Delta G_f(\text{glycine}, 3.4\ m) = \Delta G_f^\circ(\text{glycine}, aq) + RT \ln \gamma m$$

or
$$\begin{aligned} -90{,}270 &= \Delta G_f^\circ(\text{glycine}, aq) \\ &\quad + 2.303(1.987)(298.16) \log (0.73)(3.4) \\ &= \Delta G_f^\circ(\text{glycine}, aq) + 540 \end{aligned}$$

Thus
$$\begin{aligned} \Delta G_f^\circ(\text{glycine}, aq) &= -90{,}270 - 540 \\ &= -90{,}810 \text{ cal} \end{aligned}$$

7-7 FREE ENERGIES OF REAL SOLUTES: ELECTROLYTES

In view of the treatment of Sec. 7-6, recognition of nonideal ionic behavior would lead us to replace m_{HCl} in the expression for the concentration dependence of the free energy of aqueous HCl by a_{HCl} or, further, by $\gamma_{HCl} m_{HCl}$. Then the use of ionic terms rather than parent-electrolyte terms would lead in place of the ionic product $m_{H^+} m_{Cl^-}$ to $a_{H^+} a_{Cl^-}$ and further to $\gamma_{H^+} m_{H^+} \gamma_{Cl^-} m_{Cl^-}$ or $\gamma_{H^+} \gamma_{Cl^-} m_{H^+} m_{Cl^-}$. With these allowances for nonideal behavior we write

$$\begin{aligned} \Delta G_f(\text{HCl}, aq) &= \Delta G_f^\circ(\text{HCl}, aq) + RT \ln a_{H^+} a_{Cl^-} \\ &= \Delta G_f^\circ(\text{HCl}, aq) + RT \ln \gamma_{H^+} \gamma_{Cl^-} m_{H^+} m_{Cl^-} \\ &= \Delta G_f^\circ(\text{HCl}, aq) + RT \ln \gamma_{H^+} \gamma_{Cl^-} m_{HCl}^2 \end{aligned} \qquad \textbf{39}$$

If the free energy is interpreted in terms of ionic contributions, this equation can be factored to give

$$\Delta G_f(\text{H}^+, aq) = \Delta G_f^\circ(\text{H}^+, aq) + RT \ln \gamma_{H^+} m_{H^+}$$

and
$$\Delta G_f(\text{Cl}^-, aq) = \Delta G_f^\circ(\text{Cl}^-, aq) + RT \ln \gamma_{Cl^-} m_{Cl^-} \qquad \textbf{40}$$

There is, however, no experimental way of obtaining individual ionic activity coefficients, just as there is no way of obtaining ionic free energies. We obtain only ionic activity-coefficient products corresponding to electrolytes, e.g., the $\gamma_{H^+} \gamma_{Cl^-}$ term in Eq. **39** for HCl. The closest we can come to ionic activity coefficients is to deal with *mean activity coefficients.*

Thus, for γ_{Cl^-} or γ_{H^+} we would use $\sqrt{\gamma_{H^+}\gamma_{Cl^-}}$. In general notation we write $\gamma_\pm$ for the mean ionic activity coefficient and write

$$\gamma_\pm = \sqrt{\gamma_+\gamma_-} \quad \mathbf{41}$$

For 1:2 or 2:1 electrolytes, the appropriate mean is obtained as

$$\gamma_\pm = (\gamma_+\gamma_-{}^2)^{1/3} \qquad \text{or} \qquad \gamma_\pm = (\gamma_+{}^2\gamma_-)^{1/3} \quad \mathbf{42}$$

If standard ionic or electrolyte free energies are available, estimates of $\gamma_\pm$ values for the ions of electrolytes make it possible to calculate electrolyte free energies in real, nonideal solutions. It is often helpful to generalize deduced activity coefficient results so that they can be extended to new systems without resorting to additional activity measurements.

Most generalizations are based on a property of solutions known as their *ionic strength*. The use of this property has an empirical basis, but the rationale that nonideality depends on the concentration of charged species and that multiple-charge species have particularly large effects can be given.

The ionic strength μ of a solution is defined by

$$\mu = \frac{1}{2}\sum_i c_i z_i{}^2 \quad \mathbf{43}$$

where c_i and z_i are the concentration and charge of the ith ion and the summation extends over all the ions in the solution. For example, the ionic strength of an 0.5 m NaCl solution, which has the ions Na^+ and Cl^- at this 0.5 m concentration, is

$$\mu = \tfrac{1}{2}[0.5(1)^2 + 0.5(1)^2] = 0.5$$

Further, the value of μ for a solution containing 0.5 m $CaCl_2$ and 0.5 m NaCl, which has Na^+ at 0.5, Ca^{2+} at 0.5, and Cl^- at 1.5, is

$$\mu = \tfrac{1}{2}[0.5(1)^2 + 0.5(2)^2 + 1.5(1)^2] = 2.0$$

Electrolytes in aqueous solutions in the limit of zero ionic strength behave ideally; i.e., the ionic activity coefficients are unity, and the activities of the ions can be replaced by their molal concentrations.

For solutions with ionic strengths up to about 0.01, the activity coefficients can be calculated from a relation which comes from the Debye-Hückel theory of ionic solutions:

$$\log \gamma_\pm = 0.5091 z_+ z_- \sqrt{\mu} \quad \mathbf{44}$$

where z_+ and z_- are the charges, including sign, of the ions of the electrolyte and 25°C and an aqueous solution are assumed.

At higher ionic strengths, no such generalization is possible, and each electrolyte takes on its own characteristic activity-coefficient values. In such cases, values must be determined experimentally for the particular electrolyte and the particular solution before the tabulated standard free-energy data can be used. Some examples are shown in Fig. 7-5. Since these values correspond to the electrolyte itself dissolved in water, they cannot be carried over to natural and biological fluids with their mixtures of electrolytes, but they do indicate the general magnitude of the activity-coefficient effects.

It is often inconvenient or impossible to estimate activity coefficients. Then one can resort to using a high but constant ionic strength, hoping that the activity coefficients of the ionic species will remain constant even if unknown.

Experimental studies of equilibria in more complex solutions can lead to estimates of the activities and activity coefficients of the ions present. Examples of the ionic activity coefficients that have been suggested are shown in Table 7-3. Such data show how important the nonideal effects

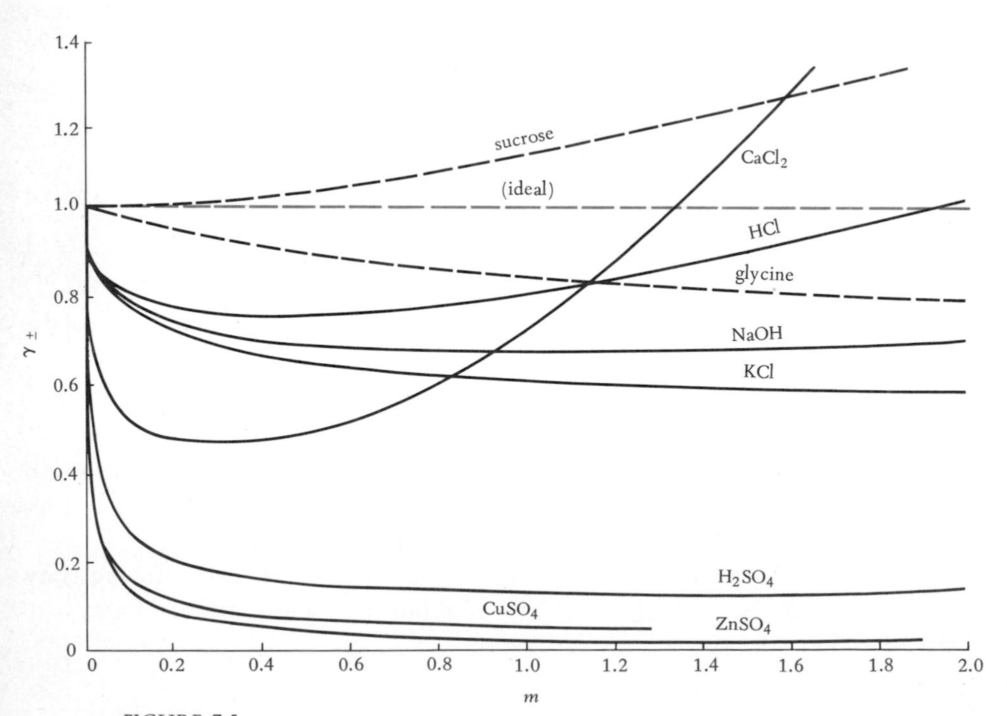

FIGURE 7-5
Mean activity coefficients of electrolytes (solid lines) and nonelectrolytes (dashed lines) in water at 25°C.

TABLE 7-3
Estimates of Ionic Activity Coefficients for Ions in Seawater

species	γ_+	species	γ_-
Na^+	0.76	Cl^-	0.64
K^+	0.64	HCO_3^-	0.68
Mg^{2+}	0.36	CO_3^{2-}	0.20
Ca^{2+}	0.28	SO_4^{2-}	0.12

†From R. M. Garrels and M. E. Thompson, *Am. J. Sci.*, **260**:57.

are in such solutions. Thus, although the sulfate-ion concentration in seawater is about 0.028 *m*, it behaves, i.e., has an activity, as if it had a concentration of about a tenth of this, 0.0033. Deductions about equilibrium concentrations would be grossly in error if such nonideal effects were ignored.

PROBLEMS

1 Which would have the greater free energy:

a Water vapor at 100°C and a pressure of $\frac{1}{10}$ atm or liquid water at this temperature and this pressure?

b 1 g of crystalline salt, NaCl, and 1 liter of water or a solution consisting of 1 g of salt dissolved in 1 liter of water?

c 1 mol of water vapor at 1 atm pressure and 25°C or 1 mol of hydrogen and $\frac{1}{2}$ mol of oxygen both as gases at the same pressure and temperature?

2 Use standard-free-energy data to confirm your conclusion in Prob. 9 of Chap. 6.

3 Verify that the standard free energy of formation of liquid ethyl alcohol as listed in Appendix II is consistent with the standard heats of formation and standard entropies given for ethyl alcohol and for the elements of which it is composed.

4 The vapor pressure of water at 25°C is 23.7 mm Hg, or 0.0312 atm. The standard-free-energy data given for $H_2O(l)$ and $H_2O(g)$ in Appendix I are for the standard-state condition of 1 atm pressure and 25°C. [This is a hypothetical state for water vapor since it cannot exist at these conditions. You can think of the value as that which $H_2O(g)$ would have if it could exist under these conditions.] With the above data, deduce the difference in free energy between water vapor at 1 atm and water vapor at 0.0312 atm. Which has the higher, more positive value? Does this seem reasonable?

5 What is the molar free energy of formation of liquid water at 20 atm pressure and 25°C?

6 Verify that the expression for the dependence of the free energy of an ideal gas on pressure gives approximately the free-energy difference for water vapor at 1 atm and 0.0312 atm deduced in Prob. 4. (In fact, under these conditions, water vapor deviates appreciably from ideal behavior.)

7 What is the free energy of formation of oxygen at its normal sea-level partial pressure of 0.21 atm?

8 The standard free energies listed for $CO_2(g)$ and $CO_2(aq)$ are $-94{,}260$ and $-92{,}260$ cal/mol, respectively. Assuming ideal-solution behavior, deduce the solubility of CO_2 in water at 25°C. Compare with the listed solubility of 0.145 g per 100 g of water at a CO_2 pressure of 1 atm.

9 From the data of Table 7-2, calculate the difference between the molar free energy of oxygen in an 0.01 *M* aqueous solution and the value that would be calculated on the assumption of ideal behavior.

10 From the free-energy data of Appendix I and the ΔG_f° value of -277.7 kcal/mol for $CaF_2(c)$, calculate the solubility product of fluorite, CaF_2. Compare with the value tabulated for the solubility product at 25°C of 3.9×10^{-11}.

11 From the standard-free-energy data of Appendix I, deduce the dissociation constant for water and compare with the value 1.0×10^{-14} at 25°C.

12 A graph of the pressure of NH_3 versus the equilibrium concentration in water at 25°C shows, as low concentrations are reached, a straight line that is fitted by $P = 0.01764\ m$, where P is in atmospheres and m is the molality.

a What gas pressure would give a 1 *m* solution if this equation were obeyed up to a concentration of 1 *m*?

b The standard free energy of formation of $NH_3(g)$ is $-3{,}960$ cal/mol. What is the free energy of formation at the pressure obtained in part (*a*)?

c Calculate the standard free energy of formation of $NH_3(aq)$ and compare with the value in Appendix I.

13 Ionic activity coefficients are often estimated for individual ionic species even if there is no rigorous way of going beyond mean, or average, ionic activity coefficients. Values estimated for the ions of seawater include the values 0.28 for Ca^{2+} and 0.20 for CO_3^{2-}. After allowance is made for the complex ions formed, the molar concentrations of these ionic species are 0.0091 for Ca^{2+} and 2.43×10^{-5} for CO_3^{2-}. Using the standard-free-energy data of Appendix I, calculate the free energy of formation of $CaCO_3$ in seawater. Compare with the value listed for solid $CaCO_3$ in Appendix I and discuss the conclusions that might be drawn regarding the degree of saturation of seawater with calcium carbonate. (A similar calculation might be attempted with body fluids, but the complex-ion formation and activity-coefficient effects are again troublesome.)

14 The principal ions and their molal concentration of surface seawater are Cl^-, 0.535; SO_4^{2-}, 0.028; Na^+, 0.456; Mg^{2+}, 0.055; and Ca^{2+}, 0.010. What is the ionic

strength of seawater? Judging from Fig. 7-5, do ions or electrolytes exhibit their own peculiar activity coefficients in such a solution?

15 The principal ions of human blood plasma and their molal concentrations are Na^+, 0.14; Cl^-, 0.10; and HCO_3^-, 0.025. What is the ionic strength of blood plasma?

16 The principal ions of the intercellular fluid of muscles and their molal concentrations are K^+, 0.15; Mg^{2+}, 0.04; and HPO_4^{2-}, 0.14. What is the ionic strength of this intercellular fluid?

17 Use the Debye-Hückel equation **44** to calculate the mean activity coefficient of the ions of (*a*) KCl and (*b*) $CaCl_2$ in water at 25°C at concentrations of 0.005 *m*.

18 The curves of Fig. 7-5 show the remarkably different shape of the γ-versus-m curves for different electrolytes. Some of this individuality results because at a given molality the ionic strength of an electrolyte with multiply charged ions is greater than that for an electrolyte with singly charged ions. For several electrolytes with ions of various charges, plot γ-versus-μ curves from the data of Fig. 7-5 in the concentration range up to about 1 *m*. Does such a plot show any greater common behavior of the electrolytes?

REFERENCES

PITZER, K. S., and L. BREWER: "Thermodynamics," 2d ed., McGraw-Hill Book Company, New York, 1961. The classic treatment of thermodynamics by G. N. Lewis and M. Randall is brought up to data in this revision. Particularly careful and complete accounts of the treatments of the free energies of the components of real solutions are given.

8 Chemical Driving Forces and Chemical Equilibria

The thermodynamic apparatus for relating the properties of compounds to the equilibrium states reached by reactions of these compounds has now been developed. We are now ready to use it. In doing so we shall treat separately the important classes of reactions that involve proton transfer (acid-base reactions), electron transfer (oxidation-reduction reactions), and phosphate-group transfer. Special terms are introduced to deal with each of these reaction types. The free-energy change provides a unifying basis for these terms.

The free energies of compounds are often used to obtain a value for the equilibrium constant, which describes the equilibrium state toward which the reacting system strives. Free energies can also be used as a measure of the driving force with which compounds move from some existing states or concentrations toward those of the equilibrium condition. In biological systems the driving-force interpretation is often the most informative. The living state frustrates the drives of many types of reactions toward their equilibrium states. How this is done can be seen by considering these free energy–related driving forces and the organization, or coupling, of reactions characteristic of the living state.

8-1 FREE ENERGIES AND THE EQUILIBRIUM-CONSTANT EXPRESSION

In the last several sections of Chap. 7, expressions were obtained for the dependence of the free energy on gas pressure and on solute concentration. If ideal behavior can be assumed and free energies are denoted by G and G° rather than by the cluttered symbols ΔG_f and ΔG_f°, these expressions can be written

$$\text{G} = \text{G}^\circ + RT \ln P \qquad \textbf{1}$$

and

$$\text{G} = \text{G}^\circ + RT \ln m \qquad \textbf{2}$$

where the G° terms are the free energies at unit pressure and unit molality. Written more completely, these expressions would be

$$\text{G} = \text{G}^\circ + RT \ln \frac{P}{1 \text{ atm}} \qquad \textbf{3}$$

and

$$\text{G} = \text{G}^\circ + RT \ln \frac{m}{1\,m} \qquad \textbf{4}$$

Thus the equations show how the free energy varies from the standard-state value as the pressure or the concentration varies from that of the standard state.

For many of the solutes we deal with, ideality cannot be assumed, but expressions of the same form can be retained by writing

$$G = G^\circ + RT \ln a \tag{5}$$

or

$$G = G^\circ + RT \ln \gamma m \tag{6}$$

The idea of Sec. 7-1 that the free-energy decrease in a reaction measures the driving force of the reaction can now be extended. We begin by calculating the concentrations or activities of the reagents that would give a reaction zero driving force, i.e., would produce the equilibrium state.

Consider a specific example, the ionization of acetic acid. The equation for the reaction and the free energies of the reagents, assuming ideal behavior and using square brackets to indicate molal concentrations, are

$$
\begin{array}{llcccc}
 & CH_3COOH(aq) & \rightarrow & CH_3COO^-(aq) & + & H^+(aq) \\
\Delta G_f,\ \text{cal:} & \Delta G_f^\circ + RT \ln [CH_3COOH] & & \Delta G_f^\circ + RT \ln [CH_3COO^-] & & \Delta G_f^\circ + RT \ln [H^+] \\
 & = -96{,}580 + RT \ln [CH_3COOH] & & = -90.08 + RT \ln [CH_3COO^-] & & = 0 + RT \ln [H^+] \\
 & & & \multicolumn{3}{c}{-90{,}080 + RT \ln [CH_3COO^-][H^+]}
\end{array}
$$

For the reaction, we now obtain

$$
\begin{aligned}
\Delta G &= (-90{,}080 + RT \ln [CH_3COO^-][H^+]) \\
&\qquad\qquad - (-96{,}580 + RT \ln [CH_3COOH]) \\
&= +6500 + 2.303(1.987)(298.16) \log \frac{[CH_3COO^-][H^+]}{[CH_3COOH]} \\
&= +6500 + 1364 \log \frac{[CH_3COO^-][H^+]}{[CH_3COOH]}
\end{aligned} \tag{7}
$$

The concentrations that would give a value of zero to ΔG are those satisfying the equation

$$0 = 6500 + 1364 \log \frac{[CH_3COO^-][H^+]}{[CH_3COOH]}$$

or

$$\log \frac{[CH_3COO^-][H^+]}{[CH_3COOH]} = -4.76 \tag{8}$$

or

$$\frac{[CH_3COO^-][H^+]}{[CH_3COOH]} = 1.75 \times 10^{-5} \tag{9}$$

When the concentrations of the reagents are such that Eq. **8** or **9** is satisfied, no driving force remains to push the reaction in either direction. We have come from considerations of free energy to the familiar equilibrium-constant expression and to a value of the equilibrium constant.

Now let us repeat the development for a more general case. Consider a schematic process

$$a\text{A} + b\text{B} \rightarrow c\text{C} \qquad \textbf{10}$$

and assume that all reagents are solutes for which activity-coefficient data are available. Then for 1 mol of each reagent, with the simplified notation G°_A for ΔG°_f (A,aq) and so forth, we can write

$$\begin{aligned} G_A &= G^\circ_A + RT \ln a_A \\ G_B &= G^\circ_B + RT \ln a_B \\ G_C &= G^\circ_C + RT \ln a_C \end{aligned} \qquad \textbf{11}$$

For a, b, and c moles of the reagents A, B, and C we have

$$\begin{aligned} aG_A &= aG^\circ_A + aRT \ln a_A \\ &= aG^\circ_A + RT \ln (a_A)^a \\ bG_B &= bG^\circ_B + bRT \ln a_B \\ &= bG^\circ_B + RT \ln (a_B)^b \\ cG_C &= cG^\circ_C + cRT \ln a_C \\ &= cG^\circ_C + RT \ln (a_C)^c \end{aligned} \qquad \textbf{12}$$

Then, the change in free energy corresponding to the reaction is

$$\begin{aligned} \Delta G &= cG_C - (aG_A + bG_B) \\ &= cG^\circ_C - (aG^\circ_A + bG^\circ_B) + RT \ln \frac{(a_C)^c}{(a_A)^a (a_B)^b} \\ &= \Delta G^\circ + RT \ln \frac{(a_C)^c}{(a_A)^a (a_B)^b} \end{aligned} \qquad \textbf{13}$$

The free-energy change is thus expressed by a term for all reagents in their standard states and one that shows the effect of the activities in the system being considered.

Of particular interest, again, is the system at equilibrium, where no driving force operates to shift the reaction. Then

$$\Delta G = 0$$

and $$\Delta G^\circ = -RT \ln \frac{(a_C)^c}{(a_A)^a (a_B)^b} \qquad \textbf{14}$$

But $\Delta G°$ is some fixed quantity at the constant temperature being considered. It follows that the activity term, for this equilibrium situation, is also a fixed constant quantity. If we add a designation to show that this is true only for the activities of the system at equilibrium, we can write

$$\left[\frac{(a_C)^c}{(a_A)^a(a_B)^b}\right]_{equil} = K \qquad \textbf{15}$$

and $$\Delta G° = -RT \ln K \qquad \textbf{16}$$

where K designates a constant. We have again come from thermodynamic considerations to the equilibrium-constant expression.

The result, Eq. **16,** is the culmination of much of the development of the two preceding chapters. It allows us to use thermodynamic data, in particular standard free energies, for compounds to deduce the value of the equilibrium constant for any real or imagined reaction in which they are involved. Conversely, from the relation of K and $\Delta G°$ of Eq. **16** we see that we can use or interpret equilibrium-constant values as measures of the driving force that tends to move a reaction system from standard conditions.

Example 8-1

The standard free energy of $NH_3(g)$ at 25°C is −3.96 kcal; as for all elements in their standard state, the values for $H_2(g)$ and $N_2(g)$ are zero. What is the equilibrium constant for the synthesis-of-ammonia reaction at 25°C?

Solution

Let us first proceed through all the details. We write

$$N_2(g) + 3H_2(g) \rightarrow 2NH_3(g)$$

For $N_2(g)$: Free energy $= \Delta G_f° + RT \ln P_{N_2} = RT \ln P_{N_2}$

For $3H_2(g)$: Free energy $= 3(\Delta G_f° + RT \ln P_{H_2}) = RT \ln P^3_{H_2}$

For $2NH_3(g)$: Free energy $= 2(\Delta G_f° + RT \ln P_{NH_3}) = -7{,}920 + RT \ln P^2_{NH_3}$

Then

$$\Delta G = -7920 + RT \ln P^2_{NH_3} - RT \ln P^3_{H_2} - RT \ln P_{N_2}$$

$$= -7920 + RT \ln \frac{P^2_{NH_3}}{P^3_{H_2} P_{N_2}}$$

For $\Delta G = 0$

$$\log \frac{P^2_{NH_3}}{P^3_{H_2} P_{N_2}} = \frac{+7920}{2.303(1.987)(298.16)} = 5.81$$

and $$\frac{P^2_{NH_3}}{P^3_{H_2} P_{N_2}} = 6.5 \times 10^5$$

Alternatively, we could proceed directly with Eq. **16** and write

$$\Delta G^\circ = -RT \ln K$$

$$\text{or} \quad \log K = -\frac{\Delta G^\circ}{2.303RT}$$

With $\Delta G^\circ = -7{,}920$ we again obtain

$$\log K = \log \frac{P^2_{NH_3}}{P^3_{H_2}P_{N_2}} = 5.81$$

Notice that, since the ΔG° value of -7920 cal is used, the equilibrium constant for the reaction is written as

$$N_2(g) + 3H_2(g) \rightarrow 2NH_3(g)$$

8-2 PROTON-TRANSFER REACTIONS AND MEASURES OF ACID STRENGTH

The tendency of a reaction to proceed from any given set of reagent concentrations, or activities, can be deduced by calculating ΔG for the reaction for these reagent concentrations. In tabulating reaction tendencies, however, it is convenient to agree on standard-state conditions and to tabulate the tendency of the system to move from these conditions to the equilibrium conditions. The more negative is ΔG°, the greater is this tendency.

There remain, for various types of processes, several other ways of reporting these reaction tendencies. The method for reporting proton-transfer tendencies in acid-base reactions will be shown here, followed in later sections by studies of electron-transfer and phosphate-group-transfer processes.

The dissociation of an acid is often written

$$HA \rightarrow H^+ + A^- \qquad \textbf{17}$$

and this is perfectly satisfactory for many acid-base calculations. But to show the relation between the measures of acid strength and other types of reactions, it is important to recognize that acid dissociation is a proton-transfer process and that acid strengths are expressed in terms of transfer to the reference substance, water. All this is indicated by the more complete description of the dissociation process

$$HA + H_2O \rightarrow H_3O^+ + A^- \qquad \textbf{18}$$

As for all reaction types, a measure of the tendency for this reaction to move the reagents from their standard states is given by the values of ΔG° for the reactions of various acids. This can be calculated for the reaction of either Eq. **17** or Eq. **18,** the same value being obtained because the thermodynamic properties of H_3O^+ are given the values for H^+ plus those for H_2O.

Let us indicate the acid-dissociation free-energy value as ΔG_a°. The few representative values are given in Table 8-1. All the acids of Table 8-1 are weak acids. All the listed ΔG_a values are positive. This means that if all reagents HA, H^+, and A^- are in standard states, approximately one molal, a driving force would operate to shift the reaction to the left. More HA

TABLE 8-1
Measures of the Acid Strength of Some Acids in Water at 25°C

acid	**formula**	ΔG_a, **kcal/mol**	K_a	pK_a
Phosphoric acid	H_3PO_4	2.92	7.3×10^{-3}	2.14
Malonic acid	$CH_2(COOH)_2$	3.85	1.5×10^{-3}	2.82
Monochloroacetic acid	$CH_2ClCOOH$	3.89	1.4×10^{-3}	2.85
Phthalic acid	$C_6H_4(COOH)_2$	3.94	1.3×10^{-3}	2.89
Nitrous acid	HNO_2	4.49	5.1×10^{-4}	3.29
Formic acid	$HCOOH$	5.12	1.8×10^{-4}	3.75
Lactic acid	$CH_3CHOHCOOH$	5.25	1.4×10^{-4}	3.85
Succinic acid	$(CH_2COOH)_2$	5.72	6.3×10^{-5}	4.19
Benzoic acid	C_6H_5COOH	5.73	6.3×10^{-5}	4.20
Anilinium ion	$C_6H_5NH_3^+$	6.28	2.5×10^{-5}	4.60
Acetic acid	CH_3COOH	6.47	1.8×10^{-5}	4.74
Butyric acid	$CH_3(CH_2)_2COOH$	6.58	1.5×10^{-5}	4.82
Pyridinium ion	$C_5H_5NH^+$	7.14	6.0×10^{-6}	5.23
Phthalic acid ion	$C_6H_4COOHCOO^-$	7.38	3.9×10^{-6}	5.41
Succinic acid ion	$HOOCCH_2CH_2COO^-$	7.67	2.4×10^{-6}	5.62
Carbonic acid	H_2CO_3	8.66	4.5×10^{-7}	6.35
Hydrogen sulfide	H_2S	9.55	1.0×10^{-7}	7.00
Dihydrogen phosphate	$H_2PO_4^-$	9.82	6.3×10^{-8}	7.20
Hydrogen cyanide	HCN	11.84	2.1×10^{-9}	8.68
Boric acid	H_3BO_3	12.61	5.8×10^{-10}	9.24
Ammonium ion	NH_4^+	12.62	5.6×10^{-10}	9.25
Phenol	C_6H_5OH	13.64	1.0×10^{-10}	10.00
Bicarbonate ion	HCO_3^-	14.09	4.7×10^{-11}	10.33
Methyl ammonium ion	$CH_3NH_3^+$	14.46	2.5×10^{-11}	10.60
Hydrogen phosphate	HPO_4^{2-}	16.92	4.0×10^{-13}	12.40
Hydrogen sulfide ion	HS^-	17.63	1.2×10^{-13}	12.92

and less H^+ and A^- would be present when equilibrium is reached. The driving force is less for the acids at the top of the table, the stronger acids, than for those at the bottom of the table. The ΔG_a values are clearly measures of acid strength.

In view of Eq. **16** one could also tabulate equilibrium constants for Eq. **17** or **18.** The values of these acid equilibrium constants K_a are related to ΔG_a values by

$$\begin{aligned}\Delta G_a^\circ &= -RT \ln K_a \\ &= -2.303RT \log K_a\end{aligned} \quad \textbf{19}$$

Tabulations of K_a for acids are in fact made, and values are included in Table 8-1. Usually, however, it is pK_a, defined as

$$pK_a = -\log K_a \quad \textbf{20}$$

that is used to report acid strengths. From writing

$$pK_a = \frac{\Delta G_a^\circ}{2.303RT} \quad \textbf{21}$$

we see that these provide an alternative way of expressing the tendency of the reaction to drive the system away from the standard states. The qualitative implications of the various ways of expressing the tendency of acids to donate protons to the reference substance are shown in Table 8-2.

Example 8-2

From the following standard free energies for aqueous solutions of the glucose 6-phosphoric acid species

TABLE 8-2
Measures of the Tendency of Acids to Transfer Protons to the Reference Substance, Water

	ΔG_a°	K_a	pK_a
Stronger acid	negative	1 to ∞	negative
No reaction tendency from standard states (at equilibrium $[HA] \cong [H_3O^+] \cong [A^-] \cong 1$)	0	1	0
Weaker acid	positive	0 to 1	positive

species	$\Delta G_f^\circ(aq)$
Glucose 6-phosphoric acid, $C_6H_{11}O_6PO_3H_2$	−429.6
Glucose 6-hydrogen phosphate, $C_6H_{11}O_6PO_3H^-$	−427.6
Glucose 6-phosphate, $C_6H_{11}O_6PO_3^{2-}$	−419.1

estimate the first and second ionization constant for glucose 6-phosphoric acid and deduce the pH of the solution that would give equal concentrations of the parent and ionized forms for these two reactions.

Solution

For the first ionization

$$C_6H_{11}O_6PO_3H_2 \rightarrow C_6H_{11}O_6PO_3H^- + H^+$$

$$\Delta G_f^\circ\text{, kcal:} \quad -429.6 \qquad \underbrace{-427.6 \qquad 0}_{-427.6}$$

$\Delta G_1^\circ = +2.0$ kcal. From $\Delta G_1^\circ = -RT \ln K_1$

$\log K_1 = -1.5 \qquad pK_1 = +1.5 \qquad K_1 = 3.2 \times 10^{-2}$

For the second ionization

$$C_6H_{11}O_6PO_3H^- \rightarrow C_6H_{11}O_6PO_3^{2-} + H^+$$

$$\Delta G_f^\circ\text{, kcal:} \quad -427.6 \qquad \underbrace{-419.1 \qquad 0}_{-419.1}$$

$\Delta G_2^\circ = +8.5$ kcal

$\log K_2 = -6.2 \qquad pK_2 = +6.2 \qquad K_2 = 6.3 \times 10^{-7}$

The concentration ratios are obtained from Eq. **29** of Sec. 4-3, the Henderson-Hasselbalch equation. For $[A^-] = [HA]$, or $[A^-]/[HA] = 1$ and $\log [A^-]/[HA] = 0$

$$pH = pK_1 = 1.5$$

and $pH = pK_2 = 6.2$

This shows that at all pH values above 6.2, the principal species is the completely ionized species, glucose 6-phosphate.

8-3 OXIDATION-REDUCTION REACTIONS: FREE ENERGIES AND EMF OF ELECTROCHEMICAL CELLS

The tendency for many types of reactions to proceed and the equilibrium state they would reach can be organized on the basis of free energies. Oxidation-reduction reactions constitute an important class of

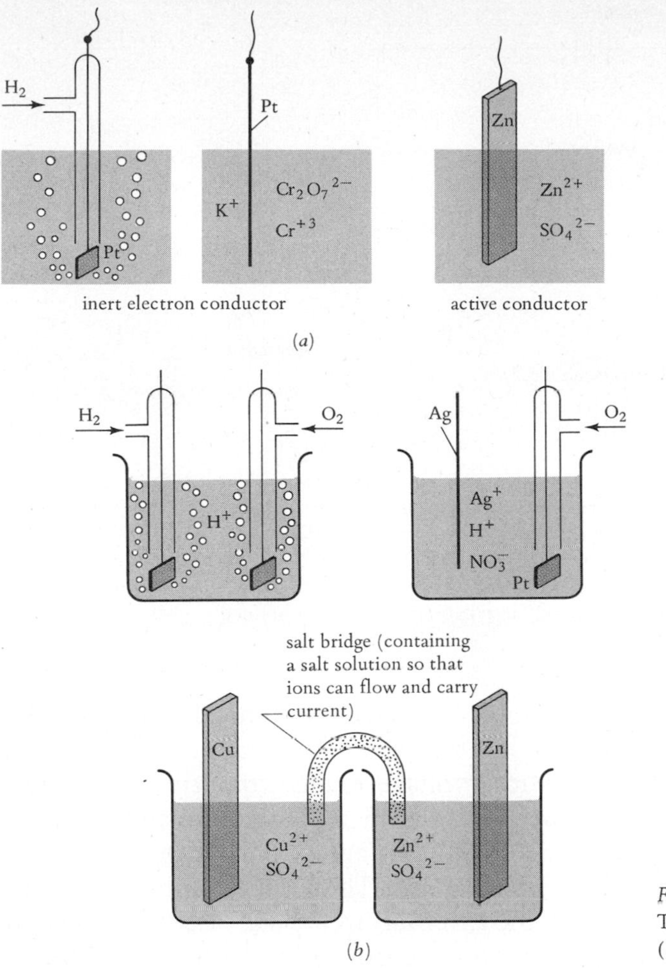

FIGURE 8-1
Types of (*a*) electrodes and (*b*) cells.

reactions that can be so organized. Our ability to make some of these reactions occur in electrochemical cells provides us with yet another way of expressing the driving force of chemical reactions.

When reagents that accept or donate electrons are arranged so that the electrons can enter or leave the reaction through a metallic conductor, an electrochemical cell can be set up. Examples are shown in Fig. 8-1. A particular advantage of such cells is that certain reactions can be made to occur at very nearly a state of balance. It is only necessary to balance the voltage generated by the cell by means of an opposing voltage. A procedure for doing this and for determining the voltage produced at this state of balance is incorporated in the potentiometer procedure of Fig. 8-2.

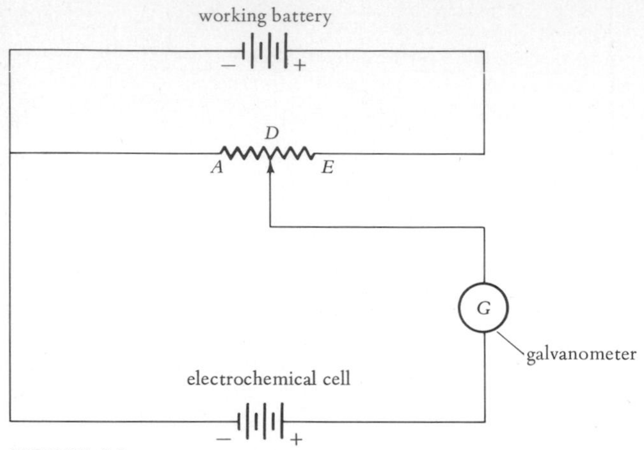

FIGURE 8-2
The potentiometric method for measuring the voltage produced by an electrochemical cell operating at a state of balance. A fraction AD/AE of the working battery can be picked off to produce a voltage that drives a current in the lower circuit in a direction opposite to that in which the electrochemical cell drives the current. When position D is such that no current flows in the lower circuit, as determined by the galvanometer, the voltage of the electrochemical cell at a state of balance is obtained as AD/AE times the voltage of the working battery.

Let us consider, to be specific, the electrochemical cell of Fig. 8-3. Potentiometric measurements show that if all the reagents are in their standard states, the cell will produce a voltage, or *electromotive force* (*emf*), of 0.771 V, tending to drive the electrons in the direction indicated. To avoid having to draw such pictures, we first agree to indicate which electrode is at the left and which at the right. Then the direction in which the electrons are driven through the external circuit will be shown by giving the emf $\mathcal{E}$ a plus sign if this direction is from the electrode on the left to the electrode on the right. A minus sign will indicate that the electrons are driven in the reverse direction. With this convention, the emf of the cell with an $H_2|H^+$ electrode at the left and an $Fe^{3+}|Fe^{2+}$ electrode at the right would be reported as $+0.771$ V.

In reactions that correspond to a positive emf, electrons are driven through the external circuit from left to right, i.e., they are released by the left electrode, and accepted by the right one. Thus we write

At right electrode:

$$Fe^{3+} + e^- \rightarrow Fe^{2+}$$

At left electrode:

$$\tfrac{1}{2}H_2 \rightarrow H^+ + e^-$$

Overall reaction:

$$Fe^{3+} + \tfrac{1}{2}H_2 \rightarrow Fe^{2+} + H^+ \qquad 22$$

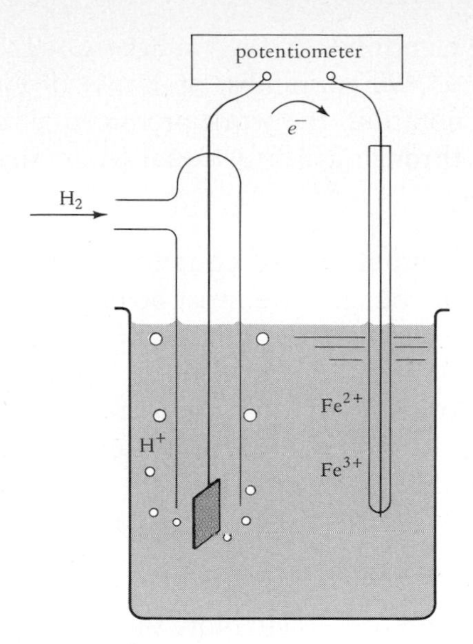

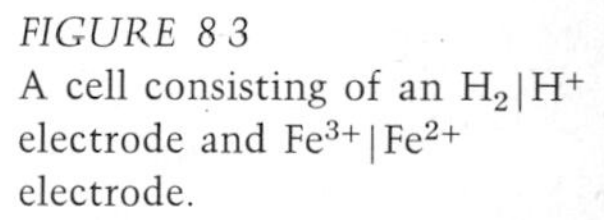
FIGURE 8 3
A cell consisting of an $H_2|H^+$ electrode and $Fe^{3+}|Fe^{2+}$ electrode.

The positive emf that is measured shows that hydrogen gas can, in fact, be effective in reducing ferric to ferrous iron, all reagents being in their standard states.

Half reactions or *electrode reactions* like those that led to Eq. **22** must be balanced in terms of both atoms and charges. Furthermore, such half reactions can be combined only when the same number n of electrons released in one reaction is picked up in the other. Thus, one cannot combine the half reactions

$$Fe^{3+} + e^- \rightarrow Fe^{2+}$$

and $$H_2 \rightarrow 2H^+ + 2e^-$$

They must be rewritten so that either one electron is released and consumed or two electrons are released and consumed, as in

$$2Fe^{3+} + 2e^- \rightarrow 2Fe^{2+}$$
$$H_2 \rightarrow 2H^+ + 2e^-$$
$$\overline{2Fe^{3+} + H_2 \rightarrow 2Fe^{2+} + 2H^+} \qquad \mathbf{23}$$

Notice that the half reactions reveal the number of transferred electrons implied by the overall reaction.

Since the measured emf indicates the driving force of the oxidation-reduction reaction in the cell, it must be related to the free-energy change for the reaction. The electrical work done by a cell as it drives an Avogadro's number of electrons through an opposing voltage, as imposed for example, by a potentiometer arrangement, is equal to the product of the total charge of these electrons and the voltage through which this charge is driven. This charge, 96,490 coulombs, of 1 mol of electrons is termed a *faraday* $\mathfrak{F}$. With this notation, the work produced as n mol of electrons driven by the cell emf through a nearly equal balancing voltage is

$$\Delta E_{\text{mech surr}} = n\mathfrak{F}\mathcal{E} \qquad \textbf{24}$$

In using Eq. **24** or other related equations that yield energies from the product of charge and voltage, care must be taken to obtain the answer in suitable units. For example, if the charge $n\mathfrak{F}$ is in coulombs and the voltage $\mathcal{E}$ in volts, the calculated energy will be in joules (J). Such a result can be converted to calories by division by 4.184 J/cal.

The mechanical energy, or maximum useful work the reaction can produce, is equal to the *decrease* in the free energy of the reacting system as a result of the process described by the cell reaction. Thus

$$\Delta G = -n\mathfrak{F}\mathcal{E} \qquad \textbf{25}$$

where $\mathcal{E}$ is the reversible emf of the cell. Note that when a reaction proceeds as written, $\mathcal{E}$, and thus $\Delta E_{\text{mech surr}}$, have positive signs. ΔG is then negative, as it must be for a spontaneous reaction.

With the relation $\Delta G = -n\mathfrak{F}\mathcal{E}$ we can use our previous result for the dependence of ΔG for a reaction on the concentrations (really activities) to show the corresponding dependence of $\mathcal{E}$. For the reaction

$$a\text{A} + b\text{B} \rightarrow c\text{C} \qquad \textbf{26}$$

we had

$$\Delta G = \Delta G^\circ + RT \ln \frac{(a_\text{C})^c}{(a_\text{A})^a(a_\text{B})^b} \qquad \textbf{27}$$

With $\Delta G = -n\mathfrak{F}\mathcal{E}$ and $\Delta G^\circ = -n\mathfrak{F}\mathcal{E}^\circ$ we obtain

$$\mathcal{E} = \mathcal{E}^\circ - \frac{RT}{n\mathfrak{F}} \ln \frac{(a_\text{C})^c}{(a_\text{A})^a(a_\text{B})^b} \qquad \textbf{28}$$

This important equation, known as the *Nernst equation,* shows how the emf of a cell can be calculated from the emf for all reagents in their standard states and the activity coefficients of the reagents. It is clear that when the activities of all the reagents are unity, the logarithmic term drops out and $\mathcal{E} = \mathcal{E}^\circ$, that is, the emf for the reagents in their standard states.

At 25°C, the temperature at which most standard electrode potentials are reported, the factor before the logarithm term can be explicitly worked out. For the general reaction of Eq. **27** it is

$$\mathcal{E} = \mathcal{E}^\circ - \frac{0.05915}{n} \log \frac{(a_\text{C})^c}{(a_\text{A})^a(a_\text{B})^b} \qquad \textbf{29}$$

Included in the numerical factor is the term 2.303 for the conversion to logarithms to the base 10. The activity term has the familiar form of the equilibrium-constant expression. It is, however, not the equilibrium constant. The activities of the reagents have the values that are determined by the solutions used to make up the cell, and these are not generally the equilibrium values.

As in the development of the equilibrium-constant expression, the activities of solid or otherwise fixed concentration reagents are not explicitly included. The contribution to the free energy and the emf of the cell of such reagents is implicit in the ΔG° and $\mathcal{E}^\circ$ terms.

It is the expression of Eq. **29** that allows us to obtain data for the emf of cells for all reagents in their standard states. In practice, one measures $\mathcal{E}$ for various concentrations of reagents and deduces $\mathcal{E}^\circ$ by an extrapolation procedure. Once a value for $\mathcal{E}^\circ$ is established, measurement of $\mathcal{E}$ for a solution with reagents in any solution leads to a value for the logarithm of the activities term. Then activities and activity coefficients of the reagents can often be deduced.

8-4 STANDARD ELECTRODE POTENTIALS AND ELECTRON-TRANSFER MEASURES

The potentials of all cells that have been studied could be listed, for the reagents in their standard states, to show the driving forces of the various cell reactions. Much more convenient is the list of the driving-force contributions each electrode makes. Such values cannot be obtained experimentally since we can measure only the emfs that result when two electrodes vie with each other for electrons. We proceed by arbitrarily assigning an electrode emf to one electrode. We again pick on hydrogen and its ion and set $\mathcal{E}^\circ = 0$ for the standard hydrogen electrode. Thus we write

$$H^+ + e^- \rightarrow \tfrac{1}{2}H_2 \qquad \mathcal{E}^\circ = 0 \tag{30}$$

Then measured cell emfs can be made to yield *standard electrode potentials.* Thus, for the example used above

$$\begin{array}{ll} Fe^{3+} + e^- \rightarrow Fe^{2+} & \mathcal{E} = \mathcal{E}^\circ_{Fe^{3+},Fe^{2+}} \\ \tfrac{1}{2}H_2 \rightarrow H^+ + e^- & \mathcal{E} = -\mathcal{E}^\circ = 0 \\ \hline Fe^{3+} + \tfrac{1}{2}H_2 \rightarrow Fe^{2+} + H^+ & \mathcal{E} = \mathcal{E}^\circ_{Fe^{3+},Fe^{2+}} + 0 \\ & \phantom{\mathcal{E}} = \mathcal{E}^\circ_{Fe^{3+},Fe^{2+}} \end{array}$$

The measured cell emf of $+0.771$ can thus be identified with the standard electrode potential of the $Fe^{3+}|Fe^{2+}$ electrode.

Electrode potentials for typical inorganic oxidation-reduction systems are shown in Table 8-3. Note that high up on the table, the more

TABLE 8-3
Some Standard Electrode Potentials for Aqueous Acid Solutions, $a_{H^+} = 1$, and 25°C

electrode	electrode reaction	$\mathcal{E}°$, V
$Pt \mid F_2 \mid F^-$	$F_2(g) + 2e^- = 2F^-$	+2.87
$Pt \mid H_2O_2 \mid H^+$	$H_2O_2 + 2H^+ + 2e^- = 2H_2O$	+1.77
$Pt \mid Mn^{2+}, MnO_4^-$	$MnO_4^- + 8H^+ + 5e^- = Mn^{2+} + 4H_2O$	+1.51
$Pt \mid Cl_2 \mid Cl^-$	$Cl_2 + 2e^- = 2Cl^-$	+1.3595
$Pt \mid Tl^+, Tl^{3+}$	$Tl^{3+} + 2e^- = Tl^+$	+1.25
$Pt \mid Br_2 \mid Br$	$Br_2 + 2e^- = 2Br^-$	+1.065
$Ag \mid Ag^+$	$Ag^+ + e^- = Ag$	+0.7991
$Pt \mid Fe^{2+}, Fe^{3+}$	$Fe^{3+} + e^- = Fe^{2+}$	+0.771
$Pt \mid O_2 \mid H_2O_2$	$O_2 + 2H^+ + 2e^- = H_2O_2$	+0.682
$Pt \mid I_2 \mid I^-$	$I_3^- + 2e^- = 3I^-$	+0.536
$Cu \mid Cu^{2+}$	$Cu^{2+} + 2e^- = Cu$	+0.337
$Pt \mid Hg \mid Hg_2Cl_2 \mid Cl^-$	$Hg_2Cl_2 + 2e^- = 2Cl^- + 2Hg$	+0.2676
$Ag \mid AgCl \mid Cl^-$	$AgCl + e^- = Ag + Cl$	+0.2225
$Pt \mid Cu^+, Cu^{2+}$	$Cu^{2+} + e^- = Cu^+$	+0.153
$Cu \mid CuCl \mid Cl^-$	$CuCl + e^- = CuCl$	+0.137
$Ag \mid AgBr \mid Br^-$	$AgBr + e^- = Ag + Br^-$	+0.0713
$Pt \mid H_2 \mid H^+$	$2H^+ + 2e^- = H_2$	0.0000
$Pb \mid Pb^{2+}$	$Pb^{2+} + 2e^- = Pb$	−0.126
$Ag \mid AgI \mid I^-$	$AgI + e^- = Ag + I^-$	−0.1518
$Cu \mid CuI \mid I^-$	$CuI + e^- = Cu + I^-$	−0.1852
$Pb \mid PbSO_4 \mid SO_4^{2-}$	$PbSO_4 + 2e^- = Pb + SO_4^{2-}$	−0.3588
$Pt \mid Ti^{2+}, Ti^{3+}$	$Ti^{3+} + e^- = Ti^{2+}$	−0.369
$Cd \mid Cd^{2+}$	$Cd^{2+} + 2e^- = Cd$	−0.403
$Fe \mid Fe^{2+}$	$Fe^{2+} + 2e^- = Fe$	−0.4402
$Cr \mid Cr^{3+}$	$Cr^{3+} + 3e^- = Cr$	−0.744
$Zn \mid Zn^{2+}$	$Zn^{2+} + 2e^- = Zn$	−0.7628
$Mn \mid Mn^{2+}$	$Mn^{2+} + 2e^- = Mn$	−1.180
$Al \mid Al^{3+}$	$Al^{3+} + 3e^- = Al$	−1.662
$Mg \mid Mg^{2+}$	$Mg^{2+} + 2e^- = Mg$	−2.363
$Na \mid Na^+$	$Na^+ + e^- = Na$	−2.7142
$Ca \mid Ca^{2+}$	$Ca^{2+} + 2e^- = Ca$	−2.866
$Ba \mid Ba^{2+}$	$Ba^{2+} + 2e^- = Ba$	−2.906
$K \mid K^+$	$K^+ + e^- = K$	−2.925
$Li \mid Li^+$	$Li^+ + e^- = Li$	−3.045

oxidized species has a great tendency to accept electrons, i.e., to be reduced and to act as an oxidizing agent. For electrodes toward the bottom of the table this tendency is minimal, and it is the less oxidized state that tends to release electrons, i.e., to act as a reducing agent. These features are reflected in the placement of some familiar oxidizing and reducing agents.

The data of Table 8-3 apply to standard states which include the unit activity, or approximately 1 *m* hydrogen-ion concentration. For inorganic electrodes that perform only in basic solutions it makes more sense to list the emfs for unit hydroxide-ion activity. Many electrode reactions show no involvement of H^+ and OH^-, and unless some other factors enter to disrupt the electrode, its emf can be taken to be that shown for the unit hydrogen-ion activity standard state regardless of the acidity of the solution.

The emfs for reactions written for basic solutions, which involve the OH^- species, can be related to those written for acid solutions, which involve H^+ species. The relation is based on the value of 10^{-14} for the ion product $[H^+][OH^-]$ or the activity product $a_{H^+} a_{OH^-}$, which holds for aqueous solutions at 25°C.

The hydrogen electrode serves as an example. The electrode reaction is

$$H^+ + e^- \rightarrow \tfrac{1}{2}H_2$$

and at a hydrogen-ion activity of a_{H^+} and a hydrogen gas pressure of 1 atm the electrode potential is

$$\mathcal{E} = \mathcal{E}° - 0.05916 \log \frac{1}{a_{H^+}} \qquad \mathbf{31}$$

When $a_{H^+} = 1$, $\mathcal{E} = \mathcal{E}°$, which has the agreed upon reference value of zero. Thus

$$\mathcal{E} = 0 - 0.05916 \log \frac{1}{a_{H^+}}$$

$$= -0.05916 \log \frac{1}{a_{H^+}} \qquad \mathbf{32}$$

To apply this expression to basic solutions, it is more convenient to exhibit a_{OH^-}, which can be done by substituting $a_{OH^-} = 10^{-14}/a_{H^+}$, to give

$$\mathcal{E} = -0.05916 \log \frac{a_{OH^-}}{10^{-14}}$$

$$= -0.828 - 0.05916 \log a_{OH^-} \qquad \mathbf{33}$$

Thus if $a_{OH^-} = 1$ is chosen as the standard state for the hydrogen electrode, we have

$$\mathcal{E}^\circ \text{ (basic solution)} = -0.828 \text{ V} \qquad \textbf{34}$$

This and other basic electrode potentials are given in Table 8-4.

But neither unit H^+ activity nor unit OH^- activity is close to biological conditions. As a result, electrode-potential data of importance for biological electron-transfer processes are often reported for the neutral condition of $a_{H^+} = a_{OH^-}$, that is, for approximately $[H^+] = [OH^-] = 10^{-7}$. Then $\mathcal{E}'$ is used in place of $\mathcal{E}^\circ$. We can correct the values of Table 8-3 or 8-4 to these conditions in a way similar to that already illustrated. In particular, from Eq. **32** with $a_{H^+} = 10^{-7}$ or from Eq. **33** with $a_{OH^-} = 10^{-7}$, the hydrogen electrode has

$$\mathcal{E}'\text{(neutral)} = -0.414 \text{ V} \qquad \textbf{35}$$

An important reactant for many processes, including biological ones, is oxygen. The oxygen electrode in basic solutions is described in Table 8-4 as

$$O_2 + 2H_2O + 4e^- \rightarrow 4OH^- \qquad \mathcal{E}^\circ \text{ (basic)} = +0.401 \text{ V} \qquad \textbf{36}$$

For unit oxygen pressure we have

$$\mathcal{E} = \mathcal{E}^\circ - \frac{0.05916}{4} \log a_{OH^-}^4$$

$$= +0.401 - 0.05916 \log a_{OH^-}$$

Then for $a_{OH^-} = [OH^-] = 10^{-7}$

$$\mathcal{E}' = +0.401 + 0.414$$

$$= +0.815 \text{ V} \qquad \textbf{37}$$

This corresponds to the value listed as $\mathcal{E}'$ (neutral) for the oxygen electrode.

Standard neutral electrode potentials of some species important in biological processes are shown in Table 8-5. As in Tables 8-3 and 8-4, the

TABLE 8-4
Some Standard Electrode Potentials for Basic Aqueous Solutions, $a_{OH^-} = 1$, and 25°C

electrode	**electrode reaction**	$\mathcal{E}^\circ$**,V**
$Pt\|MnO_2\|MnO_4^-$	$MnO_4^- + 2H_2O + 3e^- = MnO_2 + 4OH^-$	+0.588
$Pt\|O_2\|OH^-$	$O_2 + 2H_2O + 4e^- = 4OH^-$	+0.401
$Pt\|S\|S^{2-}$	$S + 2e^- = S^{2-}$	−0.447
$Pt\|H_2\|OH^-$	$2H_2O + 2e^- = H_2 + 2OH^-$	−0.828
$Pt\|SO_3^{2-}, SO_4^{2-}$	$SO_4^{2-} + H_2O + 2e^- = SO_3^- + 2OH^-$	−0.93

oxidized form of any reagent will accept electrons from the reduced form of any reagent lower down in the table if all reagents are in their standard states. Thus oxygen is the most powerful of the listed oxidizing agents, and it can withdraw electrons from any of the reduced forms of all the other species in the table.

Emf data, such as the electrode potentials of Tables 8-3 to 8-5, can be used to obtain equilibrium concentrations or equilibrium-constant values. This can be seen by combining the relations $\Delta G^\circ = -n\mathfrak{F}\mathcal{E}^\circ$ and $\Delta G^\circ = -RT \ln K$. Alternatively, one can set up the expression for the concentration dependence of the emf that would be generated by the reaction. Setting this emf equal to zero leads to information about the concentrations that would produce this equilibrium condition.

Example 8-3

What can be said about the equilibrium concentrations that result when NAD^+ oxidizes ethanol, the products being NADH and acetaldehyde? The structures of NAD^+ and NADH, the oxidized and reduced forms of nicotinamide adenine dinucleotide, are shown on the next page.

Solution

We begin by writing the two half reactions, verifying so that they show the same number of electrons. Addition then gives the desired reaction and the corresponding emf. Thus, from Table 8-5,

$$NAD^+ + H^+ + 2e^- \rightarrow NADH \qquad \mathcal{E}' = -0.32 \text{ V}$$
$$\text{Ethanol} \rightarrow \text{acctaldehyde} + 2H^+ + 2e^- \qquad \mathcal{E} = -\mathcal{E}' = +0.18$$
$$NAD^+ + \text{ethanol} \rightarrow NADH + \text{acetaldehyde} + H^+ \qquad \mathcal{E}' = -0.14 \text{ V}$$

TABLE 8-5
Standard Electrode Potentials at pH 7 for Some Reactions of Biological Importance
Values are applicable either at 25 or 37°C.

reaction	$\mathcal{E}^\circ$, V
$\frac{1}{2}O_2 + 2H^+ + 2e^- \rightarrow H_2O$	+0.82
Cytochrome c–Fe^{3+} + e^- → cytochrome c–Fe^{2+}	+0.26
Cytochrome b–Fe^{3+} + e^- → cytochrome b–Fe^{2+}	0.00
Acetaldehyde + $2H^+$ + $2e^-$ → ethanol	−0.18
Pyruvate + $2H^+$ + $2e^-$ → lactate	−0.19
NAD^+ + H^+ + $2e^-$ → NADH	−0.32
α-Ketoglutaric acid + CO_2 + $2H^+$ + $2e^-$ → isocitric acid	−0.38
$2H^+ + 2e^- \rightarrow H_2$	−0.42
Acetic acid + $2H^+$ + $2e^-$ → acetaldehyde + H_2O	−0.60

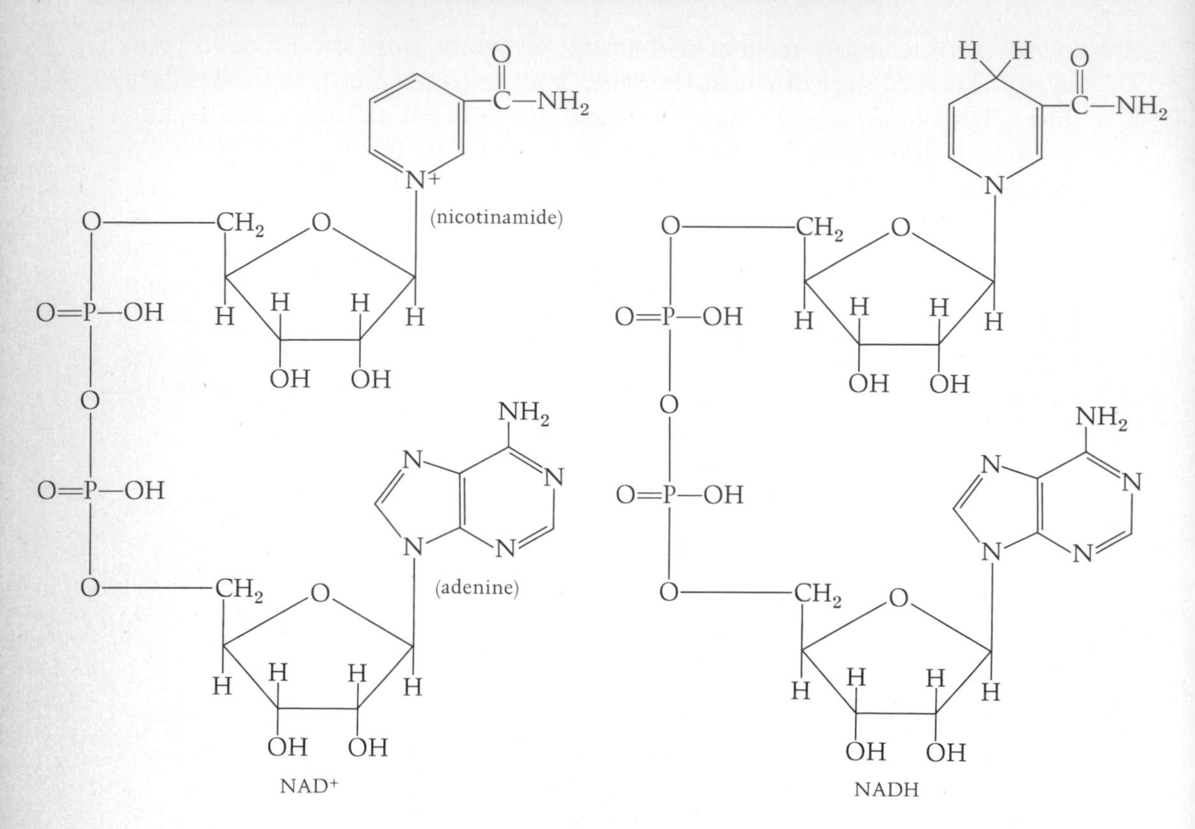

Now the Nernst equation **28** can be written, but we must remember that emf data for an $[H^+] = 10^{-7}$ standard state have been used. The hydrogen-ion concentration (more properly activity) that must be entered is the ratio of the actual H^+ concentration to this reference value. (All the other concentration, or activity terms are really ratios to unit concentration, or activity.) We thus write

$$\mathcal{E} = -0.14 - \frac{0.05915}{2} \log \frac{[\text{NADH}][\text{acetaldehyde}][\text{H}^+]/10^{-7}}{[\text{NAD}^+][\text{ethanol}]}$$

At pH 7 the $[H^+]/10^{-7}$ term equals 1, and the equilibrium concentrations that would yield $\mathcal{E} = 0$ are obtained from

$$\log \frac{[\text{NADH}][\text{acetaldehyde}]}{[\text{NAD}^+][\text{ethanol}]} = -4.74$$

or

$$\frac{[\text{NADH}][\text{acetaldehyde}]}{[\text{NAD}^+][\text{ethanol}]} = 1.8 \times 10^{-5}$$

This equilibrium-constant value applies only at pH 7.

An alternative procedure retains the [H+] term in the emf equation. Then the 10^{-7} factor must also be retained, and one arrives, as in the above steps, at

$$\frac{[\text{NADH}][\text{acetaldehyde}][\text{H}^+]}{[\text{NAD}^+][\text{ethanol}]} = 1.8 \times 10^{-12}$$

The tendency of reagents to be involved in electron transfers is indicated, as in Tables 8-3 to 8-5, by the value of $\mathcal{E}°$ or $\mathcal{E}'$ corresponding to processes of the type

$$\text{A} + e^- \rightarrow \text{A}^- \qquad \textbf{38}$$

We can deal with the tendency of electrons to be donated, and thus have an electron-donation process comparable to the hydrogen-ion donation processes treated in the preceding section, by writing

$$\text{A}^- \rightarrow \text{A} + e^- \qquad \textbf{39}$$

The driving-force measure for such reactions is $-\mathcal{E}°$. Again we must recognize that this measure involves a reference acceptor species, now the hydrogen ion of the standard hydrogen electrode. The electron-donating tendency of A^- is therefore more completely indicated by

$$\begin{array}{ll} \text{A}^- \rightarrow \text{A} + e^- & \mathcal{E} = -\mathcal{E}° \\ \text{H}^+ + e^- \rightarrow \frac{1}{2}\text{H}_2 & \mathcal{E} = 0 \\ \hline \text{A}^- + \text{H}^+ \rightarrow \text{A} + \frac{1}{2}\text{H}_2 & \mathcal{E} = -\mathcal{E}° \end{array} \qquad \textbf{40}$$

The driving force for electron transfer to the reference substance H^+ at approximately 1, 10^{-14}, or 10^{-7} m is thus the interpretation that can be put on $\mathcal{E}°$ or $\mathcal{E}'$ values from Tables 8-3, 8-4, or 8-5. This special driving force measure is related to the free-energy measure through $\Delta G° = -n\mathfrak{F}\mathcal{E}°$ or, with $\Delta G_e°$ indicating the electron-donating tendency of a substance,

$$\Delta G_e° = -n\mathfrak{F}\mathcal{E}° \qquad \text{or} \qquad \Delta G_e' = -n\mathfrak{F}\mathcal{E}'$$

and

$$-\mathcal{E}° = \frac{\Delta G_e°}{n\mathfrak{F}} \qquad \text{or} \qquad -\mathcal{E}' = \frac{\Delta G_e'}{n\mathfrak{F}} \qquad \textbf{41}$$

The qualitative features of these driving-force measures are shown in Table 8-6.

8-5 PHOSPHATE-GROUP TRANSFER POTENTIALS

Proton- and electron-transfer processes abound in both chemical and biological processes, but in biochemistry the transfer of phosphate groups is also ubiquitous. The tendency for such transfers to occur can be treated so that the process is just another example of a general group transfer.

Phosphate-transfer processes that are important in biological systems often involve the reagents adenosine triphosphate (ATP) and adenosine diphosphate (ADP). The phosphate-transfer role of these reagents can be illustrated by the reaction

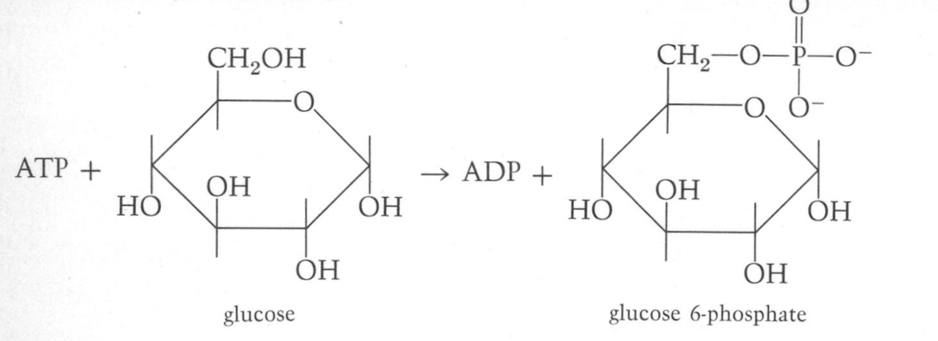

glucose glucose 6-phosphate

The value of $\Delta G'$ for this reaction at pH 7 and in the presence of Mg^{2+} at 25°C is about −4.0 kcal. (The role of the magnesium will be dealt with in Sec. 8-7.)

For each such phosphate-transfer reaction for which the equilibrium concentrations can be determined or for which $\Delta G_f'$ values for the reagents are available, the value of the free-energy driving force $\Delta G'$ can be reported. The variety of compounds and the many reactions that can be considered in phosphate-transfer processes lead us to look for an index of the phosphate-group transfer tendency that can be assigned to each compound. We do this, in a way that is parallel to that used for proton and electron transfers, by reporting the $\Delta G'$ values for phosphate transfer to a reference compound. If this compound is chosen as water, the reference phosphate-transfer reactions are the phosphate-group hydrolyses. Thus with P_i representing the inorganic phosphate, e.g., HPO_4^{2-}, the phosphate-transfer tendencies of the reagents of Eq. **42** would be described by

$$ATP + H_2O \rightarrow ADP + P_i \qquad \textbf{43}$$

TABLE 8-6
Measures of the Tendencies of Species (Such as the Reduced Forms of Tables 8-3 to 8-5) to Donate Electrons to the Reference Substance $H^+(aq)$

	ΔG_e°	$-\mathcal{E}^\circ$	$\mathcal{E}^\circ$
Stronger electron-donating agent	negative	negative	positive
No electron accepting or donating to the H_2,H^+ system (for standard states with $[H^+]$ approximately equal to 1, 10^{-7}, or 10^{-14} *m*)	0	0	0
Weaker electron-donating agent	positive	positive	negative

for which the $\Delta G'$ value is approximately -7.3 kcal and

Glucose 6-phosphate + $H_2O \rightarrow$ glucose + P_i **44**

with a $\Delta G'$ value of about -3.3 kcal.

You might expect that listing the values of $\Delta G'$ for such phosphate-to-water transfers is all that is needed to indicate the phosphate-transfer tendency of compounds. Embellishment of the symbol to $\Delta G'_P$ might be anticipated, but here too a special treatment is often used. The term *phosphate-group transfer potential* is introduced and defined as $-\Delta G'_P$, expressed in kilocalories. Some values of this transfer-tendency measure are shown in Table 8-7.

The listed values or the related $\Delta G'_P$ values can be used to determine the free-energy change for any phosphate-transfer reaction. Thus

Step 1:
ATP + $H_2O \rightarrow$ ADP + P_i $\Delta G' = -7.3$ kcal
Step 2:
Glucose 6-phosphate $\rightarrow$ glucose + P_i $\Delta G' = -3.3$ kcal
Step 3 (reverse of Step 2):
Glucose + $P_i \rightarrow$ glucose 6-phosphate $\Delta G' = +3.3$ kcal

Steps 1 and 3:
ATP + glucose $\rightarrow$ ADP + glucose 6-phosphate $\Delta G' = -4.0$ kcal

TABLE 8-7
Measures of the Tendency of Phosphate Groups to Be Transferred to the Reference Substance, Water at pH 7†

phosphate-containing substance	$\Delta G'_P$, **kcal/mol**	**phosphate-group transfer potential** $= -\Delta G'_P$
Phosphoenolpyruvate	−14.8	14.8
1,3-Diphosphoglycerate	−11.8	11.8
Phosphocreatine	−10.3	10.3
Acetyl phosphate	−10.1	10.1
Phosphoarginine	−7.7	7.7
ATP (to ADP)	−7.3	7.3
Glucose 1-phosphate	−5.0	5.0
Fructose 6-phosphate	−3.8	3.8
Glucose 6-phosphate	−3.3	3.3
Glycerol 1-phosphate	−2.2	2.2

†From A. L. Lehninger, "Biochemistry," Worth Publishers, Inc., New York, 1970.

Although the transfer of other groups besides protons, electrons, and phosphate groups is important, these three examples display the types of measures used to indicate the driving force of such reaction types. A summary of these measures is given in Table 8-8.

The term *high-energy phosphate bond* in common use in treatments of phosphate-transfer reactions has been avoided here, and it is hoped you will also try to avoid it. If you are tempted to use it, notice that ATP occupies a middle position in Table 8-7. And, in any case, no bond energy is released in phosphate-transfer processes. There is only a rearrangement of bonds and an equilibrium constant, or a related free energy, for the overall reaction.

8-6 COUPLED REACTIONS

In the chemical laboratory, reaction systems can usually be put together so that only one reaction moves the reagents from their initial concentrations toward the equilibrium values. Then the use of free-energy data to reveal the direction in which the system is driven and the state that will be reached when this force is spent is quite straightforward.

In contrast, reaction systems in living and biological systems are characterized by the *coupling* of reactions. This coupling allows a spontaneous reaction to drive a reluctant reaction. We now consider the thermodynamics of two-reaction systems in which one reaction has a forward thermodynamic driving force but is slow and does not proceed appreciably on its own. In the presence of the second reluctant reaction and perhaps other catalytic reagents, the two reactions can occur with appreciable speed. Reactions and processes with this coupled relationship abound in biological systems. Coupling is the basis of the storage and the use of chemical driving

TABLE 8-8
Summary of the Measures Used to Indicate the Tendency of Substances to Donate Protons, Electrons, and Phosphate Groups
Note that all measures are proportional to $\Delta G°$ for the transfer reactions.

transferred group	**reagent**	**reference receiver**	**reaction**	**measure of transfer tendency**
H^+	HA	H_2O	$HA + H_2O \rightarrow H_3O^+ + A^-$	$pK_a = \dfrac{\Delta G°}{2.303RT}$
e^-	A	H^+	$A + H^+ \rightarrow A^+ + \frac{1}{2}H_2$	$-\mathcal{E}° = \dfrac{\Delta G°}{n\mathfrak{F}}$ in volts
PO_4^{3-}	$A\text{-}PO_4^{3-}$	H_2O	$A\text{-}PO_4^{3-} + H_2O \rightarrow AOH + HPO_4^{2-}$	$-\Delta G°$ in kilocalories

forces in living systems. It also can be brought into laboratory systems and used to study reactions of biologically important substances.

A suitable laboratory illustration of coupled reactions is provided by the most important reaction couples in biological systems. Thus we look again to the hydrolysis of adenosine triphosphate (ATP) to form adenosine diphosphate (ADP) and phosphate ions. This reaction can be described here simply as

$$ATP + H_2O \rightarrow ADP + P_i \qquad \mathbf{45}$$

where P_i indicates the inorganic phosphate ions, principally $H_2PO_4^-$ and HPO_4^{2-}. At equilibrium the reaction has moved so far to the right that the ATP concentration is unsuitably small for reliable analysis.

This phosphate-transfer reaction couples with many other chemical reactions and transport processes. Some of these reactions can be used to modify the ratio of ATP to ADP present at equilibrium so that an analytically more satisfactory ratio is achieved. One example is the formation of glutamine from ammonium glutamate. With the abbreviations G-ine and G-ate$^-$, as in Fig. 8-4, for these reagents, this reaction, by itself, is

$$NH_4^+ + \text{G-ate}^- \rightarrow \text{G-ine} + H_2O \qquad \mathbf{46}$$

The equilibrium constant at pH 7 is about 0.003 at 25°C. Thus, we write

$$K_G = \frac{[\text{G-ine}]}{[\text{G-ate}^-][NH_4^+]} = 0.003 \qquad \mathbf{47}$$

and $$\Delta G_G^\circ = -RT \ln K_G = +3.4 \text{ kcal} \qquad \mathbf{48}$$

This glutamine-synthesis reaction can be coupled to the ATP hydrolysis reaction, a reaction which by itself is slow. But in the presence of ammonium glutamate and the enzyme *glutamine synthetase,* the entire system moves rapidly to an equilibrium state. The reaction is

$$NH_4^+ + \text{G-ate}^- + ATP \rightarrow \text{G-ine} + ADP + P_i \qquad \mathbf{49}$$

The equilibrium expression and the constant deduced from measured concentrations are

$$K_{G,\text{ATP}} = \frac{[\text{G-ine}][ADP][P_i]}{[\text{G-ate}^-][NH_4^+][ATP]} = 9600 \qquad \mathbf{50}$$

The value of the constant is at pH 7 and 25°C and in the presence of Mg^2. It can be compared to that of Eq. **47** to see the effect of the thermodynamically driven but inherently slow ATP hydrolysis reaction. We see this also by

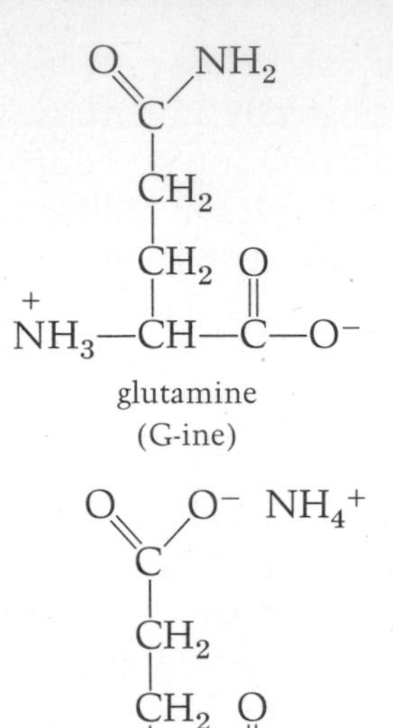

FIGURE 8-4
Structures and abbreviations for glutamine and ammonium glutamate.

comparing

$$\Delta G^\circ_{G,\mathrm{ATP}} = -RT \ln K_{G,\mathrm{ATP}}$$
$$= -5.4 \text{ kcal} \qquad \mathbf{51}$$

with $\Delta G^\circ_G = +3.4$ kcal.

From the experimentally convenient systems of the glutamine-synthesis reaction, alone and coupled with ATP, one can extract information on the ATP hydrolysis process. Thus, formation of the ratio $K_{G,\mathrm{ATP}}/K_G$, from Eqs. **47** and **50,** gives

$$\frac{K_{G,\mathrm{ATP}}}{K_G} = \frac{[\text{G-ine}][\text{ADP}][\text{P}_i]/[\text{G-ate}^-][\text{NH}_4^+][\text{ATP}]}{[\text{G-ine}]/[\text{G-ate}^-][\text{NH}_4^+]}$$
$$= \frac{[\text{ADP}][\text{P}_i]}{[\text{ATP}]} \qquad \mathbf{52}$$

Thus, from this study we deduce, for 25°C and pH 7 and in the presence of Mg^{2+},

$$K_{\mathrm{ATP}} = \frac{[\text{ADP}][\text{P}_i]}{[\text{ATP}]} = \frac{9600}{0.003} = 3 \times 10^6 \qquad \mathbf{53}$$

From this result, or more directly by subtracting the reaction equations **46** and **49** and the free energies of Eqs. **48** and **51,** we also obtain

$$\text{ATP} + \text{H}_2\text{O} \rightarrow \text{ADP} + \text{P}_i \qquad \Delta G^\circ = -9.0 \text{ kcal} \qquad \mathbf{54}$$

Some subtleties surrounding the numerical value remain to be unraveled in the following section, but how the equilibria and free energies of coupled reactions are treated should be clear.

Enzymes and catalysis are treated in Chaps. 10 and 11, but it should be pointed out that a catalyst speeds up the passage toward equilibrium. One could therefore proceed in the reaction of Eq. **49** from ATP and ammonium glutamate or from ADP, P_i, and glutamine. The coupled reactions would reach the same equilibrium state. If one starts with ATP and ammonium glutamate, it is the glutamine synthesis that must be driven up the free-energy hill by the ATP hydrolysis. If one proceeded from glutamine, ADP, and P_i, it is the formation of ATP that is pushed.

Some biological processes are equivalent to the second direction in that more ATP is generated by a driving, coupled reaction than would be present in the equilibrium ATP-ADP-P_i system itself. These are driving-force or free-energy-storage reactions. Other processes make use of this more than equilibrium concentration of ATP to drive reluctant reactions, as in the glutamine-synthesis direction of the example above. Such reactions in biological systems utilize the stored driving force or free energy.

A number of complicating factors have been ignored in keeping the focus on the coupling of the reactions. One of these, the involvement of H^+ and of Mg^{2+} ions in the equilibria, particularly those involving ATP, ADP, and the phosphate ions, will be treated in the following section.

8-7 SIMULTANEOUS EQUILIBRIA

In many chemical and biological systems the reagents involved in a reaction or in a set of coupled reactions are simultaneously involved in other reaction processes. Thus, in the example of the preceding section, each of the symbols ATP, ADP, and P_i actually represents several species interconnected by acid-base and metal-ion-complex formation reactions. The use of the single designations ATP, ADP, and P_i simplifies the formulation of the reaction process and reflects the fact that analytical procedures usually group the related species together into these three categories. Let us see the consequences of this procedure. These will be illustrated by the ATP hydrolysis reaction, but the treatment will illustrate generally applicable procedures.

In aqueous solution not too far from pH 7 the most important species included in the ATP, ADP, and P_i designations are those shown in Table 8-9, to which we restrict our attention. We then can write, with square brackets implying molal concentrations,

$$
\begin{aligned}
[\mathrm{ATP}] &= [\mathrm{HATP}^{3-}] + [\mathrm{ATP}^{4-}] \\
[\mathrm{ADP}] &= [\mathrm{HADP}^{2-}] + [\mathrm{ADP}^{3-}] \\
[\mathrm{P}_i] &= [\mathrm{H_2PO_4}^{-}] + [\mathrm{HPO_4}^{2-}]
\end{aligned} \tag{55}
$$

Note that [ATP] and [ADP] without any charge indication and $[P_i]$ imply the sum of the various individual species.

The equilibrium-constant expression formulated in terms of these summed species can be compared with any of the normal equilibrium expressions which involve particular chemical species. Let us compare the equilibrium of the summed species

$$\mathrm{ATP} + \mathrm{H_2O} \rightarrow \mathrm{ADP} + \mathrm{P}_i$$

TABLE 8-9
The Most Important ATP, ADP, and P_i Species in the Neighborhood of pH 7

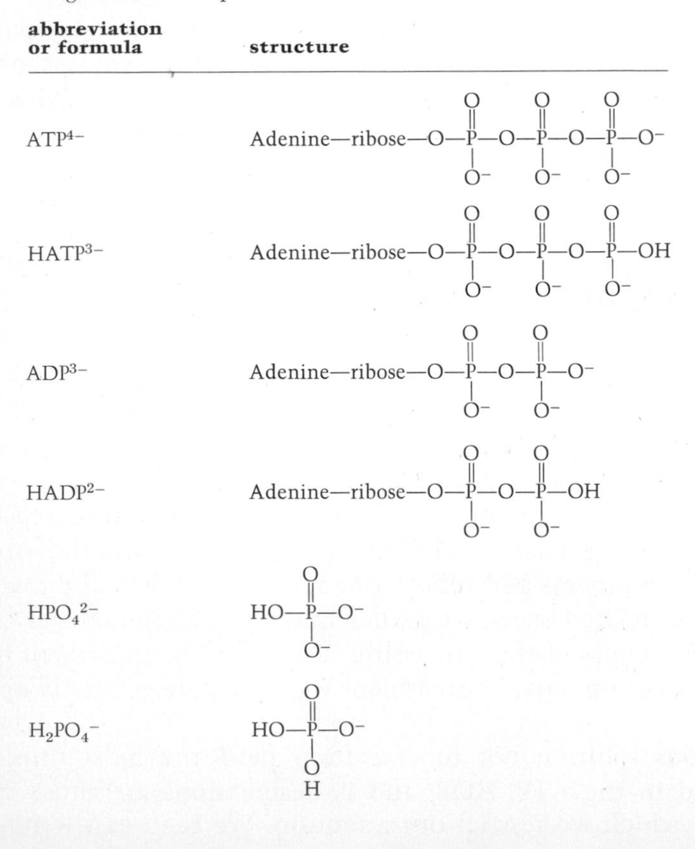

abbreviation or formula	structure
ATP^{4-}	Adenine—ribose—O—P(=O)(O⁻)—O—P(=O)(O⁻)—O—P(=O)(O⁻)—O⁻
$HATP^{3-}$	Adenine—ribose—O—P(=O)(O⁻)—O—P(=O)(O⁻)—O—P(=O)(O⁻)—OH
ADP^{3-}	Adenine—ribose—O—P(=O)(O⁻)—O—P(=O)(O⁻)—O⁻
$HADP^{2-}$	Adenine—ribose—O—P(=O)(O⁻)—O—P(=O)(O⁻)—OH
HPO_4^{2-}	HO—P(=O)(O⁻)—O⁻
$H_2PO_4^{-}$	HO—P(=O)(OH)—O⁻

and the corresponding constant

$$K_\Sigma = \frac{[\text{ADP}][\text{P}_i]}{[\text{ATP}]} \tag{56}$$

with $\text{ATP}^{4-} + \text{H}_2\text{O} \rightarrow \text{ADP}^{3-} + \text{HPO}_4{}^{2-} + \text{H}^+$

and
$$K = \frac{[\text{ADP}^{3-}][\text{HPO}_4{}^{2-}][\text{H}^+]}{[\text{ATP}^{4-}]} \tag{57}$$

Equations **56** and **57** can be related by the ionization equilibrium constants of Table 8-10 and the relations of Eq. **55.** One first obtains

$$\begin{aligned} [\text{ATP}] &= [\text{ATP}^{4-}]\left(1 + \frac{[\text{H}^+]}{1.12 \times 10^{-7}}\right) \\ [\text{ADP}] &= [\text{ADP}^{3-}]\left(1 + \frac{\text{H}^+}{1.32 \times 10^{-7}}\right) \\ [\text{P}_i] &= [\text{HPO}_4{}^{2-}]\left(1 + \frac{\text{H}^+}{1.66 \times 10^{-7}}\right) \end{aligned} \tag{58}$$

Substitution of these expressions in Eq. **56** then gives

$$K_\Sigma = \frac{K}{[\text{H}^+]}\left\{\frac{\left(1 + \frac{[\text{H}^+]}{1.32 \times 10^{-7}}\right)\left(1 + \frac{[\text{H}^+]}{1.66 \times 10^{-7}}\right)}{1 + \frac{[\text{H}^+]}{1.12 \times 10^{-7}}}\right\} \tag{59}$$

Calculation of the clumsy $[\text{H}^+]$-dependent term for various pH values leads to the curve for K_Σ/K shown in Fig. 8-5.

From Eq. **59** we can also relate the free-energy change corresponding to the standard states of the summed species to the pH-independent standard-free-energy change corresponding to Eq. **57.** The result of applying $\Delta G° = -RT \ln K$ to Eq. **59** is shown in Fig. 8-6.

Note that K is a proper equilibrium constant and that if other

TABLE 8-10
Equilibria Connecting the Most Important ATP, ADP, and P_i Species in the Neighborhood of pH 7

$\text{HATP}^{3-} \rightarrow \text{ATP}^{4-} + \text{H}^+$	$K = \frac{[\text{ATP}^{4-}][\text{H}^+]}{[\text{HATP}^{3-}]} = 1.12 \times 10^{-7}$
$\text{HADP}^{2-} \rightarrow \text{ADP}^{3-} + \text{H}^+$	$K = \frac{[\text{ADP}^{3-}][\text{H}^+]}{[\text{HADP}^{2-}]} = 1.32 \times 10^{-7}$
$\text{H}_2\text{PO}_4{}^- \rightarrow \text{HPO}_4{}^{2-} + \text{H}^+$	$K = \frac{[\text{HPO}_4{}^{2-}][\text{H}^+]}{[\text{H}_2\text{PO}_4{}^-]} = 1.66 \times 10^{-7}$

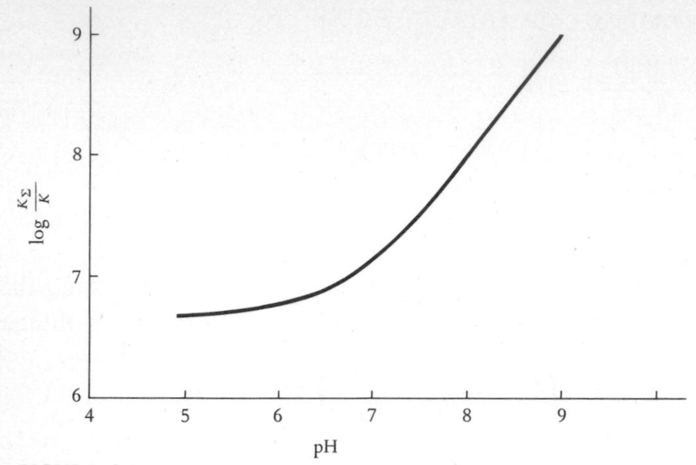

FIGURE 8-5
The dependence of the ratio of K_Σ of Eq. **56** and K of Eq. **57** on pH according to the relation shown in Eq. **59.**

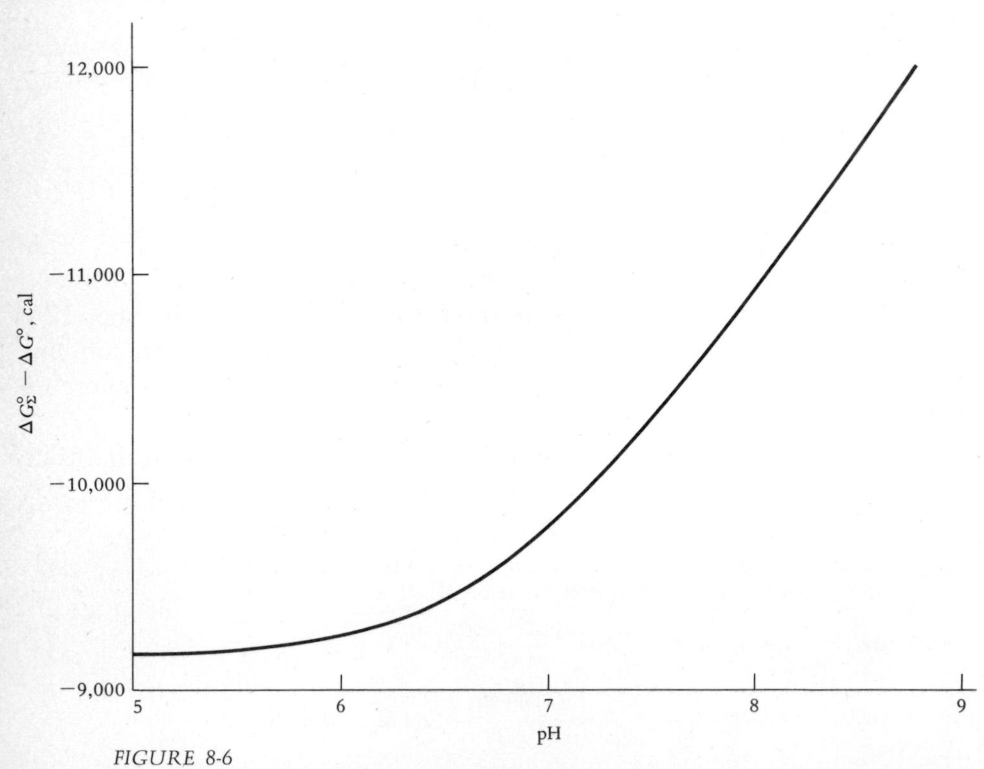

FIGURE 8-6
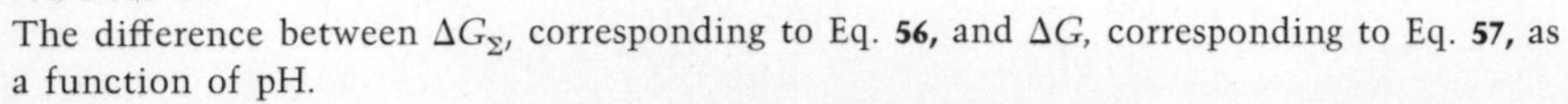
The difference between ΔG_Σ, corresponding to Eq. **56,** and ΔG, corresponding to Eq. **57,** as a function of pH.

complicating factors are set aside, K will be constant. Likewise, ΔG° is a fixed pH-independent quantity. It follows that K_Σ and ΔG_Σ° will not be constant, as can be seen by considering Figs. 8-5 and 8-6, if solutions with various pH values are dealt with. They remain, however, useful quantities.

In many biological systems all these phosphate-containing species tend to form complex ions with the metal ions of the solution. Most important are the complexes with Mg^{2+}. A set of equilibria similar to those involving H^+ attachment are then set up. The more important species and reactions are shown in Table 8-11. We now could proceed by choosing various fixed pH values and developing expressions for the dependence of K_Σ on the Mg^{2+} concentration just as we have developed the dependence of K_Σ on the H^+ concentration. Instead we refer to the results obtained from a computer-based treatment of both the pH and pMg dependence that also takes account of minor species we have omitted. At 25°C and a pH of 7 and an Mg^{2+} concentration of 0.35 or pMg = 1.45, the term that is comparable to the $[H^+]$ dependence deduced in Eq. **59** leads to

$$K_\Sigma = K(4.87 \times 10^6) \qquad \textbf{60}$$

The value of K_Σ of Sec. 8-6 of 3×10^6 applies to these conditions and thus the calculated value of K is 0.6. Then for various H^+ and Mg^{2+} concentrations, values of K_Σ can be deduced from

$$K_\Sigma = 0.6 \times ([H^+], [Mg^{2+}]) \qquad \textbf{61}$$

where the function value is like that deduced for the $[H^+]$ dependence in Eq. **59.**

The results are displayed by the contour diagram for log K_Σ and ΔG_Σ° in Figs. 8-7 and 8-8. Note again that K_Σ and ΔG_Σ° are indeed variables that depend on the two reagents H^+ and Mg^{2+}. The latter dependence is appre-

TABLE 8-11
Important Species, Equilibria, and Equilibrium Constants of the ATP, ADP, P_i System near pH 7 and Mg^{2+} Concentrations up to 1 *m*†

equilibrium	p*K*	*K*
$MgATP^{2-} \rightarrow Mg^{2+} + ATP^{4-}$	4.00	1.0×10^{-4}
$MgHATP^- \rightarrow Mg^{2+} + HATP^{3-}$	1.49	3.2×10^{-2}
$MgADP^- \rightarrow Mg^{2+} + ADP^{3-}$	3.01	9.8×10^{-4}
$MgHADP \rightarrow Mg^{2+} + HADP^{-2}$	1.45	3.6×10^{-2}
$MgHPO_4 \rightarrow Mg^{2+} + HPO_4^{2-}$	1.88	1.3×10^{-2}

†From R. A. Alberty, *J. Biol. Chem.*, **244:**3290 (1969).

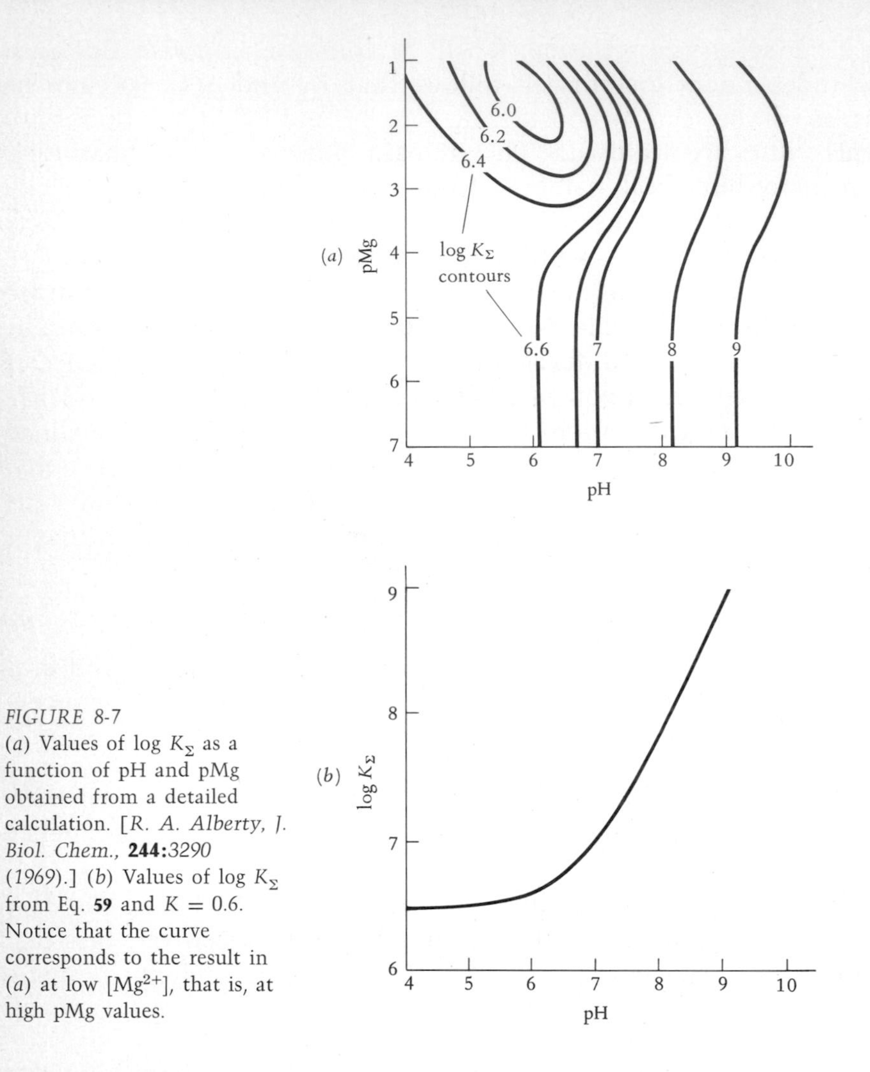

FIGURE 8-7
(*a*) Values of log K_Σ as a function of pH and pMg obtained from a detailed calculation. [*R. A. Alberty, J. Biol. Chem.*, **244:**3290 (1969).] (*b*) Values of log K_Σ from Eq. **59** and $K = 0.6$. Notice that the curve corresponds to the result in (*a*) at low $[Mg^{2+}]$, that is, at high pMg values.

ciable, however, only at Mg^{2+} concentrations of more than about 10^{-4} *M*. At lower Mg^{2+} concentrations the pH dependence shown by Eq. **59** or Fig. 8-6 is adequate, and to a valid representation of the dependence of log K_Σ and AG°_Σ on pH is given. The comparison of this result with the lower part of the contour diagram of Fig. 8-7*a* is shown by the agreement of the contour-line values with the ordinate values of Fig. 8-7*b*.

Even inclusion of the effects of H^+ and Mg^{2+} on the illustrative ATP equilibrium does not provide results applicable to all solutions of interest. The results do show how the equilibrium constant K_Σ and standard-free-energy change ΔG°_Σ depend on the hydrogen ion and the magnesium ion concentrations.

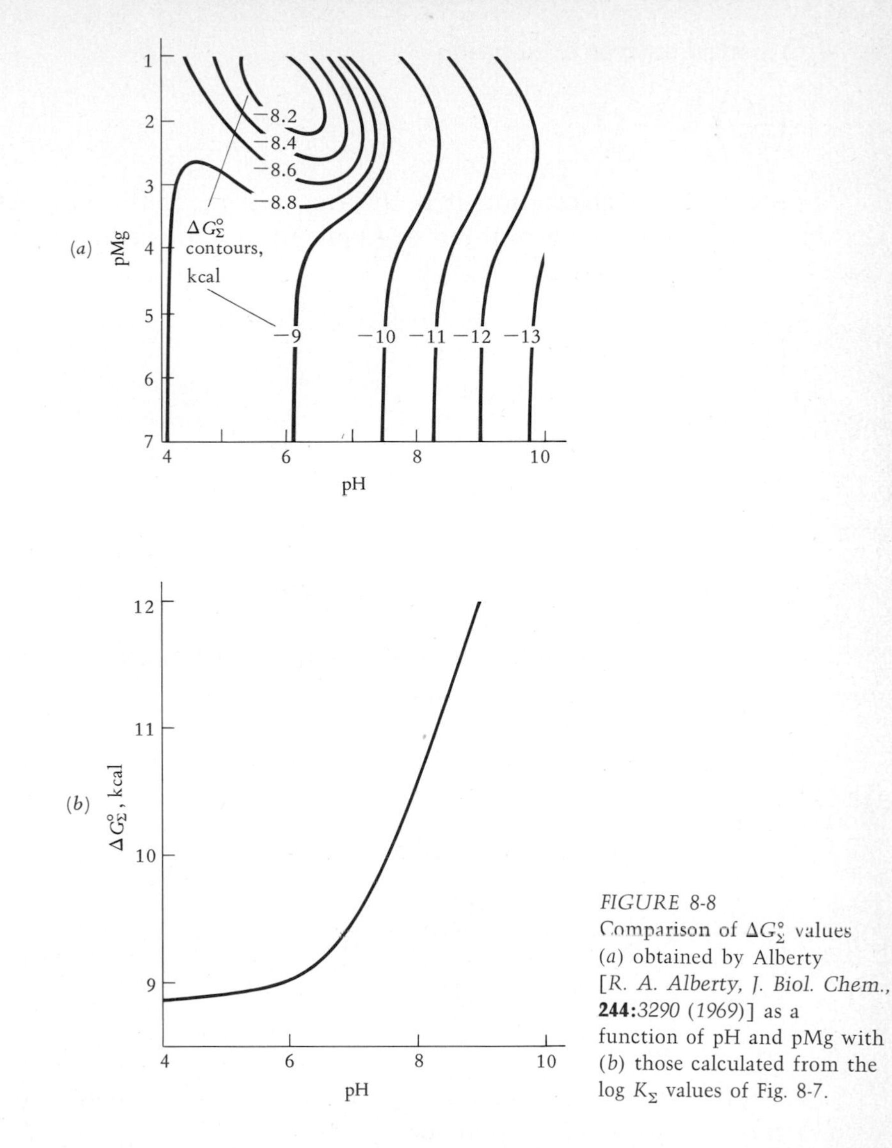

FIGURE 8-8
Comparison of ΔG_Σ° values (*a*) obtained by Alberty [*R. A. Alberty, J. Biol. Chem.*, **244:**3290 (1969)] as a function of pH and pMg with (*b*) those calculated from the log K_Σ values of Fig. 8-7.

8-8 TEMPERATURE DEPENDENCE OF THE FREE ENERGY AND THE EQUILIBRIUM CONSTANT OF A REACTION

The free energy of each compound involved in a reaction depends on the temperature according to the relation (Eq. **10** of Chap. 7), valid at constant pressure,

$$dG = -S\,dT \qquad \text{or} \qquad \frac{dG}{dT} = -S$$

or, with partial-derivative notation,

$$\left(\frac{\partial G}{\partial T}\right)_P = -S$$

For a chemical reaction it is the free energy of the products less that of the reactants that is of interest. Application of this equation to each reagent allows us to write

$$\left[\frac{\partial(\Delta G)}{\partial T}\right]_P = -\Delta S \qquad \textbf{62}$$

where

$$\Delta G = G_{\text{prod}} - G_{\text{react}} \qquad \textbf{63}$$

and $$\Delta S = S_{\text{prod}} - S_{\text{react}} \qquad \textbf{64}$$

An expression for the temperature dependence of ΔG that is easier to use results if ΔS is eliminated from Eq. **62.**

At any constant temperature the changes of free-energy, enthalpy, and entropy for any reaction are related by

$$\Delta G = \Delta H - T\,\Delta S \qquad \text{or} \qquad \Delta S = \frac{\Delta H - \Delta G}{T} \qquad [T = \text{const}] \qquad \textbf{65}$$

The second expression can be used to eliminate ΔS from Eq. **62** to give

$$\left[\frac{\partial(\Delta G)}{\partial T}\right]_P = \frac{-\Delta H + \Delta G}{T} = -\frac{\Delta H}{T} + \frac{\Delta G}{T}$$

or $$\left[\frac{\partial(\Delta G)}{\partial T}\right]_P - \frac{\Delta G}{T} = -\frac{\Delta H}{T} \qquad \textbf{66}$$

The two terms on the left side of Eq. **66** can be shown [using $d(u/v) = (vdu - udv)/v^2$] to be equivalent to

$$T\frac{\partial(\Delta G/T)}{\partial T} = T\frac{T[\partial(\Delta G)/\partial T]_P - \Delta G}{T^2} = \left[\frac{\partial(\Delta G)}{\partial T}\right]_P - \frac{\Delta G}{T} \qquad \textbf{67}$$

Now comparison of Eqs. **66** and **67** lets us write

$$T\left[\frac{\partial(\Delta G/T)}{\partial T}\right]_P = -\frac{\Delta H}{T} \qquad \textbf{68}$$

When this relation is applied to the reagents of a reaction, each at the constant pressure corresponding to the standard states, it becomes

$$T\frac{d(\Delta G^\circ/T)}{dT} = -\frac{\Delta H^\circ}{T} \qquad \begin{bmatrix} P = \text{standard-state} \\ \text{pressure} \\ T = \text{constant} \end{bmatrix} \tag{69}$$

Finally, the relation between ΔG° and the equilibrium constant, Eq. **16,** can be inserted to give, with rearrangement,

$$\frac{d(\ln K)}{dT} = \frac{\Delta H^\circ}{RT^2} \tag{70}$$

This important formula is the goal of the derivation. The rate of change of the equilibrium constant with temperature is seen to depend on the standard heat of the reaction.

The change of ln K or K over not too large a temperature range can be obtained from this derivative result by an integration of this expression with the assumption of a constant value of ΔH°. Integrations can be carried out by first rearranging Eq. **70** to

$$\frac{d(\ln K)}{d(1/T)} = -\frac{\Delta H^\circ}{R} \qquad \text{or} \qquad \frac{d(\log K)}{d(1/T)} = -\frac{\Delta H^\circ}{2.303R} \tag{71}$$

The integrated form of these equations, on the assumption that ΔH° is temperature-independent, is

$$\log K = -\frac{\Delta H^\circ}{2.303R}\frac{1}{T} + \text{const} \tag{72}$$

Both the integrated and differential forms show that a plot of log K versus $1/T$ should give a straight line with a slope equal to $-\Delta H^\circ/2.303R$. The linearity shown by good measurements can be judged by the example of Fig. 8-9. The straight line, furthermore, has been drawn with the slope $\Delta H^\circ/R$ for the reaction.

Thus a measured value of ΔH° can be used to calculate the equilibrium constant at temperatures other than that for which it is given. Conversely, it is possible to use measurements of the equilibrium constant at a number of temperatures to evaluate the standard-enthalpy change for the reaction.

In biological studies it is often necessary to interconvert data between the customary chemical reference temperature of 25°C and the common biological reference temperature of 37°C. The relation of equilibrium constants at these two temperatures for various heats of reaction is

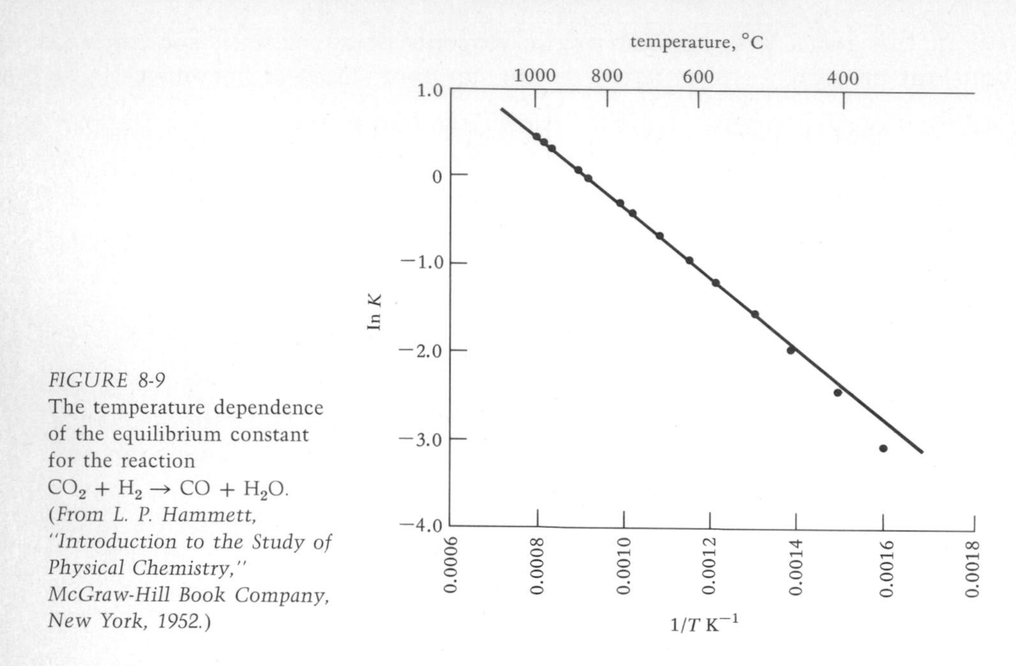

FIGURE 8-9
The temperature dependence of the equilibrium constant for the reaction $CO_2 + H_2 \rightarrow CO + H_2O$. *(From L. P. Hammett, "Introduction to the Study of Physical Chemistry," McGraw-Hill Book Company, New York, 1952.)*

shown in Fig. 8-10. The effect is that predicted by Le Châtelier's principle.

When much larger temperature ranges are considered, the basis of the dependence of the equilibrium constant on temperature can be seen more clearly by returning to the expressions

$$\Delta G^\circ = \Delta H^\circ - T\,\Delta S^\circ$$

and $$\Delta G^\circ = -RT \ln K \qquad [T = \text{const}]$$

or $$RT \ln K = -\Delta H^\circ + T\,\Delta S^\circ \tag{73}$$

Generally, the $T\,\Delta S^\circ$ term, as might be expected from the presence of the explicit T factor, is the more temperature-dependent. At high temperatures this term dominates the ΔH° term to give a $RT \ln K$ value that is increasingly positive or negative according as ΔS for the reaction is positive or negative. Thus at high temperatures the equilibrium constant generally becomes increasingly greater if ΔS is positive or smaller if ΔS is negative. Examples of this behavior are shown in Fig. 8-11.

In general, the more gas-phase molecular or atomic particles there are, the higher the entropy. This fact and the overwhelming importance of the entropy of the system at high temperatures lead to the general breakup or dissociation of species at these temperatures. One can generalize by saying that at high temperatures the side of the equation with more gas-phase species will be dominant, a generalization that is not valid unless the $T\,\Delta S^\circ$ term dominates the ΔH° term in contributing to ΔG°.

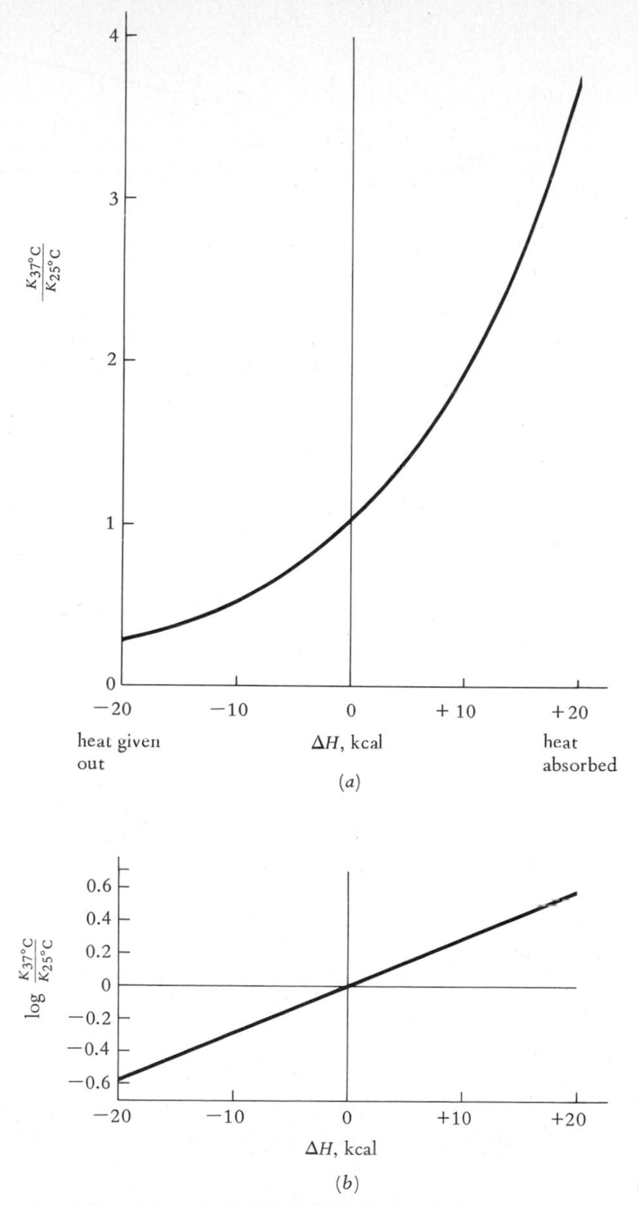

FIGURE 8-10
The effect of a temperature change from 25 to 37°C, that is, $\Delta T = 12°C$, as a function of ΔH for (*a*) the ratio $K_{37°}/K_{25°}$ and (*b*) for log $(K_{37°}/K_{25°})$.

PROBLEMS

1 From the listed standard free energies of liquid and gaseous water at 25°C deduce the vapor pressure of water at this temperature. Compare your answer with the observed vapor pressure of 23.76 mm Hg.

2 Using free-energy data in Appendix I together with the ΔG_f° values of −26.22 kcal

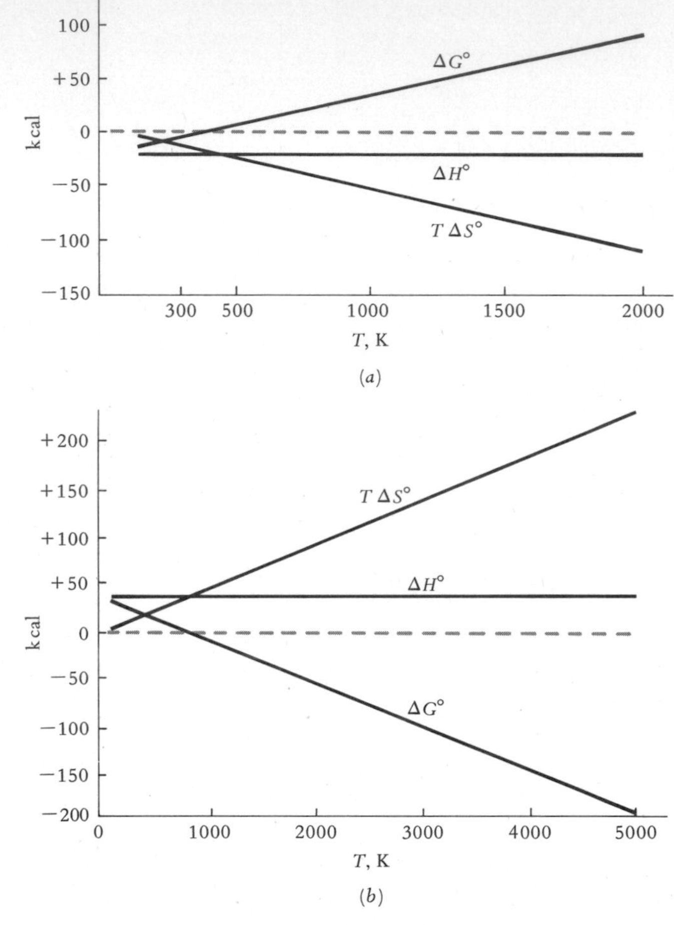

FIGURE 8-11
(*a*) The thermodynamic functions for the reaction $N_2(g) + 3H_2(g) \rightleftharpoons 2NH_3(g)$. Note that $\Delta S°$ is negative and therefore the $T\,\Delta S°$ term tends to make $\Delta G°$ positive. This favors the dissociation rather than the formation of ammonia at higher temperatures. (*b*) The thermodynamic functions for the reaction $H_2O(g) + C(s) \rightarrow CO(g) + H_2(g)$. In contrast to the reaction in (*a*), there are more product moles of gas than there are reactant moles. Thus $T\,\Delta S$ has the opposite effect and at high enough temperature makes ΔG more negative. The reaction proceeds further to the right at higher temperatures.

for AgCl(*c*) and +18.43 kcal for $Ag^+(aq)$, calculate the equilibrium constant for the reaction

$$AgCl(s) \rightarrow Ag^+(aq) + Cl^-(aq)$$

Compare with the tabulated solubility product of 1.56×10^{-10} at 25°C. The assumption of ideal behavior is suitable here because of the very low solubility of silver chloride.

3 Find the value of the equilibrium constant for the formation of lactic acid from glucose in water at 25°C, according to the equation

$$\text{Glucose}(aq) \rightarrow 2 \text{ lactic acid}(aq)$$

4 Carbon monoxide is an extremely poisonous gas. (It reacts with hemoglobin and thus prevents it from filling its oxygen-transport role.)

a Write a chemical equation to show that under equilibrium conditions in a gas such as air, if oxygen and carbon dioxide are present, carbon monoxide will also be present.

b Use the free-energy data of Appendix I to deduce the free-energy change for this reaction if all reagents are at 1 atm pressure and 25°C. What is the value of the equilibrium constant for the reaction?

c In unpolluted sea-level air the partial pressures of O_2 and CO_2 are 0.2 and 0.003 atm, respectively. What is the equilibrium partial pressure of carbon monoxide?

5 At equilibrium in aqueous solution at 25°C, what is the ratio of the activities and (if the activity coefficients are assumed to be equal) the concentrations of glucose 1-phosphate and glucose 6-phosphate?

6 What is the equilibrium-concentration ratio of ATP to ADP in an aqueous solution at 25°C that contains phosphate at unit activity? Assume a pH of 7, no Mg^{2+}, and use the data of Fig. 8-7.

7 At 37°C, 1 atm, and pH of 7, the hydrolysis of ATP to ADP and P_i occurs with a $\Delta G°$ value of about $^{-}9$ kcal.

a Calculate the equilibrium constant for the reaction.

b Calculate ΔG for the reaction when the concentrations of ATP, ADP, and P_i are 10^{-2}, 10^{-4}, and 10^{-1} M, respectively.

c How much work, muscular work for example, could be obtained by the consumption of 1 mol of ATP for the reagents at these concentrations?

d Calculate ΔG for the reaction when the concentrations of ATP, ADP, and P_i are 10^{-7}, 10^{-1}, and 10^{-1} M, respectively. How much work could be obtained per mole of ATP at these concentrations?

8 The value of $\Delta H°$ for the reaction of Prob. 7 is -4.8 kcal/mol. What will be the equilibrium constant at 0°C, 1 atm, and pH 7? (Assume that $\Delta H°$ is independent of temperature in this temperature range.)

9 The value of $\Delta V°$, the volume change for the reaction of Prob. 7 when the reagents are in their standard states, is -20 ml per mole of ATP. Assume that this value is independent of pressure. Given that $\Delta G°$ for the ATP hydrolysis reaction is -9 kcal per mole of ATP at 1 atm pressure, calculate ΔG for 40 atm pressure. Does the reaction have a greater or lesser tendency to proceed at high pressures?

10 Deduce the form of the equilibrium-constant expression and the value of the equilibrium constant for the ionization of lactic acid by making use of free-energy data and following the procedure of Sec. 8-1.

11 Use the tabulated $\Delta G_f°$ values for $H_2O(l)$, $H^+(aq)$, and $OH^-(aq)$ to deduce the value of the dissociation constant for water.

12 Using standard-free-energy data, deduce the pK_a measure of proton-transfer tendencies, i.e., acid strengths, of the parent dipolar ion form of L-glutamic acid. What will be the principal form in regions near pH 7?

13 From the standard free energies of liquid and gaseous water at 25°C the vapor pressure of water can be deduced (as in Prob. 1) at that temperature. Furthermore, this vapor pressure can be treated as the equilibrium constant for the reaction

$$H_2O(l) \rightarrow H_2O(g)$$

With this approach, take 10,300 cal/mol as a suitable average heat of vaporization of water over the temperature range from 25°C to the boiling point and see what temperature this leads to for the normal boiling point of water.

14 Use emf data from Table 8-5 and standard free energies of formation from Sec. 8-6 to deduce the standard free energy of the respiratory-chain coupled-reaction system

$$\text{NADH} + \text{H}^+ + 3\text{ADP} + 3\text{P}_i + \tfrac{1}{2}\text{O}_2 \rightarrow \text{NAD}^+ + 4\text{H}_2\text{O} + 3\text{ATP}$$

15 The reaction

$$\text{Glucose 6-phosphate} + \text{H}_2\text{O} \rightarrow \text{glucose} + \text{P}_i$$

has been studied by O. Meyerhof and H. Green [*J. Biol. Chem.*, **178**:655 (1948)]. The equilibrium constant expressed in terms of the sums of the related species was found at pH 8.5 and 38°C to be

$$K_\Sigma = \frac{[\text{glucose}][\text{P}_i]}{[\text{glucose 6-P}]} = 122$$

The following ionization constant data are available:

$$\text{Glucose—O—}\overset{\overset{\displaystyle\text{O}}{\|}}{\underset{\underset{\displaystyle\text{O}^-}{|}}{\text{P}}}\text{—OH} \rightarrow \text{glucose—O—}\overset{\overset{\displaystyle\text{O}}{\|}}{\underset{\underset{\displaystyle\text{O}^-}{|}}{\text{P}}}\text{—O}^- + \text{H}^+ \qquad K = 9.3 \times 10^{-7}$$

and $H_2PO_4^- \rightarrow HPO_4^{2-} + H^+ \qquad K = 1.9 \times 10^{-7}$

a Derive an expression that relates K_Σ and the pH to the equilibrium constant of the reaction

$$\text{Glucose—O—}\overset{\overset{\displaystyle\text{O}}{\|}}{\underset{\underset{\displaystyle\text{O}^-}{|}}{\text{P}}}\text{—O}^- + \text{H}_2\text{O} \rightarrow \text{glucose} + \text{H—O—}\overset{\overset{\displaystyle\text{O}}{\|}}{\underset{\underset{\displaystyle\text{O}^-}{|}}{\text{P}}}\text{—O}^-$$

b What is the value for the equilibrium constant for this reaction at 38°C?

c What would the value for the equilibrium constant for the summed species be at pH 7?

16 Using the thermodynamic data of Appendix II, deduce the equilibrium constant for the reaction

$$2\text{Glycine}(aq) \rightarrow \text{glycylglycine}(aq) + H_2O(l)$$

at 25°C. What value would you estimate for 37°C?

17 The ionization of water

$$H_2O \rightarrow H^+ + OH^-$$

has the well-known equilibrium constant of 1×10^{-14} at 25°C. What is its value at 37°C?

REFERENCES

General

KLOTZ, I.: "Energy Changes in Biochemical Reactions," Academic Press, Inc., New York, 1967.

LEHNINGER, A. L.: "Bioenergetics," W. A. Benjamin, New York, 1965.

Phosphate Transfer

KALEKAR, H. M.: "Biological Phosphorylations: Development of Concepts," Prentice-Hall, Inc., Englewood Cliffs, N.J., 1969. A collection of reprints and introductory notes showing the development of our modern ideas of bioenergetics.

ALBERTY, ROBERT A.: *J. Biol. Chem.*, **243:**1337 (1968) and **244:**3290 (1969).

9 Physical Equilibria and Membrane Phenomena

Chemical transformations are not the only processes that lead us to investigate equilibrium states and to set up measures of the driving force with which the system tends to move from any other state to the equilibrium state. The distribution of a substance between two phases, e.g., liquid and vapor, or between two solutions might involve no change in the molecules of the substance. Equilibrium and processes of this type can be described as physical if the word is taken to imply an absence of the molecular transformations that characterize chemical processes. The thermodynamic tools for investigating equilibria and driving forces developed in Chap. 8 will now be applied to such processes.

Reference to liquid-vapor equilibrium or even phase equilibria in general fails to suggest the great biological importance of the equilibria we now study. Some of this importance will be shown by the use of osmotic pressure measurements in the deduction of the molecular weights of macromolecules.

Of greater biological importance are processes involving solutions separated by semipermeable membranes. These provide the principal reason for this chapter. All cells and intracellular compartments are bounded by such membranes. Transport of ions and molecules into and out of these compartments are critical steps in the life process. Here we shall work up from simple physical processes and equilibria to these complex systems. You will then see that it might have been better to retain the word "chemical," with its implication of molecular intricacy for such processes, than to have allowed "physical," with its more sterile implications, to replace it.

9-1 AN EXAMPLE OF PHASE EQUILIBRIA: THE VAPOR PRESSURE OF A LIQUID

The relation between the pressure of a gas and its free energy, as given by Eq. **15** of Sec. 7-3, for example, is simple and convenient. As a result, the free-energy properties of a liquid are often deduced from those of the vapor in equilibrium with the liquid. The pressure of this equilibrium vapor is the *vapor pressure* of the liquid. It is shown as a function of temperature by vapor-pressure curves like those of Fig. 9-1.

Investigation of the basis for the temperature dependence of the liquid-vapor equilibrium shown by the curves of Fig. 9-1 introduces the approaches that are taken to study physical equilibria.

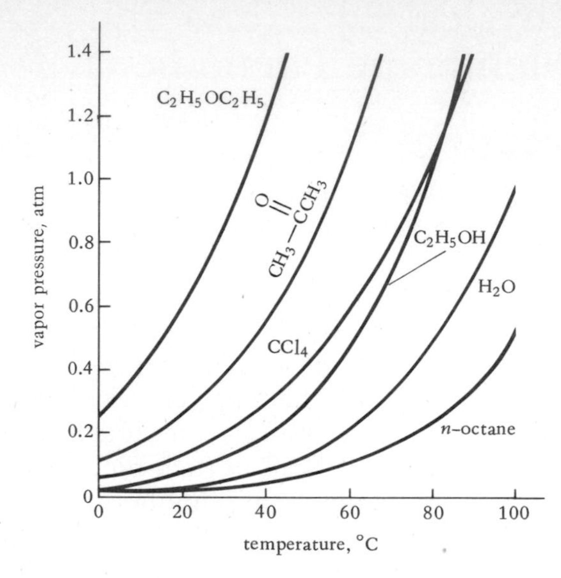

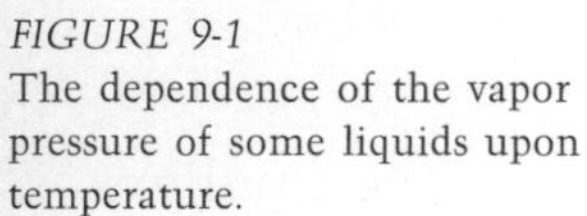
FIGURE 9-1
The dependence of the vapor pressure of some liquids upon temperature.

The free energy of 1 mol of liquid is equal to the free energy of 1 mol of the vapor that is in equilibrium with the liquid. Therefore, with subscript *l* denoting liquid and *v* denoting vapor, we can write

$$G_l = G_v \qquad \mathbf{1}$$

For an infinitesimal change in the system during which equilibrium is maintained, the differential equation

$$dG_l = dG_v \qquad \mathbf{2}$$

can be written.

Since no chemical changes occur in the liquid-vapor equilibrium we are dealing with here, only pressure and temperature changes affect the free energies.

Each free-energy change of Eq. **2** can be described by adding together the free-energy change resulting from any pressure change dP and that resulting from any temperature change dT. The rate with which the free energy changes as these two variables change is given by the partial derivatives $(\partial G/\partial P)_T$ and $(\partial G/\partial T)_P$ introduced in Sec. 7-3. The free-energy change of 1 mol of the substance in either the liquid or vapor state is thus expressed by the *total differential* as

$$dG = \left(\frac{\partial G}{\partial P}\right)_T dP + \left(\frac{\partial G}{\partial T}\right)_P dT \qquad \mathbf{3}$$

The partial derivatives are related, according to Eqs. **12** and **13** of Sec. 7-3, to the molar volume and entropy. Thus we arrive again at Eq. **9** of Sec. 7-3, and for the free-energy change in a molar amount in each phase we can write

$$dG = V\,dP - S\,dT \tag{4}$$

We recognize that although various temperatures and pressures can be considered, both phases are at the same temperature and pressure. Application of Eq. **4** to the liquid and equilibrium vapor gives

$$V_l\,dP - S_l\,dT = V_v\,dP - S_v\,dT$$

or

$$\frac{dP}{dT} = \frac{S_v - S_l}{V_v - V_l} = \frac{\Delta S_{\text{vap}}}{\Delta V_{\text{vap}}} \tag{5}$$

where the subscript "vap" indicates the vaporization process. We thus have an expression for the slope of the phase-equilibrium lines on PT diagrams like those of Fig. 9-1. The entropy of vaporization ΔS_{vap} can be expressed in terms of the heat of vaporization, as in Example 6-1, by

$$\Delta S_{\text{vap}} = S_v - S_l = \frac{\Delta H_{\text{vap}}}{T} \tag{6}$$

Substitution in Eq. **5** gives

$$\frac{dP}{dT} = \frac{\Delta H_{\text{vap}}}{T\,\Delta V_{\text{vap}}} \tag{7}$$

For liquid-vapor equilibria the liquid volume V_l can often be neglected compared with the vapor volume V_v, and with this approximation we can write

$$\frac{dP}{dT} = \frac{\Delta H_{\text{vap}}}{V_v T} \tag{8}$$

This equation is one form of the expression for the temperature-vapor pressure relation known as the *Clausius-Clapeyron equation*.

If the equilibrium vapor is treated as an ideal gas, the molar vapor volume can be expressed, from $PV = nRT$ and $n = 1$, as

$$V_v = \frac{RT}{P}$$

and substitution of this approximation in Eq. **8** gives

$$\frac{dP}{dT} = \frac{\Delta H_{\text{vap}} P}{RT^2} \tag{9}$$

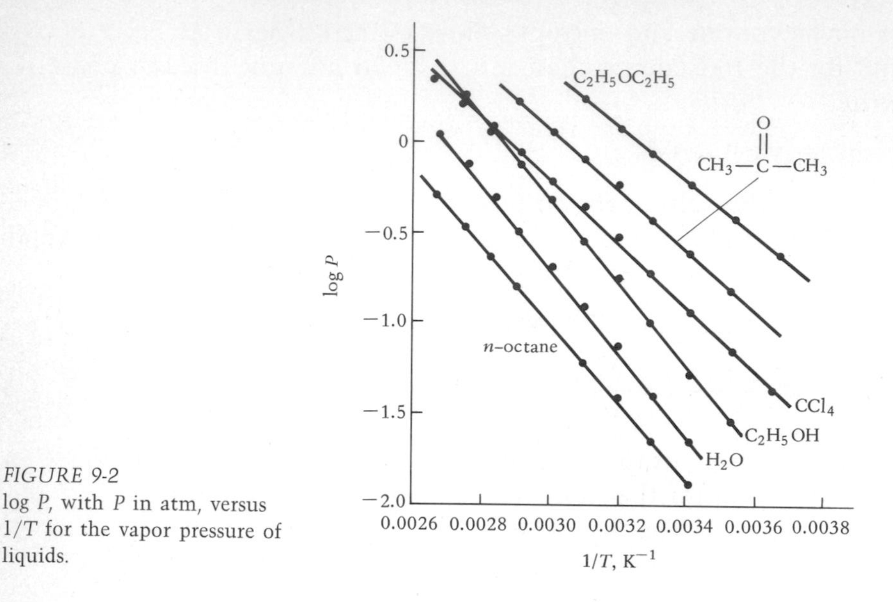

FIGURE 9-2
log P, with P in atm, versus $1/T$ for the vapor pressure of liquids.

With the relation $d \ln x = dx/x$, this result can be rearranged to the most generally used differential forms of the Clausius-Clapeyron equation,

$$\frac{d(\ln P)}{dT} = \frac{\Delta H_{vap}}{RT^2} \quad \mathbf{10}$$

and

$$\frac{d(\ln P)}{d(1/T)} = -\frac{\Delta H_{vap}}{R} \quad \mathbf{11}$$

The integrated form, assuming a constant value of ΔH_{vap} over the temperature range considered and written in terms of logarithms to the base 10, is

$$\log P = \frac{\Delta H_{vap}}{2.303R}\frac{1}{T} + \text{const} \quad \mathbf{12}$$

The preceding derivation indicates that a plot of log P versus $1/T$ should give a straight line and that the slope of such a line is to be identified with $-\Delta H_{vap}/2.303R$. As the curves of Fig. 9-2 show, essentially linear plots are obtained. A more careful look at such results, however, reveals deviations from linearity which can be attributed to the approximations introduced in obtaining Eqs. **8, 9,** and **12.**

9-2 INTRODUCTION TO PHYSICAL EQUILIBRIA INVOLVING SOLUTIONS

Now let us consider physical equilibria, i.e., equilibria set up by processes in which there is no chemical change, which involve solutions rather than one-component systems, as in Sec. 9-1. Such equilibria are involved in the *colligative properties* of solutions, the vapor-pressure lowering, the boiling-point elevation, the freezing-point depression, and the osmotic pressure. Since the osmotic pressure is of the greatest interest in biological studies, it will be used to illustrate the general treatment.

To analyze such equilibria we must be able to express the dependence of the free energy of the components of the solution not only on the pressure and temperature but also on the composition of the solution. In colligative-property studies we can simplify this step by limiting our attention to the free energy of the solvent and by considering only dilute solutions of a nonvolatile solute. We designate this solute as component B and the solvent as component A.

Colligative properties are obtained by measurements that compare the solution with the pure solvent. These properties can be related to the free-energy difference between the solvent when it is in the solution and when it is pure solvent. This free-energy difference, furthermore, can be deduced from the free-energy difference between the solvent vapor above the solution and the solvent vapor above the pure solvent. In this way we can write

$$\mathrm{G}_\mathrm{A}\,(\text{solution}) - \mathrm{G}_\mathrm{A}\,(\text{pure solvent}) = RT \ln \frac{P_\mathrm{A}(\text{solution})}{P_\mathrm{A}(\text{pure solvent})} \qquad \textbf{13}$$

In treating solvent properties we generally take the standard state to be that of pure solvent, a choice that differs from that for solutes. When the standard-state superscript is used for the solvent, Eq. **13** becomes

$$\mathrm{G}_\mathrm{A} - \mathrm{G}_\mathrm{A}^\circ = RT \ln \frac{P_\mathrm{A}}{P_\mathrm{A}^\circ} \qquad \textbf{14}$$

An empirical law, known as *Raoult's law,* can now be used to move from equilibrium vapor properties to solution concentration terms. Raoult's law, which is obeyed by ideal solutions and by all real dilute solutions, can be written

$$P_\mathrm{A} = x_\mathrm{A} P_\mathrm{A}^\circ \qquad \textbf{15}$$

where x_A is the mole fraction of solvent, i.e., number of moles of solvent divided by the number of moles of solvent plus solute. Thus the vapor pressure of the solvent falls as solute is added. Raoult's law allows us to replace P_A/P_A° in Eq. **14** by x_A, to give

$$G_A - G_A^\circ = RT \ln x_A \qquad \textbf{16}$$

Since x_A moves from unity to smaller values as solute is added, the free energy of the solvent is less in the solution than it is in the pure solvent.

The analysis of colligative properties requires an expression for the change that occurs in the solvent free energy as a result of the addition of solute to the solution. To obtain this expression we first introduce n_A as the number of moles of solvent and n_B as the number of moles of solute, so that

$$x_A = \frac{n_A}{n_A + n_B} \qquad \textbf{17}$$

and $$G_A = G_A^\circ + RT \ln \frac{n_A}{n_A + n_B} \qquad \textbf{18}$$

The quantity that will be useful is $(\partial G_A/\partial n_B)_{T,n_A}$, that is, the change in solvent free energy per mole of solute added with the temperature and the amount of solvent held constant. This derivative, which is known as the *chemical potential*, can be formed by differentiating Eq. **18** treating T and n_A as constants. We obtain

$$\begin{aligned}\left(\frac{\partial G}{\partial n_B}\right)_{T,n_A} &= RT \frac{\partial}{\partial n_B}\left(\ln \frac{n_A}{n_A + n_B}\right) \\ &= RT \frac{\partial}{\partial n_B}\left[\ln n_A - \ln (n_A + n_B)\right] \\ &= -RT \frac{\partial}{\partial n_B}\left[\ln (n_A + n_B)\right] \\ &= -\frac{RT}{n_A + n_B} \qquad \textbf{19}\end{aligned}$$

where the final step depends on the derivative relation $d \ln x = dx/x$. Thus the dependence of solvent free energy on moles of solute can be expressed as

$$dG_A = -\frac{RT}{n_A + n_B}\, dn_B \qquad \textbf{20}$$

This dependence can now be added to the temperature and pressure dependence of the free energy found in Sec. 7-3 and used for the one-component system of Sec. 9-1. For the solvent free energy we write

$$dG_A = V_A\, dP - S_A\, dT - \frac{RT}{n_A + n_B}\, dn_B \qquad \textbf{21}$$

This relation can be used to relate the pressure and temperature changes that must occur so that equilibrium is maintained when solute is added in the vapor-pressure, freezing- and boiling-point, and osmotic-pressure experiments. In the following section it will be applied to the osmotic pressure.

9-3 A COLLIGATIVE PROPERTY OF SOLUTIONS: OSMOTIC PRESSURE

Osmotic-pressure studies make use of various devices that operate on the principle shown in Fig. 9-3. The osmotic pressure is the excess pressure that must be applied to the solution so that solvent has no tendency to pass either way through a semipermeable membrane. This additional pressure increases the free energy of the solvent in the solution so that its free energy is restored to a value equal to that of the pure solvent. (The free energy of the solvent of the solution had decreased, as shown by Eq. **16,** as a result of the addition of solute.) Since the solvent in the pure solvent compartment is subject to no free-energy-changing forces, equilibrium can be maintained only if the net free-energy change of the solvent in the solution compartment is zero. Since the temperature is held constant, the $-\text{s}_\text{A}\, dT$ term of Eq. **21** drops out and we are left with the requirement

$$0 = \text{v}_\text{A}\, dP - \frac{RT}{n_\text{A} + n_\text{B}}\, dn_\text{B}$$

or

$$\text{v}_\text{A}\, dP = \frac{RT}{n_\text{A} + n_\text{B}}\, dn_\text{B} \qquad \textbf{22}$$

Integration to show the added pressure that must be applied when n_B mol of solute is added is simplified by assuming that n_B, the number of moles of solute, is small compared to n_A, the number of moles of solvent. Then since n_A is constant, the sum $n_\text{A} + n_\text{B}$ is approximately so. Further, we can assume that the molar volume of solvent v_A is a constant and has the value it has in the pure solvent. Then Eq. **22** gives

$$\text{v}_\text{A} \int_{P_\text{solvent}}^{P_\text{solution}} dP = \frac{RT}{n_\text{A} + n_\text{B}} \int_0^{n_\text{B}} dn_\text{B}$$

and

$$\text{v}_\text{A}(P_\text{solution} - P_\text{solvent}) = \frac{n_\text{B}}{n_\text{A} + n_\text{B}} RT$$

or

$$\text{v}_\text{A}\Pi = \text{x}_\text{B} RT \qquad \textbf{23}$$

where Π, the excess pressure that is applied to the solution, is the osmotic pressure.

An interesting variation of this result is obtained by multiplying

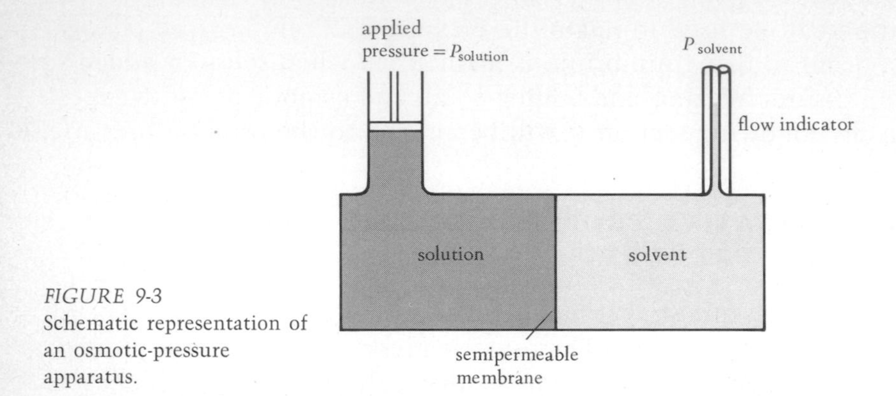

FIGURE 9-3
Schematic representation of an osmotic-pressure apparatus.

by n_A, the number of moles of solvent, which is approximately equal to $n_A + n_B$, the number of moles of solvent plus solute. This gives

$$\begin{aligned}(n_A v_A)\Pi &= n_A x_B RT \\ &= n_A \frac{n_B}{n_A + n_B} RT \\ &\cong n_B RT \qquad \mathbf{24}\end{aligned}$$

If V is introduced to represent $n_A v_A$, which is approximately the volume of the solution that contains the n_B mol of solute, we can write

$$\Pi V = n_B RT \qquad \mathbf{25}$$

The important application of Eqs. **23** and **25** to molecular-weight measurements will be dealt with in the following section.

The similarity of Eq. **25** to the ideal-gas law led van't Hoff and others to view the osmotic pressure as arising from a molecular-bombardment process. These ideas have not proved fruitful, however, and it is recognized here that Eq. **25** is merely an approximate form obtained from the thermodynamic dilute-solution result, Eq. **23.**

A comparison of the observed osmotic pressure as a function of concentration with the behavior expected on the basis of the derived expressions is shown in Table 9-1.

To this discussion of the osmotic pressure as a colligative property can be added a few comments on the mechanism by which a semipermeable membrane operates. In some cases, the membrane seems to act simply as a mechanical sieve, letting small molecules, like water, through and preventing the passage of large molecules. Other membranes do not appear to pass and reject molecules on a simple size basis. Sometimes a component appears to penetrate the membrane by dissolving in it, whereas another component that is not soluble in the membrane cannot pass through it.

Probably the clearest example of this is the passage of hydrogen through a palladium sheet. The hydrogen molecules probably dissociated into atoms on the surface of the palladium. These atoms can penetrate through the solid lattice, and on the opposite surface they can reunite into hydrogen molecules. Other molecules are not dissociated and cannot pass through the solid.

The mechanism of the process at the semipermeable membrane is a study that is quite separate from the subject of colligative properties. The expressions for the osmotic pressure derived in this section apply as long as a membrane is available that will pass solvent and will not pass solute. The procedure by which it accomplishes this is immaterial as far as the resulting osmotic pressure and the thermodynamic development are concerned. The mechanism of passage through semipermeable membranes is a valid topic for study, but it belongs in the discussion of rates and mechanisms of reactions and processes (Chaps. 10 to 12).

9-4 MOLECULAR WEIGHTS FROM OSMOTIC-PRESSURE MEASUREMENTS

The principal use of colligative-property measurements is in determining molecular weight. Osmotic-pressure measurements are particularly suitable for the study of macromolecules, both synthetic and natural.

TABLE 9-1
Osmotic Pressure of Aqueous Solutions of Sucrose at 20°C†

				calculated osmotic pressure	
molal concentration	**mole fraction of sucrose** x_B	**molar concentration**	**observed osmotic pressure, atm**	**from Eq. 23**	**from Eq. 25**
0.1	0.00180	0.098	2.59	2.40	2.36
0.2	0.00360	0.192	5.06	4.81	4.63
0.3	0.00540	0.282	7.61	7.21	6.80
0.4	0.00720	0.370	10.14	9.62	8.90
0.5	0.00901	0.453	12.75	12.0	10.9
0.6	0.0108	0.533	15.39	14.4	12.8
0.7	0.0126	0.610	18.13	16.8	14.7
0.8	0.0144	0.685	20.91	19.2	16.5
0.9	0.0162	0.757	23.72	21.6	18.2
1.0	0.0180	0.825	26.64	24.0	19.8

†Osmotic-pressure data of Morse, reported by A. Findlay, "Osmotic Pressure," Longmans, Green & Co., Inc., New York, 1919.

The result, Eq. **25,** obtained in Sec. 9-3 is valid in the limit of infinite dilution and can conveniently be used to obtain a form that allows molecular weights to be deduced. Recall that V is the volume of solution containing n_B moles of solute. We can write n_B as w_B/M_B, that is, the weight of solute divided by its molecular weight. Then Eq. **25** can be rearranged to

$$\Pi = \frac{w_B}{V}\frac{RT}{M} \qquad \textbf{26}$$

Further, w_B/V is recognized as the concentration of the solution on a weight-per-volume basis. If we introduce c for this concentration, we obtain

$$\frac{\Pi}{c} = \frac{RT}{M} \qquad \textbf{27}$$

With this equation measured values of Π/c can be extrapolated to zero concentration to give a value of RT/M that yields a valid value for M. (Generally c is used with the units of g/ml and Π with the units of atmospheres. Then R must be given the value 82.06 ml atm/deg mol.)

Some results from osmotic-pressure measurements on solutions of polyisobutylene in a "good" and a "bad" solvent are shown in Table 9-2. The graphical treatment of Fig. 9-4 shows that both solvents yield the same limiting value of Π/c, namely, 0.097 atm ml/g. Then Eq. **27** gives

$$M = \lim_{c\to 0} \frac{RT}{\Pi/c}$$

$$= \frac{82.06(298)}{0.097} = 250,000 \text{ g}$$

TABLE 9-2
Osmotic-Pressure Results at 25°C for a Polyisobutylene Fraction†

	Π, **atm**		$\frac{\Pi}{c}$, **atm ml/g**	
concentration, g/ml	**in benzene**	**in cyclohexane**	**in benzene**	**in cyclohexane**
0.0200	0.00208	0.0117	0.104	0.585
0.0150	0.00152	0.0066	0.101	0.44
0.0100	0.00099	0.0030	0.099	0.30
0.0075		0.00173		0.23
0.0050	0.00049	0.00090	0.098	0.18
0.0025		0.00035		0.14

†Calculated from the data of P. J. Flory, *J. Am. Chem. Soc.*, **65:**372 (1943).

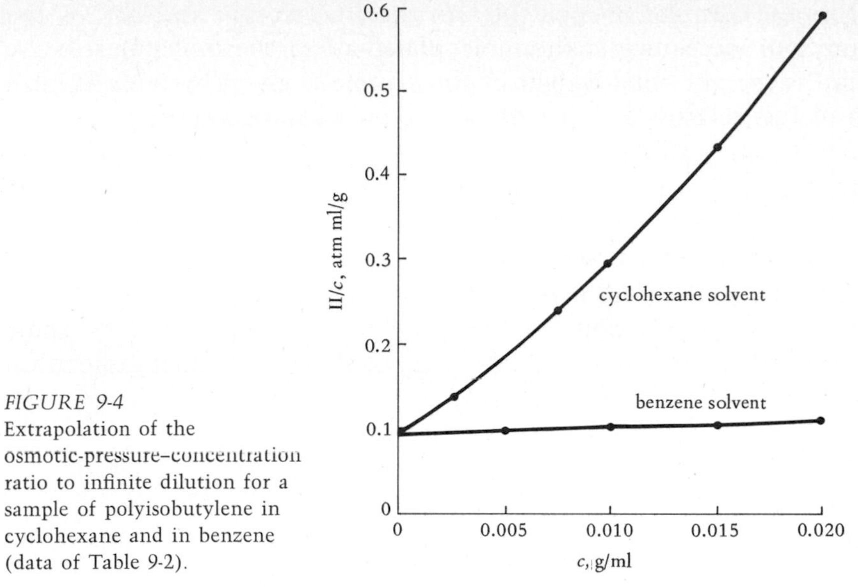

FIGURE 9-4
Extrapolation of the osmotic-pressure–concentration ratio to infinite dilution for a sample of polyisobutylene in cyclohexane and in benzene (data of Table 9-2).

Osmotic pressures provide a reliable and generally available route to the molecular weights of such high-molecular-weight materials.

Some special care must be taken in interpreting the molecular weight deduced from measurements on materials that have a distribution of molecular weights. Mixtures of molecular species, partially degraded high-molecular-weight materials, and batches of synthetic polymers fall in this category.

Osmotic pressure and other colligative studies produce information on the number of solute particles. The only solute features entering Eq. **23** or **25** are quantities that depend on the number of particles. The osmotic pressure is indifferent to the size, shape, charge, and so forth of the solute particles, at least in the dilute-solution limit where interaction effects are avoided. It follows that osmotic-pressure measurements give a value for M, the average weight of a mole, that is equal to the weight of sample divided by the total number of moles in the sample. Thus

$$M = \frac{w}{\text{No. of moles}} \qquad \text{or} \qquad \text{No. of moles} = \frac{w}{M}$$

and $$\text{No. of molecules} = \mathfrak{N}\frac{w}{M}$$

We thus can write

$$\mathfrak{N}\frac{w}{M} = \Sigma N_i \tag{28}$$

where N_i is the number of molecules in the ith fraction and ΣN_i is the total number of molecules. If the molecular mass of the molecules in the ith fraction is m_i, the total weight of the sample is given by $w = \Sigma N_i m_i$. Insertion of this relation into Eq. **28** yields, on rearrangement,

$$M = \mathfrak{N}\frac{\Sigma N_i m_i}{\Sigma N_i} \qquad \mathbf{29}$$

Such an average molecular weight is known as a *number average*. It is usually indicated by a more descriptive symbol such as $\bar{M}_n$.

If the various fractions in the sample are characterized by their molar weights M_i rather than the masses m_i of their individual molecules, Eq. **29** becomes, with $\bar{M}_n$ in place of M,

$$\bar{M}_n = \frac{\Sigma N_i M_i}{\Sigma N_i}$$

or

$$\bar{M}_n = \Sigma\left(\frac{N_i}{\Sigma N_i}\right)M_i$$

$$= x_i M_i \qquad \mathbf{30}$$

where x_i is the number fraction, or mole fraction, of the ith species in the sample.

In contrast, in other experiments, such as light scattering, each particle makes a contribution that is proportional to its size or molecular weight. Then the average molecular weight that is obtained is the *weight average*, defined as

$$\bar{M}_m = \mathfrak{N}\frac{\Sigma N_i m_i^2}{\Sigma N_i m_i}$$

$$= \frac{\Sigma (N_i m_i) M_i}{\Sigma N_i m_i}$$

$$= \sum\left(\frac{N_i m_i}{\Sigma N_i m_i}\right)M_i$$

$$= \Sigma w_i M_i \qquad \mathbf{31}$$

where w_i is the weight fraction of the ith component of the sample.

For appreciable distribution of masses, $\bar{M}_n$ and $\bar{M}_m$ can be significantly different.

9-5 DIALYSIS EQUILIBRIA

Colligative properties, e.g., osmotic pressure, depend on the modification of the solvent properties as a result of the addition of solute. More often it is the role of the solute or solutes that is of direct interest.

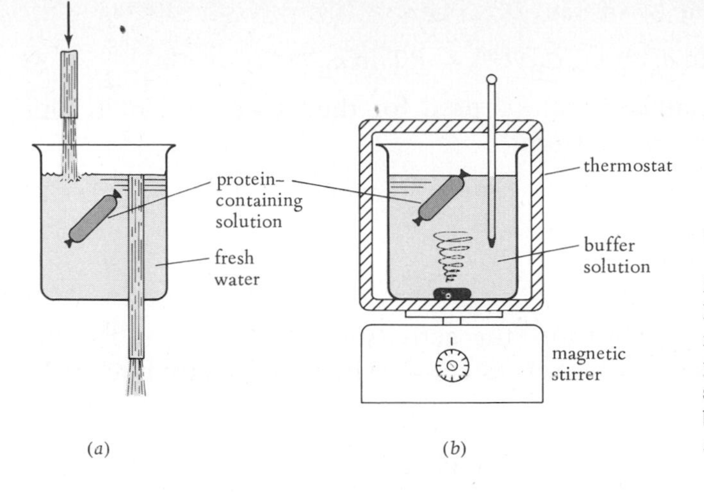

FIGURE 9-5
Dialysis arrangements for (*a*) removing small ions from a protein solution and (*b*) studying the equilibrium between a protein solution and a buffer solution.

Equilibria between solutes in different compartments are set up when the membrane dividing the compartments is permeable to the solutes. Most interesting and most complex are the membranes and the state of balance reached by various solute species in the cellular compartments of living systems. Solute equilibrium through semipermeable membranes is also important in a variety of laboratory studies. The most common practical example is the use of *dialysis* to remove small ionic species from a solution containing proteins and salts. This is simply done, as illustrated in Fig. 9-5, by placing the protein or other macromolecular material in a membrane bag that is permeable to simple ionic species. Cellophane is often a suitable material. Much of the salt can be removed by exposing the membrane bag to fresh water. Alternatively, the equilibrium between the ions in the protein compartment and those in the external fluid can be studied by allowing the two membrane-separated solutions to reach equilibrium.

In dialysis systems net transport of the species that can permeate the membrane will occur until equilibrium is established. At this stage, where there is no remaining driving force for net transport, the molar free energies of the species in the two compartments will be equal. We denote the solute species by B and the two compartments separated by a membrane permeable to B by left and right. When this solute establishes equilibrium, we can write the free-energy equality

$$G_B^{\text{left}} = G_B^{\text{right}} \qquad \textbf{32}$$

If we assume that the temperature and pressure are constant and uniform throughout the system, the free-energy terms can be described, following

Eq. **38** of Sec. 7-6 or Eq. **39** of Sec. 7-7. Then

$$(G_B^\circ)^{\text{left}} + RT \ln a_B{}^{\text{left}} = (G_B^\circ)^{\text{right}} + RT \ln a_B{}^{\text{right}} \qquad \mathbf{33}$$

If a single common standard state is used for the free energy of B, this equation shows that at equilibrium

$$a_B{}^{\text{left}} = a_B{}^{\text{right}} \qquad \mathbf{34}$$

or with activity coefficients

$$\gamma_B{}^{\text{left}} c_B{}^{\text{left}} = \gamma_B{}^{\text{right}} c_B{}^{\text{right}} \qquad \mathbf{35}$$

In the simplest application, the activity coefficients of the solute in the two compartments are taken as equal. Then Eq. **35** simplifies to

$$c_B{}^{\text{left}} = c_B{}^{\text{right}} \qquad \mathbf{36}$$

We have come by means of this thermodynamic formalism to the easily recognized result that at equilibrium the concentration of species B will be the same in the two compartments.

In a dialysis experiment one of the solutions may contain an appreciable concentration of substances that cannot penetrate the membrane. Then a solute that can move through the membrane will be exposed to the influence of these substances in one compartment but not in the other. This situation, common in dialysis studies of macromolecular substances, upsets the simple equal-concentration result in a number of ways. Two of these ways are mentioned here; a third will be treated in the following section.

First, the assumption of equal activity coefficients for species B in the two compartments may not be valid. Then we must proceed with an estimation of activity coefficients for B in the two compartments and work with the activity equality of Eq. **34** or **35** rather than the concentration equality of Eq. **36.**

Second, the nonpenetrating macromolecular substance might bind the solute species B so that the concentration of free B must be distinguished from the total concentration of B. Suppose that the right compartment contains a nonpenetrating substance that binds the penetrating solute B. Thus, if the procedure for analyzing B in the two compartments produces values of $c_{B,\text{total}}$, we have for the left compartment $c_{B,\text{total}} = c_{B,\text{free}}$, but for the right compartment $c_{B,\text{total}} = c_{B,\text{free}} + c_{B,\text{bound}}$. The concentration of the free unbound B species in the two compartments approximates the activities. Then Eq. **36** yields

$$c_{B,\text{total}}^{\text{left}} = c_{B,\text{total}}^{\text{right}} - c_{B,\text{bound}}^{\text{right}} \qquad \mathbf{37}$$

Measurements of $c_{B,\text{total}}^{\text{left}}$ and $c_{B,\text{total}}^{\text{right}}$ can then yield a value for the amount of B bound to the nonmigrating species in the right compartment. Such studies are used to investigate the binding of various species to proteins.

9-6 DONNAN MEMBRANE EQUILIBRIA

We have seen that laboratory studies of macromolecule solutions, as in osmotic-pressure and dialysis studies, confine the macromolecules to one compartment while allowing passage of small ions or solvent in or out of that compartment. Much of the transport occurring in cells and cell compartments in living systems can be similarly described. In all such cases the equilibrium state that would be reached as a result of the net transport of the small ions can be markedly affected if the macromolecule carries a charge, as is generally the case.

The studies of Chap. 4 showed that, except at the isoionic pH, proteins and nucleic acids carry a charge as a result of a net gain or loss of protons. Additional charges are acquired by the binding of other species, e.g., the binding of Mg^{2+} ions by phosphate groups. Thus, macromolecules in laboratory or biological systems generally carry a charge. The overall electrical neutrality of the solution results from a corresponding opposite charge contributed by ions, called *counterions,* included in the remaining ionic makeup of the solution.

Suppose such a macromolecule or, more specifically, a protein solution is separated from pure water by a semipermeable membrane that allows passage of small ions but prohibits the passage of protein molecules. Such a situation could arise in an osmotic-pressure study or in the dialysis of the protein solution. Suppose the protein carries a net negative charge and that Na^+ ions are the counterions. The Na^+ ions will tend to diffuse to the low-concentration region of initially pure water. Electrical neutrality would be lost and this process prevented if it were not for the dissociation of water. This occurs, and H^+ tend to accumulate on the protein side of the membrane while the corresponding OH^- accumulate, along with the Na^+ ions, on the side of the initially pure water. Unless both solutions are buffered, pH changes will occur to upset the osmotic-pressure or dialysis experiment.

In such ways we are led to deal with the equilibrium between protein solutions, which are often themselves buffered, and buffer solutions. The complications that then arise can be illustrated by considering the simplest situation of the protein–sodium-ion solution separated by a semipermeable membrane from a sodium chloride solution.

Suppose the protein species P carries a negative charge of $-z$. The neutrality of the solution is achieved by the presence of z positive charges,

Na^+ ions for example, for each protein molecule. At the start of a dialysis or osmotic-pressure study, if the protein concentration is c_P, as in Fig. 9-6, the initial Na^+ concentration in the protein compartment is zc_P.

Separated from this protein–sodium-ion solution, as in Fig. 9-6, is a salt solution with Na^+ and Cl^- ions at an initial salt concentration c_S.

Let us suppose that when equilibrium has been reached, an amount of chloride has diffused into the protein–sodium-ion compartment so that the chloride-ion concentration there has reached the value x. To maintain charge neutrality, an equal amount of sodium ions must accompany the chloride ions. (We shall assume that this effect overwhelms any H^+ or OH^- buildup that could result from water dissociation.) As a result, the concentrations of the species in the two compartments are those shown in Fig. 9-6.

The value of the unknown x in the expressions of Fig. 9-6 must now be deduced. When equilibrium is reached, the species that can penetrate the membrane will have equal activities in the two compartments, according to Eq. **34.** Here it is sodium chloride that establishes equilibrium, and thus the activity a_{NaCl} must be the same in the two compartments. Further, according to Sec. 7-7, a_{NaCl} can be interpreted as $a_{Na^+}a_{Cl^-}$ or as $(\gamma_\pm)^2 c_{Na^+}c_{Cl^-}$. If such expressions are written for the two compartments, and if the mean-activity term is assumed to be equal in the two compartments, the condition for equilibrium can be described by

$$(zc_P + x)(x) = (c_S - x)(c_S - x) \qquad \textbf{38}$$

Rearrangement leads to x, the concentration of chloride that develops in the protein compartment

$$x = \frac{c_S^2}{zc_P + 2c_S} \qquad \textbf{39}$$

At large salt concentrations the effect of the protein is overwhelmed and $x \simeq \frac{1}{2}c_S$. The two compartments achieve equal salt concentrations. On

	salt compartment		protein compartment		
species	Na^+	Cl^-	Na^+	Cl^-	P^{z-}
initial concentration	c_S	c_S	zc_P	0	c_P
change in concentration to reach equilibrium	$-x$	$-x$	$+x$	$+x$	
equilibrium concentration	$c_S - x$	$c_S - x$	$zc_P + x$	x	c_P

FIGURE 9-6
Species concentrations in a Donnan membrane equilibrium study.

the other hand, at large protein concentrations, the passage of salt into the protein compartment is prevented, even though this entails the rejection of the chloride ion by a solution that contains none of that ion. Intermediate situations are illustrated by values calculated from Eq. **39** and shown in Table 9-3. Some chloride migrates into the protein compartment, along with the corresponding Na^+ that must accompany it. The idea that no NaCl would diffuse into the protein compartment and that the NaCl would diffuse until the inside and outside concentrations were equal is incorrect. The equilibrium, an example of the Donnan membrane equilibrium, is deduced from Eq. **39.**

The effect of various concentrations of protein and electrolyte are shown in Table 9-3. Only at high electrolyte concentrations relative to the protein concentration is the effect of the confined charged protein small. Therefore many studies of proteins or other polyelectrolytes in solution are made at high electrolyte concentration and at a pH near the isoionic point.

Many extensions and refinements of the development that led to Eq. **39** are possible, and often necessary. Recognition of water dissociation and pH changes introduces additional complications. Even more difficult are the problems that arise from the nonideality of the two solutions and its effect on the activities or activity coefficients of the charged species.

9-7 INDIVIDUAL ION TRANSPORT AND MEMBRANE POTENTIALS

Biological membranes exist, and inorganic electrode materials can be produced that allow the passage of a single ionic species. In biological systems, we often consider the transport of K^+ ions and the resulting electric potential across a membrane, e.g., that of a nerve cell. Synthetic ion-

TABLE 9-3
Donnan Membrane Equilibrium Results Calculated from Eq. **39** for $z = 1$

initially		**at equilibrium**				
salt compartment	**protein compartment**	**salt compartment**		**protein compartment**		
$c_S = [Na^+] = [Cl^-]$	$c_P = [P^-] = [Na^+]$	$[Na^+] = c_S - x$	$[Cl^-] = c_S - x$	$[Na^+] = c_P + x$	$[Cl^-] = x$	$[P^-] = c_P$
1.0	0	0.5	0.5	0.5	0.5	0
1.0	0.01	0.502	0.502	0.508	0.498	0.001
1.0	0.1	0.52	0.52	0.58	0.48	0.1
1.0	1.0	0.67	0.67	1.33	0.33	1.0

permeable membranes find their applications in *ion-selective electrodes.* With these electrodes the concentration of a particular ionic species can be deduced from the measured electric potential that results when the solution to be analyzed and a reference solution are separated by the membrane. pH meters depend on a H^+-permeable glass. Electrodes are now available that are permeable to, and therefore sensitive to, many other types of ions.

The relation between ion migration and the development of an electric potential can be seen by considering a membrane permeable to K^+ ions. If this membrane separates solutions with different K^+ activities or, approximately, concentrations, there will be a tendency for K^+ ions to diffuse from the high-concentration compartment to the low-concentration compartment. Balance is brought about, and the equilibrium distribution is reached as a result of the opposing forces of the charge imbalance that develops. (This situation is in contrast to that treated in the preceding section, where the entire electrolyte could penetrate the membrane and equilibrium was established as a result of the contributions of both ions of the electrolyte to the activity of the electrolyte.)

The charge imbalance can be expressed in terms of a difference $\Delta\Phi$ in electric potential on the two sides of the membrane.

At equilibrium the activity or, more simply, concentration-driving force is balanced by the opposing electric potential. In terms of free energies, this balance can be expressed, for an ion of charge z for which the molar charge is $z\mathfrak{F}$, by

$$RT \ln \frac{c_2}{c_1} = z\mathfrak{F}\,\Delta\Phi \qquad \textbf{40}$$

A typical K^+ ratio concentration for intra- and extracellular fluids is 20:1. A representative membrane potential value can then be calculated from Eq. **40** at the physiological temperature of 37°C as

$$\begin{aligned}\Delta\Phi &= \frac{RT}{z\mathfrak{F}} \ln \frac{c_2}{c_1}\\ &= \frac{(1.987\ \text{cal/deg mol})(4.184\ \text{J/cal})(310\ \text{deg})}{1 \times 96{,}500\ \text{J}} \times 2.303 \log 20\\ &= 0.080\ \text{V}\\ &= 80\ \text{mV} \qquad \textbf{41}\end{aligned}$$

This result is similar to the electric-potential differences observed in biological systems.

It must be pointed out that many difficulties have been ignored in the development of Eq. **40** and that agreement with directly measured potentials is, at best, approximate. Activity effects must be expected to be appreciable, and pH changes may also turn out to be significant. Some penetration of most membranes by more than a single ion type is also likely. Furthermore, the experimentally measured potential, obtained by inserting inert electrodes in the two membrane-separated solutions, is not clearly or simply related to the potential difference experienced by the principal permeating ion. Still, even the qualitative idea that membranes permeable to particular ionic species can lead to potential differences is of value, and the semiquantitative results from Eq. **40** are guides to the magnitudes of the effects.

9-8 ACTIVE TRANSPORT

The analyses of the preceding sections have shown some of the complications arising when membrane action and the membrane-separated solutions exhibit the variety of which they are capable. These complications stand in the way of a complete and precise treatment of the crucial biological processes in which reagents are moved through membranes into or out of cells and cell compartments. But even incomplete and approximate treatments are revealing.

Many important biological processes require the movement of an ion, the ions of an electrolyte, or a molecule through a membrane in the direction opposite that in which the species would move spontaneously. When free energy is used to treat the driving force of the processes, we say that the movement is against the free-energy change, or gradient, for that process. An example is the movement of amino acids from the relatively dilute solution of blood plasma into solutions within some cells which are already concentrated with respect to that amino acid. Another is the formation of the approximately 1 *M* HCl solution, which ultimately is secreted into gastric juice, by driving HCl from blood plasma, at pH about 7, into *parietal cells.* A final example is the transport of K^+ ions into a great variety of cells from the K^+-poor, Na^+-rich extracellular fluids.

Calculation of the actual free-energy gradient against which such processes are driven would require detailed information on the activity coefficients of the species, the extent to which they are bound to other species, the Donnan effect, and any electric-potential effects. Ignoring these effects leaves us with the qualitatively useful free-energy–concentration relation

$$G_2 - G_1 = \Delta G = RT \ln \frac{c_2}{c_1} \qquad \textbf{42}$$

For 37°C, insertion of numerical values converts this to

$$\Delta G = 1.987 \times 310 \times 2.303 \log \frac{c_2}{c_1}$$

$$= 1420 \log \frac{c_2}{c_1} \qquad \text{cal}$$

$$= 1.42 \log \frac{c_2}{c_1} \qquad \text{kcal} \qquad \textbf{43}$$

Thus, for a concentration gradient of 10:1 to be overcome, a free-energy driving force of about 1.4 kcal/mol is needed. For a 100:1 concentration gradient, a 2.8-kcal free-energy driving force is needed, and so forth.

Driving forces of this magnitude are generally provided by coupling to the same ATP-to-ADP reaction (Sec. 8-6) that drives many of the reluctant chemical reactions of biological systems. The equilibrium position for the ATP hydrolysis reaction is far in the direction of ADP and phosphate ion, as shown in Figs. 8-7 and 8-8. Thus, a suitable driving force is available if a free-energy storage process has led to the accumulation of more than the equilibrium amount of ATP and if the membrane can provide a coupling mechanism between an ATP phosphate-donation reaction and the reluctant transport process. Such a transport process, which requires the development of a free-energy driving force by the membrane, is said to be *active transport*.

PROBLEMS

1 If the osmotic pressures of Table 9-2 were measured with a mercury manometer, what accuracy of mercury-column-height measurement would be necessary to give data good to about 10 percent? What accuracy in reading the height would be necessary if a water manometer were used?

2 a Assuming that Raoult's law is applicable, calculate the vapor pressure of a 1 *m* solution of sucrose in water, i.e., a solution made up by dissolving 1 mol of sucrose in 1000 g of water at 30°C. The vapor pressure of water at this temperature is 31.82 mm Hg. Compare the calculated solution vapor pressure with the observed value of 31.20 mm Hg.

b A 2 *m* sucrose solution shows, at 30°C, a water vapor pressure of 30.50 mm Hg. Compare the error that results from assuming Raoult's law for 2 *m* solution with the error for a 1 *m* solution and thus illustrate the fact that Raoult's law holds as infinite dilution is approached.

3 At 100°C the vapor pressure of an aqueous solution of 0.5 mol of ammonium nitrate in 1 liter of water is 12.8 mm Hg less than that of pure water. Assuming

Raoult's law, calculate the mole fraction of the solute in the solution. To how many moles of solute per liter of water does this correspond? What is the basis for the major part of the discrepancy from the 0.5-mol value?

4 The following osmotic-pressure values have been reported for sucrose in water at 10°C:

sucrose, g per gram of water	osmotic pressure, atm
0.103	7.18
0.171	12.30
0.342	25.92

Using an extrapolation procedure like that of Fig. 9-4, calculate the molecular weight of sucrose.

5 The osmotic pressure of human blood versus water varies from about 7.2 atm, which it may be early in the morning, to about 8.0 atm, to which it may rise after a big meal. These values are for a blood temperature of 37°C. Using an average value, calculate the mole fraction of the solutes and the number of moles of solutes per 1000 g of water. Compare with information you can find on the composition of blood plasma. (Discrepancies will enter because blood plasma is too concentrated for the dilute-solution osmotic pressure relations to hold.)

6 A normal saline solution contains 9 g of NaCl per liter of water. Estimate its osmotic pressure at 37°C and compare with the osmotic pressures of blood given in Prob. 5. (Again the dilute-solution equations are not completely applicable.)

7 The process of *reverse osmosis* for preparing fresh water from salt water (seawater for example) depends on applying a pressure greater than the osmotic pressure to the solution. In this way the solvent is driven to flow from the solution to the pure solvent. The concentrations in moles per kilogram of seawater of the principal solutes of seawater are Cl^-, 0.546; Na^+, 0.456; Mg^{2+}, 0.053; SO_4^{2-}, 0.028; and Ca^{2+}, 0.010. What pressure on a seawater sample confined by a membrane permeable to water would have to be exceeded to produce reverse osmosis?

8 What are the number-average and weight-average molecular weights of (*a*) a macromolecule sample that contains equal numbers of molecules with mass 100,000 and 200,000 and (*b*) a macromolecule sample that contains equal masses of molecules with mass 100,000 and 200,000?

9 What are the number-average and weight-average molecular weights of a sample that contains equal numbers of molecules with masses 10,000, 50,000, and 100,000 amu, or daltons?

REFERENCES

Phase Equilibria and Colligative Properties

CASTELLAN, G. W.: "Physical Chemistry," chaps. 12 to 16, Addison-Wesley Publishing Company, Inc., Reading, Mass., 1971.

Membrane Equilibria

VAN HOLDE, K. E.: "Physical Biochemistry," pp. 37–47, Prentice-Hall, Inc., Englewood Cliffs, N.J., 1971.

Transport across Membranes

STEIN, W. D.: "Movement of Molecules across Cell Membranes," Academic Press, Inc., New York, 1967.

ALBERS, R. S.: Biochemical Aspects of Active Transport, *Ann. Rev. Biochem,* **36:**727 (1967). A review, emphasizing Na^+ and K^+ transport.

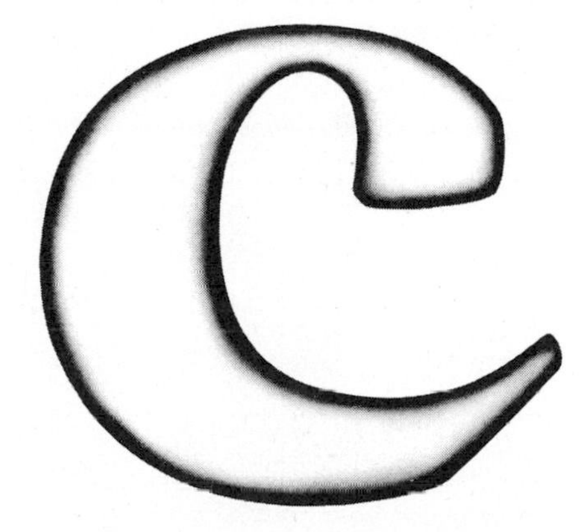

Kinetics

Thermodynamics organizes our knowledge about what can and what cannot happen. But what does happen, how does it happen, and how long does it take?

We usually have to depend on direct observation to see what happens and how long it takes. The few generalizations that stem from these *rate,* or *kinetics,* studies are not nearly as powerful as those of thermodynamics. Nevertheless, results from such studies often guide us to the mechanism by which the process occurs.

The major area we shall consider is that of chemical reactions. The treatments that are developed to answer what happens and how

long it takes are illustrated in Chap. 10. Attempts to answer how it happens are the subject of Chap. 11, where the value and limitation of rate-of-reaction studies in leading us to detailed mechanisms of chemical reactions will become apparent.

Parallel rate studies and molecular-level interpretations are applied in the third and final chapter to some physical processes. These processes involve the movement of macromolecules in solutions. The most familiar example of such processes is diffusion. Transport processes are appropriate to a biophysical chemistry study on two counts: (1) living systems depend on extensive and complex transportation of substances through body and cellular fluids and through the membranes that separate these fluids, and (2) laboratory studies of transport in solutions of macromolecules provide us with additional information on the size and shape of these molecules in solution. Much of our knowledge of proteins and their behavior in solution comes from such studies.

10 Rates of Chemical Reactions

Our study of the molecular world has so far been concerned with only two of the broad areas which interest chemists, structure and energetics, or equilibria. Now, the details of chemical reactions and physical processes are investigated. Our attention is focused not only on the reactants and products, i.e., the initial and final states, but also on the details of the transformation. That this aspect has previously been avoided is seen by the absence of the time variable in the preceding studies. It will now play a major role.

How does a process occur? How are reactants converted into products in some particular chemical reaction? These questions call for answers in terms of molecular-level happenings. The answers are given in part by a statement of the sequence of simple, or elementary, steps that are thought to constitute most overall reactions or processes. The principal measurements on which our ideas of such reaction sequences are based are those showing the dependence of the rate at which the reactions or processes occur on the concentrations of the reagents. Results of such measurements are usually summarized in a *rate equation* for the reaction. The development of rate equations for a variety of reaction types is the primary goal of this chapter.

In Chap. 11 the relation of these experimental results to ideas about the sequence of reactions (known as the *mechanism*) will be investigated.

10-1 MEASUREMENT OF THE RATES OF CHEMICAL REACTIONS

Just as there are a great variety of chemical reactions and a correspondingly great range of reaction times, from apparently instantaneous to imperceptibly slow, so there are a great variety of ways of obtaining quantitative data on the rate of reactions.

For a reaction that is slow relative to the time it takes to remove a sample from the reaction mixture and to perform an analysis, the data from which the rate equation is deduced can be obtained by any analytical method, physical or chemical, that is applicable to the particular system. When the reaction is relatively rapid, the time of sampling and analysis is appreciable and the analytical results are not easily related to any well-defined reaction time. One procedure that is then used is the rapid cooling of a sample, a step which generally slows the reaction. In other cases one can quickly dilute a sample of the reaction system. Reactions which proceed rapidly only in the presence of a catalyst can be slowed down by the removal of the catalyst. A common example is an acid-catalyzed reaction that is frozen by dilution or neutralization of the acid.

Often more convenient than the withdrawal and analysis of samples is some physical measurement which is applicable to the reaction mixture and from which the concentrations of a reactant or product can be deduced. Any physical property can be used so long as it changes as the reaction mixture changes from reactants to products. Thus, one might use the volume change in a constant-pressure system, the conductivity change that accompanies many ionic-solution reactions, or the optical rotation when the reactants or products are optically active.

The most satisfactory physical properties, however, are more definitely related to a reactant or a product. If one species in the reaction system, for example, absorbs radiation at a particular wavelength in the infrared, visible, or ultraviolet region, the amount of absorption at that wavelength can be used to measure the concentration of that species. Measurements of such quantities are less susceptible to interference from side reactions than the physical properties listed above.

If a suitably rapid physical-measurement method is available, it is often the time required for the mixing process itself that is limiting. Much progress has been made by using a flow system in which the two reagents flow together in a T arrangement and a physical property, such as electric conductivity or ultraviolet absorption, is measured at various positions along the united stream. In this way, reactions that proceed appreciably in times as short as a thousandth of a second can be studied.

Even more rapid reactions, however, interest the chemist. The reaction of an acid and a base, e.g., the combination of H^+ and OH^-, is often said to be instantaneous. A number of methods for studying such fast reactions have been developed. Those known as *relaxation methods* depend upon disturbing a reaction system from equilibrium, as by the sudden imposition of an electric field or a high pressure, and determining the time required for the system to relax, i.e., return to a new equilibrium state. It appears that reactions that occur almost as rapidly as molecules vibrate, i.e., in times down to 10^{-13} sec, are not beyond the realm in which the modern kineticist can work.

How the data obtained from chemical or physical measurements on a reacting system at various time intervals are used to deduce the rate of the reaction and its rate equation will be illustrated by specific examples in the following sections.

10-2 INTRODUCTION TO RATE EQUATIONS

The rate of a reaction is usually expressed in terms of the decrease in the amount of one of the reactants as a function of time. Alternatively, the increase in the amount of a product can be used. If the reaction system is one of constant or near-constant volume, the change in the amount of reagent will correspond to a change in the concentration of that reagent. For liquid systems one usually states the rate of a reaction in terms of the rate of change of the molar concentration of a reagent or of the number of moles of a reagent. The time units that enter into the statement of the rate of a reaction may be seconds, minutes, hours, and so forth.

The rate of a reaction is always considered to be a positive quantity. Thus, for the reaction

$$A + B \rightarrow C + D \qquad \mathbf{1}$$

the rate, which generally is a function of concentrations and temperature, can be given in terms of the decrease in A by writing

$$\text{Rate} = -\frac{d[A]}{dt} \qquad \mathbf{2}$$

where [A] is the molar concentration of A. Alternatively, if one measures the increase in the concentration of C, one might write

$$\text{Rate} = +\frac{d[C]}{dt} \qquad \mathbf{3}$$

Some care must be taken with other reactions such as those of the type

$$A + 2B \rightarrow \text{products} \qquad \mathbf{4}$$

Depending on whether A or B is followed experimentally, one might write the rate as

$$-\frac{d[A]}{dt} \quad \text{or} \quad -\frac{d[B]}{dt} \qquad \mathbf{5}$$

Since two molecules of B are used up for every molecule of A, the second derivative will be twice as large as the first.

Many reactions have rates that at a given temperature are proportional to the concentration of one or two of the reactants with each reactant raised to a small integral power. If reactions are considered in which A and B represent possible reactants, the rate equations for reactions with such concentration dependence would be of the form

$$\text{Rate} = k[\text{A}] \qquad\qquad \text{first order}$$

$$\text{Rate} = k[\text{A}]^2 \quad \text{or} \quad k[\text{A}][\text{B}] \qquad \text{second order} \tag{6}$$

The proportionality constants k which appear in such rate laws are *rate constants.*

Reactions that proceed according to such simple rate equations are said to be of the *first or second order,* as indicated. As we shall see, not all reactions have such simple rate laws. Some involve concentrations raised to nonintegral powers; others consist of more elaborate algebraic expressions. But enough reactions under certain conditions are simple first or second order to make the idea of the *order of a reaction* useful.

It should be pointed out immediately that the rate law has no necessary relation to the form of the equation for the overall reaction. The simplified derivation for the form of the equilibrium-constant expression that depends on equating the rate of the forward reaction to the rate of the reverse reaction, each written in terms of the overall equation, should not be taken as implying anything about the actual rate equations.

The units of the rate constants for first- and second-order reactions can be deduced from the form of Eqs. **6** and the fact that the rate has the units of $-d[\text{A}]/dt$. For a first-order reaction, therefore, the rate constant has the units of the reciprocal of time, i.e., reciprocal seconds, reciprocal minutes, and so forth, and is independent of the concentration units. The second-order rate constant, on the other hand, has units such as liters per mole second. The units of the rate constants for more complicated rate equations must be determined by inspection of the experimentally deduced rate equation.

Since rate constants always involve reciprocal seconds, reciprocal minutes, and so forth, a quick and approximate idea of how fast a reaction proceeds can be obtained from a reported rate constant by taking the reciprocal of the rate constant. As the units indicate, this gives a quantity that can be interpreted as the time required for the reaction to proceed appreciably when the reagents have unit concentrations. A more quantitative interpretation will be given when individual reactions are taken up.

It is important to realize that the order of a reaction and the rate equation are summaries of experimental results. In the next chapter, attempts will be made to devise mechanisms that are consistent with the rate equations. Then it will be necessary to keep in mind that mechanisms are theoretical but that the rate equation is an analytical portrayal of the experimental data.

10-3 FITTING RATE DATA TO A FIRST-ORDER RATE EQUATION

A first-order reaction is one for which, at a given temperature, the rate of the reaction depends only on the first power of the concentration of a single reacting species. Let the concentration of this species be represented by c and suppose that the volume of the system remains essentially constant during the course of the reaction. The first-order rate law can then be written

$$-\frac{dc}{dt} = kc \qquad \mathbf{7}$$

The rate constant k is then a positive quantity and has the units of the reciprocal of time.

The experimental results obtained in a study of the rate of a reaction are usually values of c, or some quantity related to c, at various times. Such data can best be compared with the integrated form of the first-order rate law. If the initial concentration, at time $t = 0$, is c_0, and if at some later time t the concentration has fallen to c, the integration gives

$$-\int_{c_0}^{c} \frac{dc}{c} = k \int_0^t dt$$

and $$-\ln\frac{c}{c_0} = \ln\frac{c_0}{c} = kt$$

or $$\log\frac{c_0}{c} = \frac{k}{2.303}t \qquad \mathbf{8}$$

A form that is sometimes more convenient is

$$\log c = -\frac{k}{2.303}t + \log c_0 \qquad \mathbf{9}$$

A reaction can therefore be said to be a first-order reaction if a plot of $\log (c_0/c)$ or of $\log c$ against t gives a straight line. If a straight line results, the slope of the line can be used to obtain the value of the rate constant k. An alternative to this graphical procedure is to calculate values of k from the individual measurements of c at the various times t, from Eq. **8** for example. The reaction is classified as first order if all the data lead to essentially the same values for k, that is, if Eq. **8** is satisfied with a constant value of k.

A classic example is given by the study of the rate of formation of glucose and fructose from sucrose, i.e., the inversion of sucrose, using measurements of the optical rotation of the solution. A solution of sucrose rotates the plane of polarized light to the right; i.e., sucrose is *dextrorotatory*. The mixture of monosaccharides formed as a result of hydrolysis rotates the plane to the left; i.e., their solution is *levorotatory*. As first shown more than 100 years ago, measurement of the degree of rotation as a function of time allows the rate of the inversion reaction to be followed.

Let α_0 be the initial angle of rotation for a given solution of sucrose. Let α_∞ be the angle of rotation when the inversion reaction is complete. The difference $\alpha_0 - \alpha_\infty$ is proportional to the concentration c_0 of sucrose present initially. If α is the angle of rotation at some time t, the concentration c of sucrose present at that time will be proportional to the amount of rotation that has yet to occur; that is, c is proportional to $\alpha - \alpha_\infty$. Thus we can write

$$c_0 = (\text{const})(\alpha_0 - \alpha_\infty)$$
$$c = (\text{const})(\alpha - \alpha_\infty)$$

Then if we suspect the reaction rate to be described by a rate equation that is first order with respect to the sucrose concentration, we can use Eq. **8** in the form

$$\log \frac{c_0}{c} = \log \frac{\alpha_0 - \alpha_\infty}{\alpha - \alpha_\infty} = \frac{k}{2.303} t \qquad \textbf{10}$$

or $$\log (\alpha - \alpha_\infty) = -\frac{k}{2.303} t + \log (\alpha_0 - \alpha_\infty) \qquad \textbf{11}$$

A plot of $\log (\alpha - \alpha_\infty)$ versus t that gives a straight line, confirms the first-order rate law for the reaction and yields, from its slope, a value of k.

Sucrose inversion, as the data of Table 10-1 and Fig. 10-1 illustrate, follows such a first-order rate law, but the rate depends on other reagents, in particular, acids, bases, or the enzyme *invertase*. In any single inversion rate study, however, these reagents are unchanged, and the reaction is first order with respect to sucrose. Reagents that affect the reaction rate but are not chemically changed in form or amount as a result of the reaction are called *catalysts*. Here the details of their role are avoided, and only their effect on the value of the rate constant is recognized.

For first-order reactions it is customary to use not only the rate constant k for the reaction but also the related quantity, the *half-life* of the reaction. The half-life is defined as the time required for the concentration or amount of the reagent to decrease to half its initial value. For a first-order reaction the relation of the half-life $t_{1/2}$ to the rate constant can be found from Eq. **8** by inserting the requirement that at $t = t_{1/2}$ the

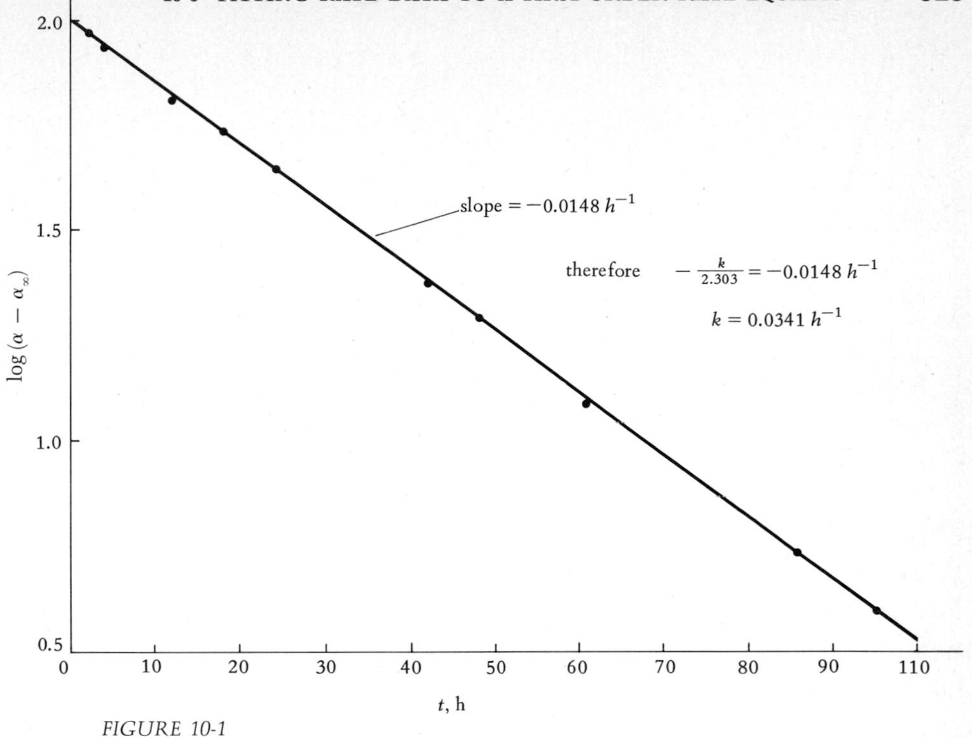

FIGURE 10-1
A plot of $\log(\alpha - \alpha_\infty)$ versus t for the data on the inversion of sucrose in Table 10-1.

TABLE 10-1
Polarimeter Readings versus Time for a Sucrose Solution at 30°C Containing an Acid Catalyst†

t, **h**	α	$\alpha - \alpha_\infty$	**log** $(\alpha - \alpha_\infty)$
0	78.00($= \alpha_0$)	99.15	1.996
2	71.00	92.15	1.964
4	65.10	86.25	1.936
12	44.60	65.75	1.804
18	32.40	53.55	1.729
24	22.45	43.60	1.640
42	2.30	23.45	1.370
48	−1.96	19.19	1.283
61	−9.00	12.15	1.085
86	−15.80	5.35	0.728
95	−17.25	3.90	0.591
∞	−21.15($= \alpha_\infty$)	0	

†From data of Rosanoff, Clark, and Sibley, *J. Am. Chem. Soc.*, **33:**1911 (1911).

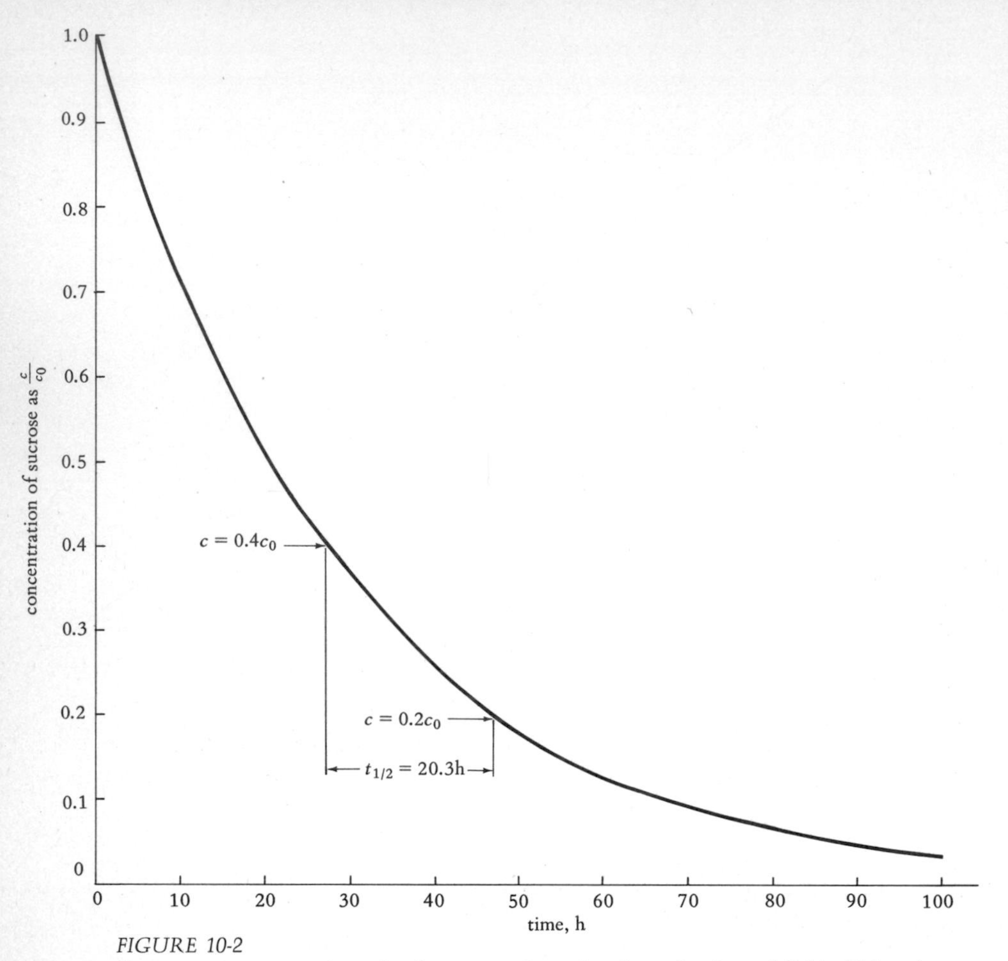

FIGURE 10-2
Concentration-versus-time plot for sucrose inversion from the data of Table 10-1 and an illustration of the half-life $t_{1/2} = 0.639/k = 20.3$ h.

concentration is $c = \frac{1}{2}c_0$. In this way one obtains

$$\log \frac{c_0}{\frac{1}{2}c_0} = \frac{k}{2.303} t_{1/2}$$

or

$$t_{1/2} = \frac{0.693}{k} \tag{12}$$

This result shows that for a first-order reaction there is a simple reciprocal relation between k and $t_{1/2}$. Furthermore, since the expression involves no term for the concentration or amount of material, the time required for half the reactant to be used up is independent of the initial concentration or amount of the reactant. The half life of sucrose, according to the data of Table 10-1 is shown graphically in Fig. 10-2. It is the simple relation of

Eq. **12** that makes the half-life a useful quantity for first-order reactions. The half-life of higher-order reactions, as we shall see, is a function of the initial concentration as well as of the rate constant, and the concept of half-life is then of little value.

A type of reaction somewhat outside the realm of chemistry which conforms beautifully to first-order kinetics is that of radioactive decay. The rate at which a radioactive species decays is proportional to the amount of that species. The decay is therefore first order, and a half-life is invariably used to characterize the decay rate.

10-4 FITTING RATE DATA TO A SECOND-ORDER RATE EQUATION

A reaction is classified as *second order* if the rate of the reaction is proportional to the square of the concentration of one of the reagents or to the product of the concentrations of two species of the reagents. The second situation leads to the same equations as the first if the two reactants are used up at the same rate and if their initial concentrations are equal. For these situations, the rate law is

$$-\frac{dc}{dt} = kc^2 \qquad \mathbf{13}$$

where c is the concentration of the single reagent or of one of the two reagents. Again, the kinetic data are usually compared with the integrated form of the equation. One has

$$-\int_{c_0}^{c} \frac{dc}{c^2} = k\int_0^t dt \qquad \mathbf{14}$$

and

$$\frac{1}{c} - \frac{1}{c_0} = kt \qquad \text{or} \qquad \frac{c_0 - c}{c_0 c} = kt \qquad \mathbf{15}$$

A reaction of the types considered so far is second order; therefore, a plot of $1/c$ versus t gives a straight line. The slope of the straight line is equal to the rate constant. An example is given in Fig. 10-3.

As Eq. **15** shows, this constant involves the units of concentration and in this respect differs from the first-order rate constant that involves only the units of time. Furthermore, the time for the concentration to drop to half its initial value is deduced from Eq. **15** to be

$$t_{1/2} = \frac{1}{kc_0} \qquad \mathbf{16}$$

The half-life depends, therefore, on the initial concentration and is not a convenient way of expressing the rate constant of second-order reactions.

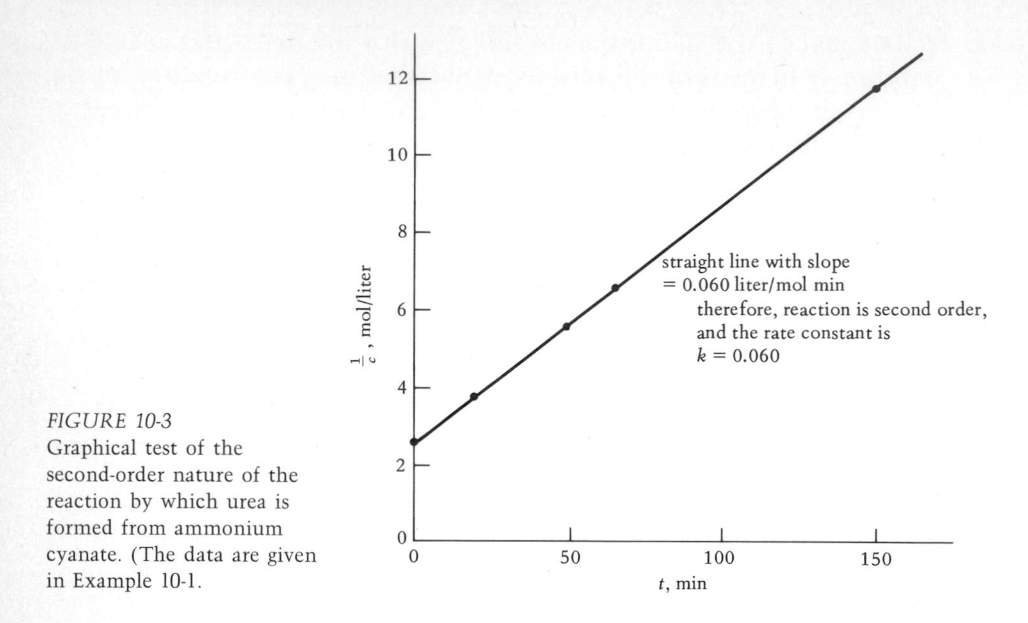

FIGURE 10-3
Graphical test of the second-order nature of the reaction by which urea is formed from ammonium cyanate. (The data are given in Example 10-1.

An illustration of a second-order rate law appropriate here is provided by the reaction of the inorganic ions ammonium and cyanate to form the organic compound urea. This was Wöhler's famous reaction that drove organic compounds out of the confines of biology and into the grasp of chemists.

Example 10-1

The tabulated data have been reported for the formation of urea from a solution containing 22.9 g per liter of ammonium cyanate at 65°C by the reaction

$$NH_4CNO \rightarrow NH_2\overset{\displaystyle O}{\overset{\|}{C}}NH_2$$

time, min	urea formed, g
0	0
20	7
50	12.1
65	13.8
150	17.7

Verify that the reaction is second order and deduce a value for the rate constant.

Solution

Although we could work with the weights of the reagents, which are approximately proportional to concentrations, it is customary to deal with concentrations. We have

$$c_0 = \text{initial concentration of } NH_4CNO$$
$$= \frac{22.9}{60.1} = 0.381 \text{ mol/liter}$$

Also, since the weight of urea formed is equal to the weight of ammonium cyanate consumed:

time, min	**concentration** c **of** NH_4CNO
0	0.381
20	0.264
50	0.180
65	0.151
150	0.086

We now can either form a plot of $1/c$ versus $(c_0 - c)/c_0c$ versus t and investigate its linearity, or we can calculate

$$\frac{c_0 - c}{c_0c} \frac{1}{t}$$

and investigate the constancy of the results. The second approach gives the following values:

time, min	c, **mol/liter**	$\frac{c_0 - c}{c_0c} \frac{1}{t}$, **liters/mol min**
0	$0.381(= c_0)$	
20	0.264	0.0585
50	0.180	0.0586
65	0.151	0.0615
150	0.085	0.0601
	Avg.	0.060

The near constancy of the calculated values in the final column confirms conformity to Eq. **15** and yields

$$k = 0.060 \text{ liter/mol min}$$

The rate equation at 65°C is thus

$$\text{Rate} = (0.060 \text{ liter/mol min}) [NH_4CNO]^2$$

where the square brackets imply molar concentration.

10-5 MORE COMPLEX RATE EQUATIONS: ENZYME-CATALYZED REACTIONS

Chemical-rate studies turn up many rate equations that are more complex than first- and second-order equations. Rate equations of this sort can also be illustrated by reactions that occur in biological systems or at least are affected by enzymes that occur in such systems.

The effect of enzymes on the rate with which chemical reactions move toward their equilibrium position provides some of the most dramatic illustrations of catalytic effects. Much of the current interest in the subject is centered on the details of the interaction between the *enzyme*, which is the catalyst, and the *substrate*, which is the material whose reaction it affects. But it is also important to understand how an enzyme-catalyzed reaction proceeds in time. Furthermore, from the measurement of the progress of such reactions, the catalytic activity of the enzyme-substrate pair is evaluated.

The experimental data for enzyme-catalyzed reactions show a variety of concentration dependencies related to the enzyme, the substrate, the temperature, the presence of interfering substances, and so forth. Much of the behavior found can be seen as a variation from the "ideal" curves of Fig. 10-4. We now develop a rate equation for such rate curves in a form that is conveniently related to the quantities measured in enzyme studies.

Inspection of the curves of Fig. 10-4 shows that at high substrate concentrations the rate of the reaction is independent of the substrate concentration and proportional to the total amount or concentration of enzyme $[E_{\text{total}}]$ in the system. Furthermore, at low substrate concentrations, as shown by the initial straight-line sections of the curves of Fig. 10-4, the rate is proportional to the substrate concentration and, as shown by the slopes of the different lines, proportional also to the total enzyme concentration. You can verify that these features can be accounted for by a rate equation of the form

$$R = \frac{(\text{const})[E_{\text{total}}][S]}{\text{const}' + [S]} = \frac{(\text{const})[E_{\text{total}}]}{1 + \text{const}'/[S]} \qquad \mathbf{17}$$

where R denotes the rate of the reaction and const and const′ are two different constants. To anticipate the notation that will be introduced when the mechanism of enzyme-catalyzed reactions is dealt with, we use k_2 and K_M for the two constants and thus write the rate equation in the form

$$R = \frac{k_2[E_{\text{total}}]}{1 + K_M/[S]} \qquad \mathbf{18}$$

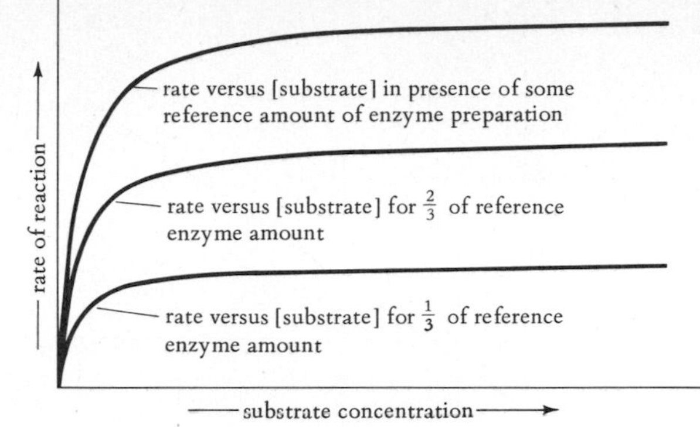

FIGURE 10-4
Rate-versus-substrate-concentration curves followed by some enzyme-substrate systems; these curves provide a basis for the interpretation of the kinetics of many enzyme-substrate reactions.

Although the parameters k_2 and K_M could be determined so that a function corresponding to the experimental curve on an R-versus-S plot is obtained, another procedure is more convenient.

First, the reciprocal of Eq. **18** is formed as

$$\frac{1}{R} = \frac{1}{k_2[E_{\text{total}}]} + \frac{K_M}{k_2[E_{\text{total}}]}\frac{1}{[S]} \qquad \textbf{19}$$

Now one sees that for a reaction containing a fixed amount of enzyme, data for R and $[S]$ can be used to form a $1/R$-versus-$1/[S]$ plot. If the reaction does, in fact, follow a rate law of the form of Eq. **19,** a straight-line plot will be obtained. Such a graph is known as a *Lineweaver-Burk plot.* (There are, in addition, other modifications of this graphical treatment that have certain advantages.) The intercept on the $1/[S]$ coordinate that occurs when $1/R$ is extrapolated to zero can be identified with $-1/K_M$. This can be seen by rearranging Eq. **19** for $1/R = 0$ to give

$$0 = \frac{1}{k_2[E_{\text{total}}]} + \frac{K_M}{k_2[E_{\text{total}}]}\frac{1}{[S]}$$

$$\frac{1}{[S]} = -\frac{1}{K_M} \qquad \left[\frac{1}{R} = 0\right]$$

The slope, as can again be seen from Eq. **19,** is identified with $K_M/k_2[\text{E}_{\text{total}}]$. If K_M has been obtained from the intercept, $k_2[E_{\text{total}}]$ can be extracted from the value of the slope. One can also recognize that at $1/[S] = 0$, the value of $1/R$ is $1/k_2[E_{\text{total}}]$. It follows that the intersection on the $1/[S] = 0$ axis can also be used to obtain $k_2[E_{\text{total}}]$. The relation of the Lineweaver-Burk plot to the parameters of Eqs. **18** and **19** is shown in Fig. 10-5.

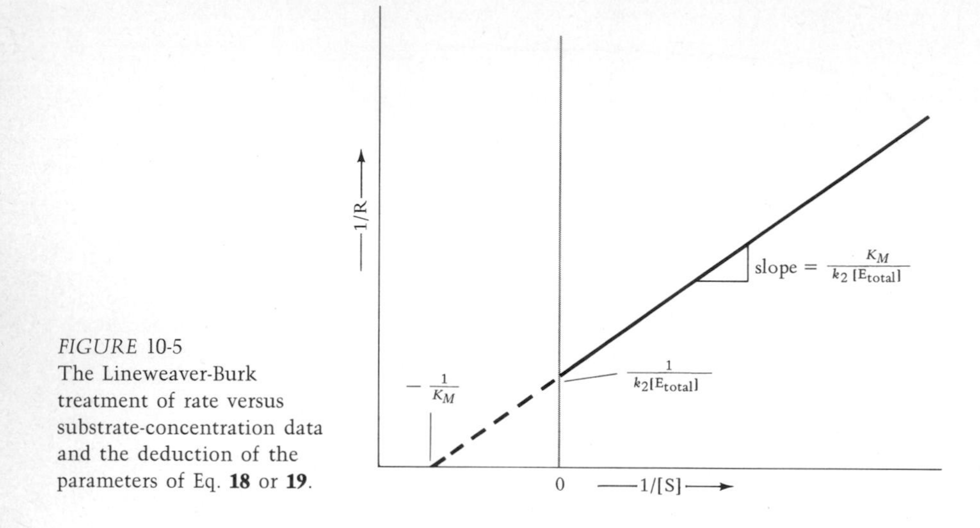

FIGURE 10-5
The Lineweaver-Burk treatment of rate versus substrate-concentration data and the deduction of the parameters of Eq. **18** or **19**.

As a specific example, consider the rate of conversion of starch to sugar as influenced by the enzyme *amylase*. The data for initial rate versus starch concentration are shown in Table 10-2. All runs were carried out with the solutions containing the same amount of the same preparation of barley *amylase*.

A plot of rate versus substrate concentration (Fig. 10-6) confirms the qualitative relation of these data to the ideal curve of Fig. 10-4.

TABLE 10-2
The Formation of Sugar (Mostly Maltose) from Starch in Solutions Containing Equal Amounts of the Enzyme *Amylase*†

starch concentration, %	initial reaction rate, mg of maltose per 5 ml of solution per minute
0.0300	0.140
0.0400	0.165
0.0500	0.180
0.0863	0.260
0.129	0.305
0.216	0.345
0.431	0.400
0.647	0.435
1.078	0.445

†Data from C. S. Hanes, *Biochem. J.*, **26:**1406 (1932).

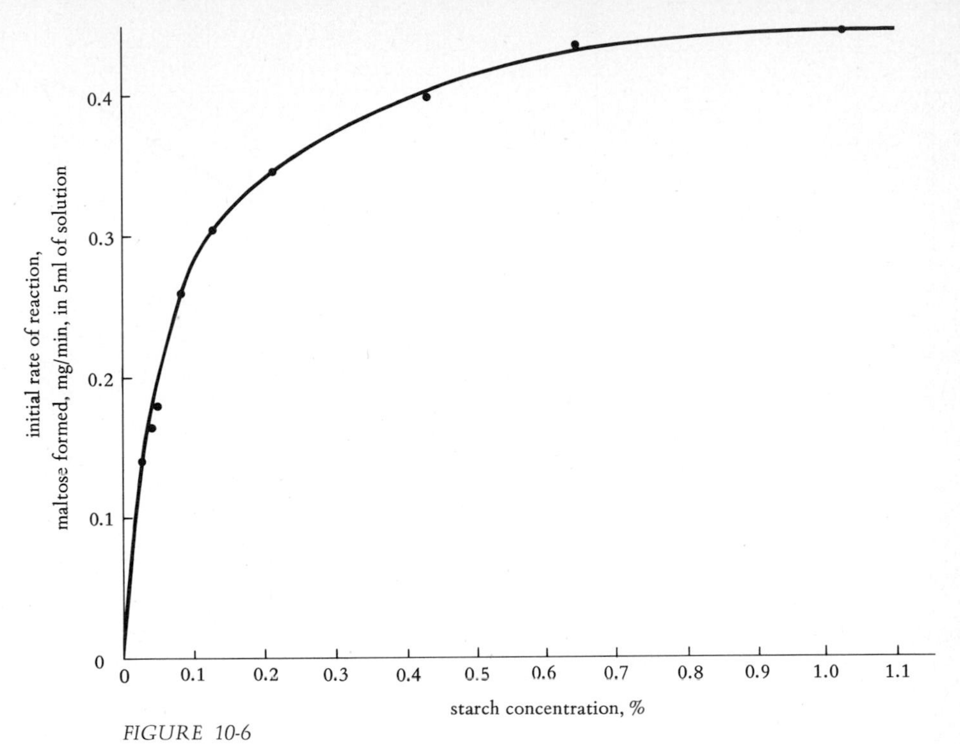

FIGURE 10-6
Plot of the data of Table 10-2 for the amylase-catalyzed hydrolysis of starch.

The reciprocal plot suggested by Eq. **19** is shown in Fig. 10-7. The linearity confirms the obedience to the form of Eq. **19** and thus to the form of Eq. **18.** Furthermore, we obtain from the slope and intercepts of Fig. 10-7 the parameter values

$$K_M = 0.073$$
$$k_2[E_{\text{total}}] = 0.47$$

With these numerical values it can be seen that $k_2[E_{\text{total}}]$ is, according to Eq. **18,** the maximum value that the rate approaches when $[S]$ becomes large and that Fig. 10-6 bears out this expectation.

Further, it is clear that measurement of the dependence of the rate on the substrate concentration does not provide any means of sorting out the two terms of the $k_2[E_{\text{total}}]$ parameter. A value can be assigned to k_2 itself only if a separate determination of the total enzyme concentration is available.

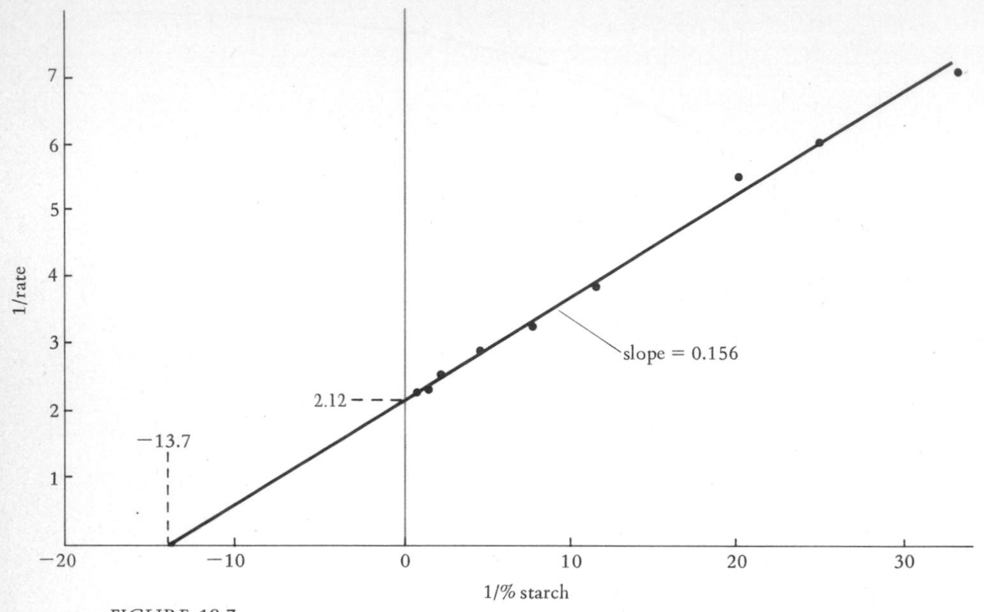

FIGURE 10-7
Reciprocal-reciprocal, or Lineweaver-Burk, plot of the data for the amylase-catalyzed hydrolysis as given in Table 10-2 and Fig. 10-6.

10-6 INHIBITED ENZYME ACTION

Many reagents can slow down reactions that are enzyme-catalyzed. Analysis of the effect of these reagents, known as *inhibitors*, illustrates how added complexities can provide additional insight into reaction mechanisms. Here the rate results which form a basis for such reaction-mechanism studies will be developed.

The action of inhibitors can be seen most clearly from the $1/R$ versus $1/[S]$ plots. If such plots are made for reaction systems containing various amounts of inhibitor, two classes of inhibitors are discovered (Fig. 10-8). One sees from these representative curves that for all inhibitors the greater the concentration of inhibitor, the greater the slowing-down effect; moreover, the smaller the concentration of the substrate, i.e., to the right on a $1/[S]$ plot, the greater the effect. But for one type of inhibitor, at large substrate concentrations, i.e., as $1/[S]$ approaches zero, the inhibitor effect is washed out, and the rate of reaction is, in the limit, unaffected by inhibitor. The other type of inhibitor remains effective even in the limit of large substrate concentrations.

The rate expression in the form of Eq. **19** for uninhibited reactions can be modified to encompass these two types of inhibitors. In cases where

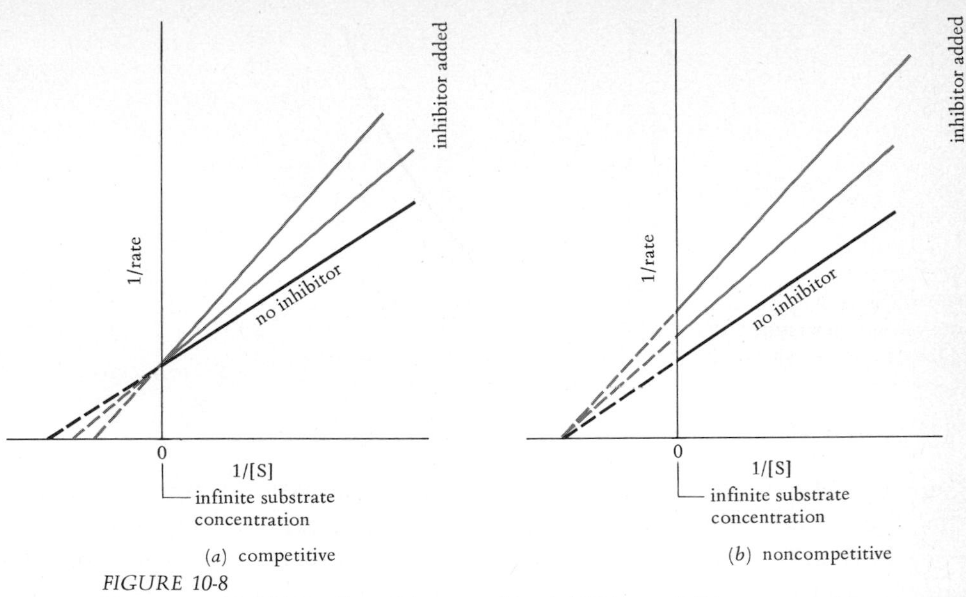

FIGURE 10-8
Two types of inhibitor effect: (*a*) competitive inhibition and (*b*) noncompetitive inhibition.

the slope but not the intercept is affected, we can write

$$\frac{1}{R} = \frac{1}{k_2[E_{\text{total}}]} + \frac{K_M}{k_2[E_{\text{total}}][S]} f([I]) \qquad \mathbf{20}$$

where $f([I])$ is a function of the inhibitor concentration. When both the slope and the intercept are altered, the rate expression

$$\frac{1}{R} = \left(\frac{1}{k_2[E_{\text{total}}]} + \frac{K_M}{k_2[E_{\text{total}}][S]}\right) f'([I]) \qquad \mathbf{21}$$

is satisfactory. Detailed studies of the rate–substrate-concentration data for various amounts of enzyme and inhibitor could lead to forms for the inhibitor terms, but we avoid this step and postpone further investigation of the role of inhibitors until we see what effects are suggested by the mechanisms that can be proposed.

10-7 TEMPERATURE DEPENDENCE OF THE RATES OF CHEMICAL REACTIONS

The rate equation and the value of the rate constant for a reaction are deduced from measurements of the rate of reaction at a fixed temperature. Experiments at several different temperatures generally show that the

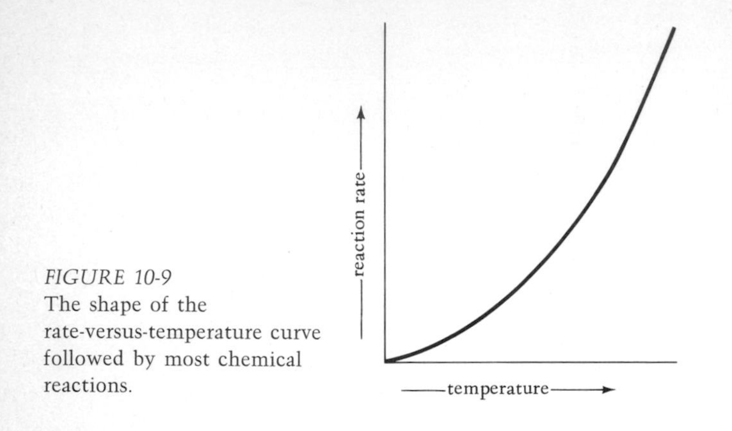

FIGURE 10-9
The shape of the rate-versus-temperature curve followed by most chemical reactions.

concentration dependence exhibited in the rate equation is unchanged but that the value of the rate constant is very temperature-dependent.

In most chemical reactions the temperature dependence of the rate "constant," which appears in the rate equation

$$\text{Rate} = k(T)(\text{concentration-dependent term}) \qquad \textbf{22}$$

is of the form shown in Fig. 10-9.

Before dwelling on this typical temperature dependence, we should recognize that other types of behavior are also observed. Figure 10-10 shows that two other types of temperature dependence are found in reactions which reach an explosive stage and in enzyme-catalyzed reactions. Other temperature dependences also show up in special circumstances, but the majority of chemical reactions have the temperature dependence illustrated in Fig. 10-9.

In 1889 Arrhenius recognized that this typical temperature dependence indicates an exponential increase of the rate, or rate constant, with temperature.

This can be confirmed empirically, as by plotting $\ln k$ versus $1/T$ or $\log k$ versus $1/T$. The data on the rate constant of a particular reaction as a function of temperature, as given in Table 10-3, are shown in such a plot in Fig. 10-11.

Such linear plots imply the relation

$$\log k \propto \frac{1}{T}$$

or $$\ln k \propto \frac{1}{T} \qquad \textbf{23}$$

or, finally $k \propto e^{\text{const}/T}$

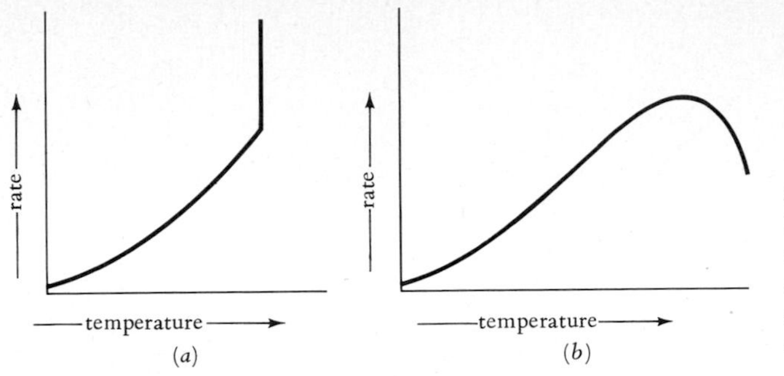

FIGURE 10-10
The rate-versus-temperature curve exhibited by some reactions that (*a*) reach an explosive stage or (*b*) are enzyme-catalyzed.

In view of interpretations of this temperature dependence to be given later and in line with the treatment of Arrhenius, this empirical relation is more conveniently written

$$k = Ae^{-E_a/RT} \qquad \textbf{24}$$

where A is called the *preexponential factor* and E_a is known as the *activation energy*. With this notation one writes the logarithmic form of Eq. **24** as

$$\ln k = -\frac{E_a}{RT} + \ln A$$

or

$$\log k = -\frac{E_a}{2.303R}\frac{1}{T} + \log A \qquad \textbf{25}$$

TABLE 10-3
The rate constant as a function of temperature for the reaction
$CH_3I + C_2H_5ONa \rightarrow CH_3OC_2H_5 + NaI$
in ethyl alcohol*

t, °C	k_2 (liters/mole sec)
0	5.60×10^{-5}
6	11.8
12	24.5
18	48.8
24	100
30	208

*From W. Hecht and M. Conrad, Z. Physik.

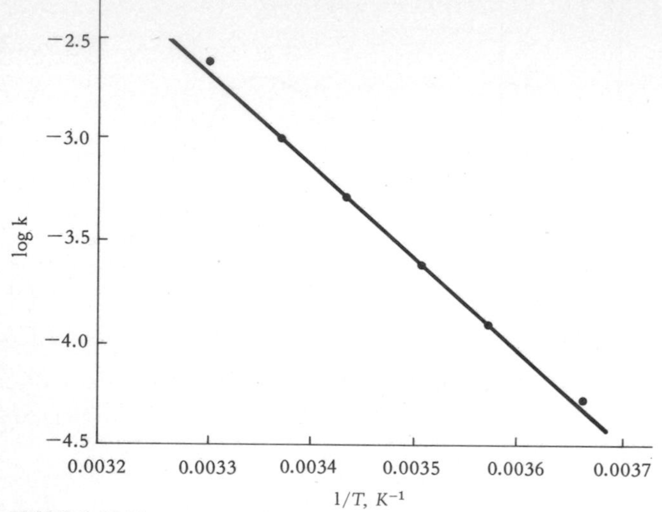

FIGURE 10-11
The Arrhenius plot for the data of Table 10-2. The straight line is represented by log $k = -4250/T + 11.38$ or log $k = 19{,}490 + \log 2.4 \times 10^{11}$ or $2.303RT$. Thus $E_a = 19{,}490$ cal and $A = 2.4 \times 10^{11}$ mol/liter sec.

The empirical constants E_a and A can therefore be deduced from the slope and intercept of the appropriate plot of the values of k at different temperatures.

Although these expressions are empirical correlations of rate data and the terms E_a and A are, for the present, to be treated as empirical parameters, the form of the expression for the rate constant might have been anticipated from the previously derived relation for the temperature dependence of the equilibrium constant. The thermodynamic equation

$$\frac{d(\ln K)}{dT} = \frac{\Delta H}{RT^2} \tag{26}$$

can be written

$$K = (\text{const})e^{-\Delta H/RT} \tag{27}$$

For an elementary reaction, K can be interpreted as k_1/k_{-1}. The simplest temperature dependence of the rate constants which is consistent with Eq. **27** is

$$k_1 = A_1 e^{-(E_a)_1/RT} \qquad \text{and} \qquad k_{-1} = A_{-1}e^{-(E_a)_{-1}/RT}$$

giving

$$\frac{k_1}{k_{-1}} = \frac{A_1}{A_{-1}} e^{-[(E_a)_1-(E_a)_{-1}]/RT} \tag{28}$$

and agreement with the form of Eq. **27**.

Example 10-2

A rule of thumb used by photographers, chemists, and biologists is that an increase in temperature of 10°C doubles the rate of reaction. According to the development of this section, the temperature dependence of rate depends on the activation energy. What activation energy must a reaction have to obey the 10-deg doubling rule?

Solution

Return to Eq. **25** and write it for two temperatures, T_1 and T_2,

$$\log\frac{k_{T_2}}{k_{T_1}} = -\frac{E_a}{2.303R}\left(\frac{1}{T_2} - \frac{1}{T_1}\right)$$

$$= -\frac{E_a}{2.303R}\frac{T_1 - T_2}{T_1 T_2}$$

Let us suppose that the 10°C increment raises the temperature from 300 to 310 K. Then, if this leads to $k_{T_2}/k_{T_1} = 2$, we have

$$\log 2 = -\frac{E_a}{2.303R}\left(\frac{-10}{300 \times 310}\right)$$

$$\text{or}\quad E_a = \frac{2.303\ R\log 2}{(10/300)(310)} = 13\ \text{kcal/mol}$$

PROBLEMS

1 Calculate a few values of k for the inversion of sucrose according to the data of Table 10-1 and Eq. **11.** Confirm that a constant value is obtained and that it is equal to the value deduced by the graphical treatment of Fig. 10-1.

2 The hydration of ethylene oxide in aqueous solutions proceeds according to the overall equation

$$\underset{\diagdown\ \text{O}\ \diagup}{CH_2\text{—}CH_2} + H_2O \rightarrow CH_2OHCH_2OH$$

The rate of the reaction has been studied [Brønsted, Kilpatrick, and Kilpatrick, *J. Am. Chem. Soc.*, **51:**428 (1929)]. Measurements were made of the height h of the liquid in the capillary tube of a dilatometer, a large thermometerlike reaction cell. At 20°C, a solution containing 0.12 M ethylene oxide and 0.007574 $HClO_4$ gave the data shown in the table on the next page. Confirm that the reaction is first order with respect to ethylene oxide. What is the rate constant?

3 The rate constant for the inversion of sucrose in the studies that led to the data of Table 10-1 is 0.034 h^{-1}. What fraction of sucrose would remain after exposure to these conditions for 10 h?

time, min	h, arbitrary units
0	18.48
30	18.05
60	17.62
90	17.25
120	16.89
240	15.70
300	15.22
360	14.80
390	14.62
∞	12.30

4 The isotope ^{32}P is radioactive and decays by the emission of a nuclear electron, or β^- particle, to the stable isotope ^{32}S. The nuclear equation is

$$^{32}P \rightarrow {}^{32}S + \beta^-$$

Such radioactive decays follow first-order rate laws. The amount of ^{32}P can be followed by the amount of β activity shown by the sample. If it is found that after 20 days the activity of a sample is 37.9 percent of what it was originally, what is the rate constant and the half-life of this decay?

5 Two of the troublesome radioactive waste products from nuclear fission power plants are ^{90}Sr and ^{137}Cs, or strontium 90 and cesium 137. Their half-lives are about 30 years. The amounts produced and the damaging effect of their radiations are such that safe storage of waste for about 600 years would be necessary. By what fraction would the radioactivity of these wastes be reduced in that length of time?

6 The decay of the radioactive isotope ^{14}C is used in the determination of the time for which carbon-containing materials have been removed from active contact with atmospheric carbon dioxide. The half-life of ^{14}C is 5670 years, and the amount of ^{14}C present in a sample is followed by measuring the β activity that results from the decay process $^{14}C \rightarrow {}^{14}N + \beta^-$. A sample in active contact with the atmosphere, growing plants for example, has a ^{14}C activity of 15.3 counts per minute per gram of carbon. Very nearly the same value appears to apply to quite ancient atmospheres.

a Plot the ^{14}C activity as a function of years for a time span over which the ^{14}C activity would fall to about 1 percent of its maximum value.

b A more convenient graph is that in which the logarithm of the activity is plotted against time. Make such a graph and cite the equation that explains its nature.

c A tree overturned in the eruption that formed Crater Lake in Oregon shows an activity of 6.6 counts per minute per gram of carbon. Use your graphs to estimate the age of Crater Lake.

7 The following data for the hydrolysis of ethyl acetate at 25°C were reported in an early study [J. Walker, *Proc. R. Soc.*, **A78:**157 (1906)]. (The concentration of reagents was deduced from measurements of the conductance of the solution.)

time, min	concentration of ethyl acetate and of hydroxide, mol/liter
0	0.01000
5	0.00755
9	0.00637
13	0.00541
20	0.00434
33	0.00320

The reaction can be described by the equation

$$CH_3-\overset{\overset{O}{\|}}{C}-OC_2H_5 + OH^- \rightarrow CH_3\overset{\overset{O}{\|}}{C}O^- + C_2H_5OH + H_2O$$

Verify that the reaction follows a second-order rate law and deduce the value of the rate constant. Use both the graphical and analytical treatments suggested by Eq. **15** and compare the results.

8 Tertiary butyl chloride decomposes in the gas phase at high temperatures to give gaseous products according to the equation

$$(CH_3)_3CCl \rightarrow (CH_3)_2C{=}CH_2 + HCl$$

The following data are for 295°C [Brearley, Kistiakowsky, and Stauffer, *J. Am. Chem. Soc.*, **58:**42 (1936)]:

time, min	pressure $(CH_3)_3CCl$, mm Hg
30	28.20
50	18.10
60	14.20
80	9.13

a What is the order of the reaction?

b What is the rate constant for the reaction?

c Plot the pressure of $(CH_3)_3CCl$ from $t = 0$ to $t = 80$ min. Include on the same graph the total pressure over this time.

9 Hypochlorite ion oxides iodide in a basic solution according to the reaction

$$OCl^- + I^- \rightarrow Cl^- + OI^-$$

At a hydroxide concentration of 1 *M*, a value which remains unchanged during the course of the reaction, the following data have been given [Y. T. Chia and R. E. Connick, *J. Phys. Chem.*, **63:**1518 (1959)]:

time, sec	$[I^-] = [OCl^-]$, **mol/liter**
0	0.00200
5	0.00127
10	0.00088
20	0.00056
30	0.00041

Deduce the rate equation, with a numerical value for the rate constant, for this reaction under the specified conditions, i.e., 25°C in a 1 *M* hydroxide aqueous solution.

10 Studies of the iodide-hypochlorite reaction of Prob. 9 show that the rate law maintains the same form but the value of the rate constant varies for solutions with various hydroxide concentrations. The following results† are representative of the rate constants obtained at different $[OH^-]$ values.

$[OH^-]$, **mol/liter**	*k*, **mol/liter sec**
1.00	61
0.50	120
0.25	230

Deduce from these results and the answer to the preceding problem a rate equation that encompasses the affects of the hydroxide-ion concentration as well as the effects of the iodide and the hypochlorite concentrations.

11 The fumarase-catalyzed dehydration of *l*-malate to yield fumarate has been extensively studied. Data for one set of runs at 25°C in the presence of 5 mmol of phosphate [R. A. Alberty, V. Massey, C. Frieden, and A. R. Fuhlbrigge, *J. Am. Chem. Soc.*, **76:**2485 (1954)] were as follows for a fixed enzyme concentration:

concentration of ***l*****-malate, mmol/liter**	**initial rate**
0.100	1.9
0.333	4.2
1.000	6.1
3.33	6.5
10.0	7.2
33.3	7.4
100	6.9

† From E. L. King, "How Chemical Reactions Occur," W. A. Benjamin, Inc., New York, 1963, and Y. T. Chia and R. E. Connick, *J. Phys. Chem.*, **63:**1518 (1959).

Do these data fit a rate equation of the form of Eq. **18**? To the extent that they do, what are the values of the parameters of that equation?

12 In the study of amylase-catalyzed conversion of starch to sugar that produced the data of Table 10-2 and Figs. 10-6 and 10-7, investigations of the effect of enzyme concentrations were also made. Since the enzyme was not isolated and added to the reaction mixture in known amounts, only the effect of relative amounts of enzyme could be reported. These showed the following values:

relative enzyme concentration	initial velocity
12.5	0.167
25	0.36
37.5	0.495
50	0.70
75	1.01
100	1.35

Is this dependence of the reaction rate on enzyme concentration consistent with the rate law of Eq. **18**?

13 On diagrams like those of Fig. 10-8, indicate, in view of Eqs. **20** and **21,** the meaning that can be given to the various intercepts and the slopes as was done for the noninhibited enzyme curves of Fig. 10-5.

14 What will be the ratios of the rate constants for a reaction run at temperatures 10°C apart if the activation energy is (*a*) 26 kcal and (*b*) 6.5 kcal?

15 The combination of myoglobin, Mgb, with oxygen gas in homogeneous aqueous solution to form oxymyoglobin, O_2Mgb, may be assumed to follow the rate law

$$\frac{d[\mathrm{O_2Mgb}]}{dt} = k[\mathrm{O_2}][\mathrm{Mgb}]$$

At 1°C, k is equal to 5×10^6 liter/mol sec for a 0.1 *M* phosphate buffer solution of pH 7.1. At 11°C, in the same buffer, k is 7×10^6.

a If the initial concentration of each reactant is 3.64×10^{-6} mol/liter, how long will it take before half of them are converted to the combined, O_2Mgb, form?

b What is the Arrhenius energy of activation and the preexponential *A* factor?

REFERENCES

General Chemical Kinetics

LAIDLER, K. J.: "Chemical Kinetics," 2d ed., McGraw-Hill Book Company, New York, 1965.

FROST, A. A., and R. G. PEARSON: "Kinetics and Mechanisms," John Wiley & Sons, Inc., New York, 1961.

CALDIN, E. F.: "Fast Reactions in Solution," Blackwell, Oxford, 1964.

EYRING, H., and E. M. EYRING, "Modern Chemical Kinetics," Reinhold Publishing Corporation, New York, 1963.

Chemical Kinetics and Application to Biochemical Systems

WILLIAMS, V. R., and H. B. WILLIAMS: "Basic Physical Chemistry for the Life Sciences," chap. 6, W. H. Freeman and Company, San Francisco, 1967.

WOLD, FINN: "Macromolecules: Structure and Function," Prentice-Hall, Inc., Englewood Cliffs, N.J., 1971. Enzyme kinetics is treated on pages 30 to 51, and applications to this material are then made throughout the book.

AMDUR, I., and G. HAMMES: "Chemical Kinetics: Principles and Selected Topics," McGraw-Hill Book Company, New York, 1966.

11 Reaction Mechanisms

Chapter 10 showed that there are a variety of rate equations and that generally there is no relation between the concentration terms in these equations and the chemical-reaction equation. This suggests that reactions do not proceed by a coming together of all the molecules shown as reactants in the reaction equation.

Reactions generally proceed by a series of simple, elementary steps. In each step two molecules come together and react or one molecule falls apart to produce one or two new molecular species.

After a look at some of the characteristics of elementary reactions in solutions, we shall be in a position to see what sequences of elementary steps can account for rate equations like those determined in Chap. 10.

11-1 ELEMENTARY REACTIONS IN LIQUID SOLUTIONS

Here we investigate the factors that determine the rate of an elementary reaction occurring in a liquid solution. By elementary we mean here a single-step reaction that depends upon two molecules' coming together or a single molecule's breaking up.

Consider a liquid system consisting of molecular species A and B, which can be identical or different, dissolved in a relatively inert solvent. Let there be N_A^* molecules of A and N_B^* molecules of B per unit volume of solution. What is the rate with which A and B molecules will encounter one another?

To proceed, we must borrow some of the concepts and results of the following chapter. In particular we must use the concept of a *diffusion coefficient D*, which is a measure of the rate with which a material diffuses across a unit cross-sectional area as a result of a unit concentration gradient.

The relation of the liquid-phase reaction process to diffusion can be recognized by first focusing on a particular A molecule and finding the rate with which B molecules would diffuse to it. The result, which seems reasonable and will not be derived here, is that the rate depends on N_B^*, the number of molecules of B per unit volume; D_B, the diffusion coefficient of B; and $r_A + r_B$, the distance between the center of B and that of A for the molecules to be in contact. The derived expression is

$$\text{Rate of B molecules diffusing to an A molecule} = 4\pi D_B(r_A + r_B)N_B^* \quad \mathbf{1}$$

For the total rate of A-B encounters per unit volume, on this basis, we would multiply by the number of A molecules per unit volume and thus get

$$\text{Rate of A-B encounters per unit volume} = 4\pi D_B(r_A + r_B)N_B^*N_A^* \quad \textbf{2}$$

Similarly focusing on a B molecule to find the rate with which A molecules diffuse toward it would give

$$\text{Rate of encounters per unit volume} = 4\pi D_A(r_A + r_B)N_A^*N_B^* \quad \textbf{3}$$

Since in general A and B will have appreciable diffusion tendencies, the encounter rate can be written as the mean of these results, namely,

$$\text{Rate of encounters} = 2\pi(D_A + D_B)(r_A + r_B)N_A^*N_B^* \quad \textbf{4}$$

We thus have an expression that corresponds to the maximum rate of reaction that could occur in liquid solutions. An important feature is the proportionality of the rate to the product of the concentration of the reacting species.

A representative value for this maximum rate can be calculated from a representative value of 10^{-5} cm^2/sec for D and 4 Å, or 4×10^{-8} cm, for r_A and r_B. Then we obtain

$$\text{Number of encounters} = 10^{-11}\ N_A^*N_B^* \text{ per sec per cm}^3 \quad \textbf{5}$$

For comparison with rate equations with customary units we need to convert to moles of encounters per liter per second and to moles per liter of reagents A and B. We obtain

$$\text{Moles of encounters, per sec per liter} = 6 \times 10^9[\text{A}][\text{B}] \quad \textbf{6}$$

This result corresponds to the rates of some of the fastest known simple reactions in solution, e.g., formation of I_2 from I atoms in nonpolar solvents. Such reactions, which have rates approaching that implied by Eq. **6** and for which the only obstacle is apparently that of one reagent species finding the other, are said to be *diffusion-controlled*.

Reaction rates even exceeding the value deduced in Eq. **6** are found for some ionic reactions in solution. Thus the acid-base neutralization reaction

$$H_3O^+ + OH^- \rightleftarrows 2H_2O$$

has been deduced to have a rate given by

$$\text{Rate} = 150 \times 10^9[H_3O^+][OH^-]$$

The increase over the representative diffusion-controlled rate constant of Eq. **6** can be attributed to the attraction between ions of opposite charge and the special ease of diffusion of a proton in an aqueous system.

The rapidity of such reactions can be appreciated by assuming that A and B are at concentrations of 1 M. The reaction would then begin with a product being formed at a rate of 6×10^9 mol/liter sec. Clearly the reaction would have progressed appreciably in a time as short as 10^{-9} sec. Few reactions proceed at such rates. We must expect that the elementary reactions require more than the coming together of the reacting species and that a *sequence* of such steps is necessary to produce products.

The coming together of molecules in the liquid has been called an encounter, implying some distinction between an encounter and a simple collision of gas-phase molecules. The principal distinction arises from the difficulty with which molecules of a liquid, once adjacent to each other, separate again. A similar difficulty stands in the way of the fragments of a molecular breakup escaping from each other. The special character of a liquid system is treated in terms of a *cage effect*.

Consider a pair of molecules (Fig. 11-1) that have just encountered each other or, what in this regard is equivalent, a pair of molecules that have just arisen from the breakup of a parent molecule. The molecules can separate from each other only as a result of their diffusional motion through the liquid. Since, on the molecular level, diffusion is pictured as the result of repeated jostlings of the neighboring molecules, we can picture the AB molecular pair as remaining for some time together in a cage formed by the surrounding solvent. The whole process of A and B coming together and remaining together for a number of subsequent collisions is what we have been calling an encounter. Now let us try to estimate the number of collisions (which, in fact, are not easily defined in a liquid system) that occur during an encounter. We must use a result from the analysis (to be carried out in Sec. 12-1) of the net distance x a particle moves as a consequence of a series of random steps taken in some time t. This result from the *random-walk problem* is that the average squared net distance is related to the diffusion coefficient by the relation (shown by Eq. **10** of Chap. 12)

$$\overline{x^2} = 6Dt$$

If we say that a pair of colliding molecules have escaped from each other when they are separated by an additional amount $r_A + r_B$, that is, their average diameter, and if we use 8 Å as a representative value for this diameter and 10^{-5} cm²/sec as a representative diffusion coefficient, we calculate

$$\text{Duration of encounter} = \frac{\overline{x^2}}{6D} = \frac{(8 \times 10^{-8})^2}{6 \times 10^{-5}} = 10^{-10} \text{ sec} \qquad \mathbf{7}$$

By contrast, the time required for representative gas-phase molecules subject to no retarding force to become separated by this same 8 Å distance is about 0.02×10^{-10} sec.

Thus, the encounter lasts some 50 times longer than would take

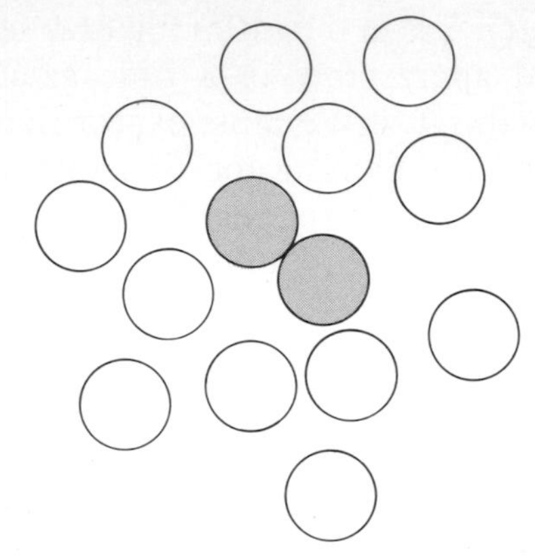

FIGURE 11-1
Two molecules, which could be the result of the breaking up of an intermediate, in a liquid cage.

the molecules to separate under gas-phase conditions. It follows that in the liquid phase each encounter can be looked upon as consisting of, say, 10 to 100 collisions.

The idea of a cage implies also that a molecular breakup is not as abrupt as it is in the gas phase. The fragments have considerable opportunity to react with each other or with the adjacent solvent before they move freely into the bulk of the solution to react with other entities.

This analysis of elementary reactions in solution is perhaps most revealing in pointing up the difficulties we have in constructing a molecular-level picture of the process. But the analysis of elementary reactions in solution does lead us to the important conclusion that the rate of any single step is proportional to the product of the concentrations of the two reacting species if the step is bimolecular and is proportional to the concentration of the single species if the step is a unimolecular breakup. This is all we need for exploring the relations between the overall rate of a reaction and the sequence of elementary steps proposed for a reaction.

11-2 REACTION MECHANISMS AND RATE LAWS: THE STATIONARY-STATE METHOD

We now investigate the rate laws expected on the basis of a sequence of the elementary reaction steps that appear to enter into the mechanism of many organic, inorganic, and biochemical reactions. To do so, we use the conclusions from Sec. 11-1 that the rate of an elementary step is proportional to the product of the concentrations of the reactants in that step and that various intermediates may enter into the mechanism even if they are not obtained as products of the overall reaction.

Let us consider some mechanisms in which the first step is a *reversible* reaction that produces an intermediate and may produce one or more of the final product species. The second step consists of the decay of the intermediate or the reaction of the intermediate with one of the reactant species. The result of the second step may be final products or additional intermediates. In the latter case, for simplicity, we shall assume that all further reactions are relatively very rapid. Two examples of the many variations are

$$\begin{array}{ll} A \rightleftarrows M + C & A + B \rightleftarrows M \\ M + B \rightarrow D & M \rightarrow C + D \\ \hline A + B \rightarrow C + D & A + B \rightarrow C + D \end{array}$$

where A and B are reactants, M is an intermediate, and C and D are products. Other variations are illustrated by the compilation of suggested mechanisms for a variety of reaction types in Table 11-1.

Deducing the rate equation for a reaction that proceeds by one of the mechanisms of this general type can be carried out most easily if we

TABLE 11-1
Some Reactions for Which Mechanisms Consisting of an Initial Reversible Step Followed by Reaction of One of the Products of This Step Have Been Suggested
Additional mechanistic steps beyond the first two are required to account for the overall reaction.

First step	$Br_2 \rightleftarrows 2Br$
Second step	$Br + H_2 \rightarrow HBr + H$
Overall reaction	$Br_2 + H_2 \rightarrow 2HBr$
First step	$[Co(CN)_5OH_2]^{2-} \rightleftarrows [Co(CN)_5]^{2-} + H_2O$
Second step	$[Co(CN)_5]^{2-} + I^- \rightarrow [Co(CN)_5\,I]^{3-}$
Overall reaction	$[Co(CN)_5OH_2]^{2-} + I^- \rightarrow [Co(CN)_5\,I]^{3-} + H_2O$
First step	$CHCl_3 + OH^- \rightleftarrows :CCl_3^- + H_2O$
Second step	$:CCl_3^- \rightarrow :CCl_2 + Cl^-$
Overall reaction	$2CHCl_3 + 7OH^- \rightarrow CO + HCOO^- + 6Cl^- + 4H_2O$
First step	$CH_3\overset{O}{\overset{\parallel}{C}}CH_3 + OH^- \rightleftarrows CH_3\overset{O}{\overset{\parallel}{C}}CH_2^- + H_2O$
Second step	$CH_3\overset{O}{\overset{\parallel}{C}}CH_2^- + Br_2 \rightarrow CH_3\overset{O}{\overset{\parallel}{C}}CH_2Br + Br^-$
Overall reaction	$CH_3\overset{O}{\overset{\parallel}{C}}CH_3 + 3Br_2 + 4OH^- \rightarrow CH_3\overset{O}{\overset{\parallel}{C}}O^- + CHBr_3 + Br^- + 3H_2O$

can assume that for most of the period over which the reaction is studied the concentration of the intermediates is small compared to that of the reactants or products. Failure to detect intermediates lends support to this idea. If the concentration of the intermediate fails to build up to an appreciable value, it follows that except at the initial and final stages of the reaction, the rate of change of the intermediate can be set equal to zero. This result, a feature of the *stationary-state* method, is very helpful in deducing the rate equation corresponding to a postulated mechanism.

Consider, to be specific,

$$\mathrm{A} \underset{k_{-1}}{\overset{k_1}{\rightleftharpoons}} \mathrm{M} + \mathrm{C} \qquad \mathbf{8}$$

$$\mathrm{M} + \mathrm{B} \xrightarrow{k_2} \text{products} \qquad \mathbf{9}$$

The value of [M] increases as a result of the elementary reaction associated with k_1. The value decreases as a result of the reverse reaction, for which the rate constant is k_{-1}, and as a result of the second reaction, for which the rate constant is k_2. Then the steady-state assumption applied to M lets us write

$$k_1[\mathrm{A}] = k_{-1}[\mathrm{M}][\mathrm{C}] + k_2[\mathrm{M}][\mathrm{B}]$$

or

$$[\mathrm{M}] = \frac{k_1[\mathrm{A}]}{k_{-1}[\mathrm{C}] + k_2[\mathrm{B}]} \qquad \mathbf{10}$$

According to the mechanism, the rate of formation of products, or the rate of consumption of B, which from the overall stoichiometry is also that for consumption of A, is equal to the rate of the second reaction, i.e.,

$$\text{Rate} = -\frac{d[\mathrm{A}]}{dt} = -\frac{d[\mathrm{B}]}{dt} = k_2[\mathrm{M}][\mathrm{B}] \qquad \mathbf{11}$$

The steady-state expression obtained for [M] allows this to be written as

$$\text{Rate} = \frac{k_1 k_2[\mathrm{A}][\mathrm{B}]}{k_{-1}[\mathrm{C}] + k_2[\mathrm{B}]} \qquad \mathbf{12}$$

Different but related rate equations can be worked out in the same way for the variations of this mechanism illustrated in Table 11-1.

That a single mechanism, such as that of Eqs. **8** and **9,** can lead to different rate equations can be illustrated by considering two extremes and an intermediate case in the values of relative rate constants.

CASE 1: $k_{-1}[\mathrm{C}] \gg k_2[\mathrm{B}]$

For this case, Eq. **12** reduces to

$$\text{Rate} = \frac{k_1 k_2}{k_{-1}} \frac{[\text{A}][\text{B}]}{[\text{C}]} \qquad \mathbf{13}$$

A similar result could have been obtained more directly by assuming that an equilibrium is established by the first reaction and that the second reaction does not use up B fast enough to upset this equilibrium. Then one would have written

$$k_1[\text{A}] = k_{-1}[\text{M}][\text{C}] \qquad \text{or} \qquad \frac{[\text{M}][\text{C}]}{[\text{A}]} = \frac{k_1}{k_{-1}} = K$$

and

$$\text{Rate} = k_2[\text{M}][\text{B}] = Kk_2 \frac{[\text{A}][\text{B}]}{[\text{C}]} \qquad \mathbf{14}$$

An illustration of this situation is the reaction of the iodide ion with hypochlorite

$$\text{I}^- + \text{OCl}^- \rightarrow \text{OI}^- + \text{Cl}^-$$

The rate of the reaction is influenced by the pH such that the rate law deduced from rate data is

$$\text{Rate} = \frac{k[\text{I}^-][\text{OCl}^-]}{[\text{OH}^-]} \qquad \mathbf{15}$$

A suitable mechanism for this reaction has an initial rapid equilibrium step

$$\text{OCl}^- + \text{H}_2\text{O} \rightleftharpoons \text{HOCl} + \text{OH}^-$$

followed by a slow step involving the reagents I^- and HOCl.

At least two mechanisms can be written beyond this point. Thus, we could postulate

$$\text{I}^- + \text{HOCl} \rightarrow \text{HOI} + \text{Cl}^-$$
$$\text{OH}^- + \text{HOI} \rightarrow \text{H}_2\text{O} + \text{OI}^-$$

or

$$\text{I}^- + \text{HOCl} \rightarrow \text{ICl} + \text{OH}^-$$
$$\text{ICl} + 2\text{OH}^- \rightarrow \text{OI}^- + \text{Cl}^- + \text{H}_2\text{O}$$

Rate data do not distinguish between these presumed fast final steps.

CASE 2: $k_{-1}[\text{C}] \ll k_2[\text{B}]$

Equation **12** now reduces to

$$\text{Rate} = k_1[\text{A}] \qquad \mathbf{16}$$

That is, the rate law would simply be first order in A even though the reaction is still assumed to proceed through the same sequence of steps.

One can recognize that the assumed inequality makes the initial dissociation of A the rate-determining step and the later steps in the mechanism of no consequence.

CASE 3: $k_{-1}[C]$ and $k_2[B]$ of comparable magnitudes
An illustration of this situation is provided by the reaction

$$Co(CN)_5OH_2^{2-} + I^- \rightarrow Co(CN)_5I^{3-} + H_2O$$

The observed rate law has the form

$$\text{Rate} = \frac{a[Co(CN)_5OH_2^{2-}][I^-]}{b + c[I^-]} \qquad \mathbf{17}$$

where a, b, and c represent empirically determined constants. A mechanism that is reasonable and leads to an expression of this form is given in the second example of Table 11-1. For this example, the analysis that led to Eq. **12** would lead to the rate equation

$$\text{Rate} = \frac{k_1k_2[Co(CN)_5OH_2^{2-}][I^-]}{k_{-1}[H_2O] + k_2[I^-]} \qquad \mathbf{18}$$

Since the reaction is run in aqueous solution with a large and essentially constant concentration of water, we see that the mechanism is consistent with the empirical rate law.

11-3 A MECHANISM FOR ENZYME-CATALYZED REACTIONS

Many different rate laws are required to portray the rates of enzyme-catalyzed reactions for the great variety of possible enzymes, substrates, and rate-influencing reagents and physical conditions. The rate law, Eq. **18** of Sec. 10-5, however, is a guide to the mechanism of this section, which is the basis for describing many enzyme-catalyzed reactions.

With S representing substrate, E the enzyme, and E • S an enzyme-substrate complex, the mechanism of the reaction is presumed to be adequately represented by

$$E + S \rightleftarrows E \cdot S \qquad \mathbf{19}$$

$$E \cdot S \rightarrow E + \text{products} \qquad \mathbf{20}$$

The net rate of forming products, or using up substrate, is given by the rate of the second step. Thus

$$R = k_2[\mathrm{E \cdot S}] \qquad \mathbf{21}$$

We need now to express [E · S] in measurable quantities.

As in the preceding section the steady-state assumption applied to E · S allows its rates of formation and consumption to be equated as

$$k_1[\mathrm{E}][\mathrm{S}] = k_{-1}[\mathrm{E \cdot S}] + k_2[\mathrm{E \cdot S}]$$

giving

$$[\mathrm{E \cdot S}] = \frac{k_1}{k_{-1} + k_2}[\mathrm{E}][\mathrm{S}] \qquad \mathbf{22}$$

It is customary to replace the terms $(k_{-1} + k_2)/k$, by the single constant K_M so that

$$[\mathrm{E \cdot S}] = \frac{1}{K_M}[\mathrm{E}][\mathrm{S}] \qquad \mathbf{23}$$

Now the rate of the reaction is given by

$$R = \frac{k_2}{K_M}[\mathrm{E}][\mathrm{S}] \qquad \mathbf{24}$$

We have not yet come to the desired result because generally [E] and [S], the concentrations of free enzyme and free substrate, are not measurable. We must develop a rate equation in terms of measurable quantities. One is $[\mathrm{E_{total}}]$, the total concentration of enzyme, both free and bound. Further, since the enzyme concentration is relatively small, [E · S] is negligible compared to the total substrate concentration. Then [S], which is really the concentration of free substrate, can be assumed to be the same as the total enzyme concentration. We can proceed by first writing

$$[\mathrm{E_{total}}] = [\mathrm{E}] + [\mathrm{E \cdot S}] \qquad \mathbf{25}$$

Then use of Eq. **23** gives, with rearrangement,

$$[\mathrm{E}] = \frac{K_M[\mathrm{E_{total}}]}{K_M + [\mathrm{S}]} \qquad \mathbf{26}$$

Now the rate expression, Eq. **24,** becomes

$$R = \frac{k_2[\mathrm{E_{total}}][\mathrm{S}]}{K_M + [\mathrm{S}]}$$

or

$$R = \frac{k_2[\mathrm{E_{total}}]}{1 + K_M/[\mathrm{S}]} \qquad \mathbf{27}$$

We have come to the form of the empirical rate law obtained in Sec. 10-4.

We now are in a position to interpret the values of the parameters K_M and $k_2[E_{total}]$ in terms of their roles in the steps of the mechanism.

Reference to Eq. **23** shows that as the reaction is proceeding,

$$\frac{[E][S]}{[E \cdot S]} = K_M \qquad \textbf{28}$$

Thus K_M is related to species concentrations, as is the dissociation constant for the species E · S. The value of K_M, however, is given by $(k_{-1} + k_2)/k_1$ and this is equal to the value of the dissociation constant for E · S, which is k_{-1}/k_1, only to the extent that k_2 is small and can be neglected compared to k_{-1}. Thus, when the breakup of the E · S complex to form original E and S species dominates the process whereby the complex forms products, the value of K_M approaches the dissociation constant for the E · S complex. Therefore, except when k_2 is large, K_M can be taken as a measure of the enzyme-substrate binding. Some values are shown in Table 11-2.

TABLE 11-2
Some Values of K_M for Enzyme-Substrate Pairs†

enzyme and substrate	K_M, **mmol/liter**
Catalase:	
H_2O_2	25
Hexokinase:	
Glucose	0.15
Fructose	1.5
Chymotrypsin:	
N-Benzoyltyrosinamide	2.5
N-Formyltyrosinamide	12.0
N-Acetyltyrosinamide	32
Glycyltyrosinamide	122
Carbonic anhydrase:	
HCO_3^-	9.0
Glutamate dehydrogenase:	
Glutamate	0.12
α-Ketoglutarate	2.0
NH_4^+	57
NAD_{ox}	0.025
NAD_{red}	0.018

†From A. L. Lehninger, "Biochemistry," Worth Publishers, Inc., New York, 1970.

What is the significance of the term $k_2[\mathrm{E}_{\mathrm{total}}]$? One first notes that the rate of the overall reaction is

$$R = k_2[\mathrm{E} \cdot \mathrm{S}] \qquad \mathbf{29}$$

It follows that $k_2[\mathrm{E}_{\mathrm{total}}]$ is the rate that the reaction would have if all the enzyme were in the form of the enzyme-substrate complex; that is, $k_2[\mathrm{E}_{\mathrm{total}}]$ is the *maximum rate* for a given value of $\mathrm{E}_{\mathrm{total}}$. The *turnover* number of an enzyme in a particular enzyme-catalyzed reaction is the maximum rate per mole, or per molecule, of enzyme. The turnover rate is thus equal to the value of k_2, and this can be calculated from $k_2[\mathrm{E}_{\mathrm{total}}]$ if the total enzyme concentration is known. Some values of k_2, the turnover number, are given in Table 11-3.

11-4 MECHANISMS FOR INHIBITED ENZYME-CATALYZED REACTIONS

The idea that enzymes act by forming an enzyme-substrate complex that leads rapidly to the reaction products suggests that inhibitors form complexes that interfere with this process. Two roles can be imagined, and we now see how they lead to rate expressions of the types found in Sec. 10-6 from empirical studies of reaction rates.

First we develop a rate law based on a mechanism in which the inhibitor binds to the same active site of the enzyme that is used for substrate binding. This situation, in which both the substrate and the inhibitor compete for the active enzyme site, is known as *competitive*

TABLE 11-3
Some Representative Turnover Numbers Under Optimun Conditions Per Minute†
Turnover numbers represent molecules of product per molecule of enzyme.

enzyme	**turnover number, min^{-1}**
Carbonic anhydrase C	36,000,000
Δ^5-3-Ketosteroid isomerase	17,100,000
β-Amylase	1,100,000
β-Galactosidase	12,500
Phosphoglucomutase	1,240
Succinate dehydrogenase	1,150

† From A. L. Lehninger, "Biochemistry," Worth Publishers, Inc., New York, 1970.

inhibition. It is shown schematically in Fig. 11-2. The reaction system is now described by

$$E + S \rightleftarrows E \cdot S \tag{30}$$

$$E + I \rightleftarrows E \cdot I \tag{31}$$

and $$E \cdot S \rightarrow E + \text{products} \tag{32}$$

As before, the steady-state assumption for the first reaction leads to

$$[E \cdot S] = \frac{k_1}{k_{-1} + k_2}[E][S] = \frac{1}{K_M}[E][S] \tag{33}$$

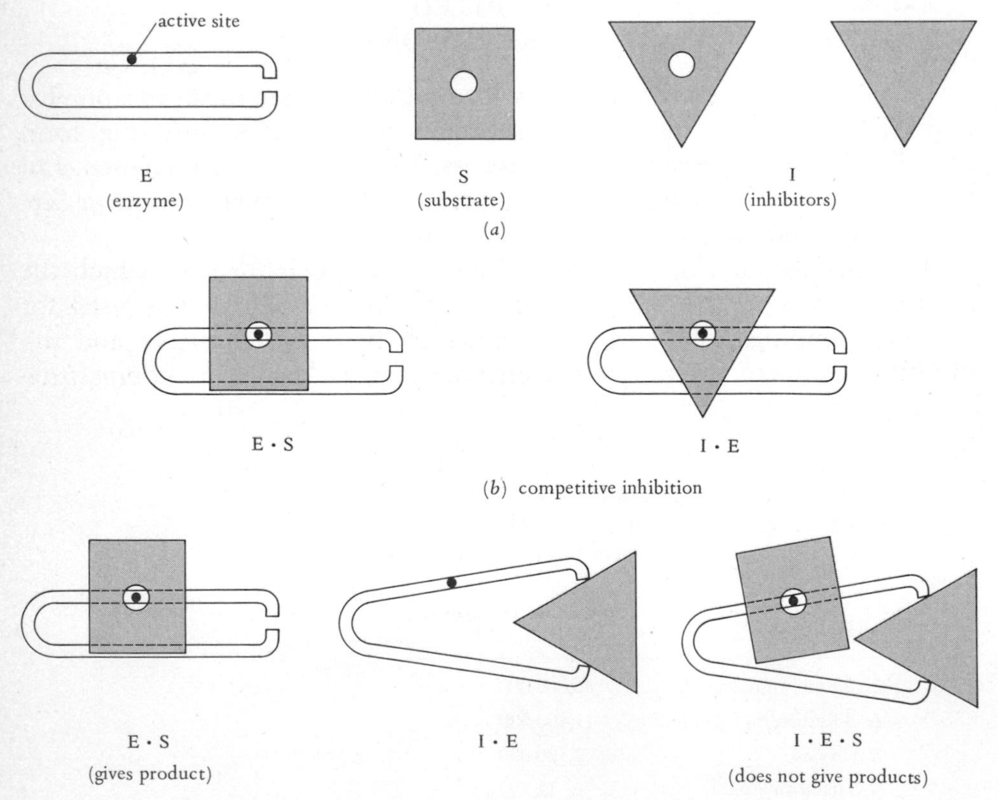

FIGURE 11-2
Models suggesting molecular behavior that could result in inhibition of an enzyme: (*a*) schematic enzyme (*E*), substrate (*S*), and inhibitors (*I*); (*b*) competitive inhibition, in which the inhibitor competes for the active site; (*c*) noncompetitive inhibition, in which the inhibitor associates with the enzyme but not at the active site.

and thus to the rate expression

$$R = k_2[\mathrm{E \cdot S}] = \frac{k_2}{K_M}[\mathrm{E}][\mathrm{S}] \qquad \mathbf{34}$$

Now, however, the total enzyme concentration is not $[\mathrm{E}] + [\mathrm{E \cdot S}]$ but

$$[\mathrm{E_{total}}] = [\mathrm{E}] + [\mathrm{E \cdot S}] + [\mathrm{I \cdot E}] \qquad \mathbf{35}$$

If the enzyme-inhibitor complex is taken as being in equilibrium with the enzyme and with the inhibitor, we can introduce $K_{\mathrm{I \cdot E}}$ as the dissociation constant for the inhibitor-enzyme complex according to

$$\mathrm{I \cdot E} \rightleftarrows \mathrm{I} + \mathrm{E} \qquad K_{\mathrm{I \cdot E}} = \frac{[\mathrm{I}][\mathrm{E}]}{[\mathrm{I \cdot E}]} \qquad \mathbf{36}$$

(The custom in biochemistry is to use dissociation constants rather than formation constants.) Now $[\mathrm{I \cdot E}]$ can be expressed as

$$[\mathrm{I \cdot E}] = \frac{[\mathrm{I}][\mathrm{E}]}{K_{\mathrm{I \cdot E}}} \qquad \mathbf{37}$$

With Eq. **33** for $[\mathrm{E \cdot S}]$ and Eq. **37** for $[\mathrm{I \cdot E}]$, we can write

$$\begin{aligned}[\mathrm{E_{total}}] &= [\mathrm{E}] + \frac{1}{K_M}[\mathrm{E}][\mathrm{S}] + \frac{1}{K_{\mathrm{I \cdot E}}}[\mathrm{I}][\mathrm{E}] \\ &= [\mathrm{E}]\left(1 + \frac{[\mathrm{S}]}{K_M} + \frac{[\mathrm{I}]}{K_{\mathrm{I \cdot E}}}\right) \\ &= [\mathrm{E}]\frac{[\mathrm{S}] + K_M(1 + [\mathrm{I}]/K_{\mathrm{I \cdot E}})}{K_M}\end{aligned} \qquad \mathbf{38}$$

Replacement of the unmeasurable quantity [E] in the expression of Eq. **34** for the rate of the reaction now gives

$$R = \frac{k_2[\mathrm{E_{total}}][\mathrm{S}]}{K_M(1 + [\mathrm{I}]/K_{\mathrm{I \cdot E}}) + [\mathrm{S}]} \qquad \mathbf{39}$$

In reciprocal form this becomes

$$\frac{1}{R} = \frac{1}{k_2[\mathrm{E_{total}}]} + \frac{K_M(1 + [\mathrm{I}]/K_{\mathrm{I \cdot E}})}{k_2[\mathrm{E_{total}}]}\frac{1}{[\mathrm{S}]} \qquad \mathbf{40}$$

Comparing Eq. **40** with Eq. **19** and **20** of Chap. 10, we see how the idea of competitive inhibition leads to a rate expression corresponding to those empirical results in Fig. 10-5, where the slope but not the intercept on a reciprocal plot is affected. This mechanism leads us to expect that at very large substrate concentrations there will be such an excess of substrate that the inhibitor will not be able to compete successfully for the binding site on the enzyme. The reaction rate will then be unaffected by the inhibitor.

A different mechanism, known as *noncompetitive inhibition,* is based on the idea that the inhibitor attaches itself to the enzyme at a site removed from that used by the substrate. The inhibitor can thus form complexes with the enzyme whether it is attached to the substrate or not. The further assumption is made that the inhibitor-enzyme-substrate complex cannot break down to yield product species. The presence of the inhibitor can be visualized as disrupting the structure of the enzyme enough to prevent this critical breakdown step, as is suggested in Fig. 11-2. Now the elementary reactions that must be considered are

$$\mathrm{E + S \rightleftarrows E \cdot S} \quad \textbf{41}$$

$$\mathrm{I + E \rightleftarrows I \cdot E} \quad \textbf{42}$$

$$\mathrm{I + E \cdot S \rightleftarrows I \cdot E \cdot S} \quad \textbf{43}$$

$$\mathrm{E \cdot S \rightarrow E + products} \quad \textbf{44}$$

We proceed as before to the stage

$$R = k_2[\mathrm{E \cdot S}] = \frac{k_2}{K_M}[\mathrm{E}][\mathrm{S}] \quad \textbf{45}$$

Now, however, the expression for [E] in terms of measurable quantities stems from

$$[\mathrm{E_{total}}] = [\mathrm{E}] + [\mathrm{E \cdot S}] + [\mathrm{I \cdot E}] + [\mathrm{I \cdot E \cdot S}] \quad \textbf{46}$$

The assumption of equilibria for the formation of the I · E and I · E · S species allows the introduction of $K_{\mathrm{I \cdot E}}$ and $K_{\mathrm{I \cdot E \cdot S}}$ for the dissociation constants for these species as

$$K_{\mathrm{I \cdot E}} = \frac{[\mathrm{I}][\mathrm{E}]}{[\mathrm{I \cdot E}]} \qquad \text{and} \qquad K_{\mathrm{I \cdot E \cdot S}} = \frac{[\mathrm{I}][\mathrm{E \cdot S}]}{[\mathrm{I \cdot E \cdot S}]} \quad \textbf{47}$$

These expressions, and that of Eq. **33** allow the last three terms of Eq. **46** to be interpreted so that after some rearrangement, [E] can be expressed as

$$[\mathrm{E}] = \frac{K_M[\mathrm{E_{total}}]}{K_M(1 + [\mathrm{I}]/K_{\mathrm{I \cdot E}}) + (1 + [\mathrm{I}]/K_{\mathrm{I \cdot E \cdot S}})[\mathrm{S}]} \quad \textbf{48}$$

With this result the rate expression becomes

$$R = \frac{k_2[\mathrm{E_{total}}][\mathrm{S}]}{(1 + [\mathrm{I}]/K_{\mathrm{I \cdot E \cdot S}})[\mathrm{S}] + K_M(1 + [\mathrm{I}]/K_{\mathrm{I \cdot E}})} \quad \textbf{49}$$

or, in reciprocal form,

$$\frac{1}{R} = \frac{1 + [\mathrm{I}]/K_{\mathrm{I\cdot E\cdot S}}}{k_2[\mathrm{E}_{\mathrm{total}}]} + \frac{K_M(1 + [\mathrm{I}]/K_{\mathrm{I\cdot E}})}{k_2[\mathrm{E}_{\mathrm{total}}]}\frac{1}{[\mathrm{S}]} \qquad \mathbf{50}$$

Further, if we make the reasonable assumption that the dissociation constant for the inhibitor-enzyme complex is little different from that of the inhibitor-enzyme-substrate complex, that is, $K_{\mathrm{I\cdot E\cdot S}} = K_{\mathrm{I\cdot E}}$, this becomes

$$\frac{1}{R} = \left(1 + \frac{[\mathrm{I}]}{K_{\mathrm{I\cdot E}}}\right)\left(\frac{1}{k_2[\mathrm{E}_{\mathrm{total}}]} + \frac{K_M}{k_2[\mathrm{E}_{\mathrm{total}}][\mathrm{S}]}\right) \qquad \mathbf{51}$$

This is the form of the rate equation for inhibitor action of the type shown in Fig. 10-5, where both the slope and intercept are affected.

11-5 THEORIES OF THE ELEMENTARY REACTION PROCESS

Now that we have seen examples of the mechanisms used to explain the observed dependence of the rate of reaction on the concentration of reagents, let us turn to the rate constant itself and its temperature dependence. To do so, we must tackle the details of elementary reaction processes. An introduction is provided by the interpretation given by Arrhenius.

Either for a decay or for a collision that is effective in producing product species from reactants, some amount of energy E_a must be available to allow for the necessary bond breakings and rearrangements. The probability of such an activated complex's occurring will be proportional, among other things, to a Boltzmann factor

$$e^{-E_a/RT}$$

where E_a is the energy necessary for whatever molecular transformation must occur. Thus we can write

$$\frac{\text{Conc. of activated complexes}}{\text{Conc. of reagents}} \propto e^{-E_a/RT} \qquad \mathbf{52}$$

Then if it is assumed that the rate of the reaction is proportional to the concentration of activated complexes, with the introduction of the proportionality factor A we can write

$$\text{Rate} = (Ae^{-E_a/RT})(\text{conc. of reagents}) \qquad \mathbf{53}$$

If the reagent-concentration term is treated in more detail, this agrees with the observed rate equations since it leads to the relation

$$k = Ae^{-E_a/RT} \qquad \mathbf{54}$$

This approach says that the empirical constant E_a is to be interpreted as the energy of the activated-complex molecules compared with the reagent molecules.

The idea of an activated complex can be presented on a plot of the energy of the system as ordinate versus the *reaction coordinate* as abscissa. The reaction coordinate is not any single internuclear distance but depends on all the internuclear distances that change as the reactant molecules are converted into product molecules. In general, it is impossible (and for the present purpose unnecessary) to give a quantitative description of the reaction coordinate. It consists merely of a qualitative description of the extent of the transformation from reactants to produce. The diagram that can be constructed to represent the reaction is shown in Fig. 11-3.

The Arrhenius theory suggests that the energy of the system, when it is partially transformed from reactants to products, is given by the activation-energy term in the empirical expression for the rate constant. This information is added to the elementary reaction curve of Fig. 11-3, and a smooth but otherwise undetermined curve is drawn to pass through the three known energies.

The Arrhenius theory leads to a considerable improvement in our understanding of the reaction process, but it is still a very qualitative theory in that it does not show how the preexponential factor A depends on the molecular properties of the reaction system; nor does it attempt to predict the value of E_a.

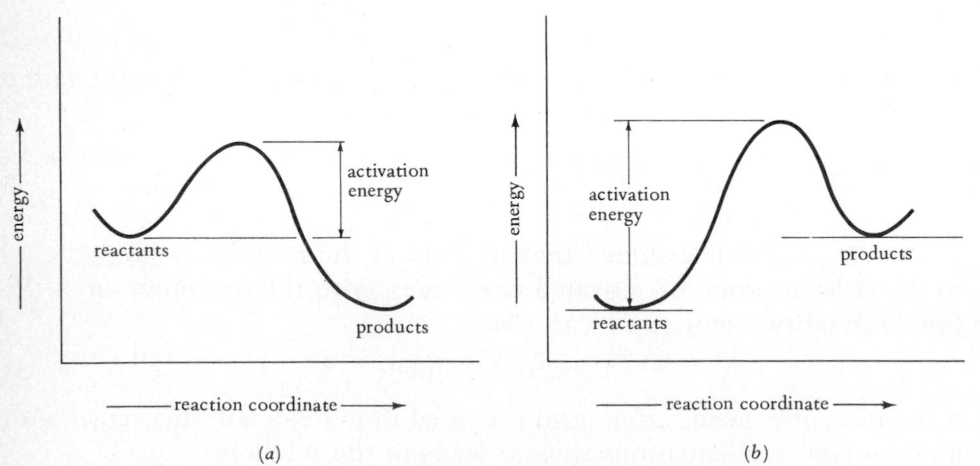

FIGURE 11-3
Forms of the energy change required to transform reactants into products, the activation energy, for (a) an exothermic and (b) an endothermic reaction.

The *transition-state* theory, which attempts to refine these ideas, focuses attention on the species in the reaction process that corresponds to the maximum-energy stage in the reaction. In this theory, this species, called the *activated complex*, or *transition state*, is treated formally as a molecule in spite of its ill-defined nature and transitory existence. More specifically, the theory assumes that this species can be treated as a thermodynamic entity.

To be specific, let us consider a bimolecular elementary reaction that leads to products. As such a reaction proceeds, at all times a small fraction of the reacting species A and B must be in the process of undergoing the bond rearrangements that ultimately convert them to product species. These transition-state species will be denoted by $(AB)^{\ddagger}$, and this species will be treated as though it were a well-defined molecule worthy of thermodynamic consideration. The assumption is satisfactory as long as the reacting system does not lurch so violently from reactants to products that the transition-state species cannot establish an equilibrium with the reactants and products.

When the molecules A and B react to give products, one now can suggest, in an approach similar to that originated by Arrhenius, that A and B establish an equilibrium concentration of the transition-state species and that this species reacts further to form products. Thus

$$\mathrm{A + B \rightleftarrows (AB)^{\ddagger} \rightarrow products}$$

The rate of the reaction depends on two factors: the concentration of the transition-state species and the rate with which it breaks up to give products.

The concentration of the activated complex can be written, at least formally, in terms of the equilibrium expression

$$K^{\ddagger} = \frac{[(\mathrm{AB})^{\ddagger}]}{[\mathrm{A}][\mathrm{B}]}$$

or

$$[(\mathrm{AB})^{\ddagger}] = K^{\ddagger}[\mathrm{A}][\mathrm{B}] \tag{55}$$

Although no value is given for $K^{\ddagger}$, it will be seen that its thermodynamic interpretation is profitable and justifies its introduction.

The rate with which the complex breaks up can be estimated by recognizing that it can fly apart into product molecules when a suitable vibration happens to have a large enough amplitude to break open the complex. The frequency of such a vibration will therefore be something like the rate with which the complex breaks up. If the binding is weak along this reaction coordinate, this frequency will be relatively low. As a result, it can be estimated by placing the vibrational energy $h\nu$ equal to the thermal-energy term kT. Thus we obtain for the frequency of vibration along the reaction coordinate the expression kT/h. (At room temperature this implies a frequency of 0.6×10^{13} Hz, and this corresponds, as we should expect, to the far-infrared spectral region.)

Now the reaction rate is expressed as

$$\text{Rate} = -\frac{d[\text{A}]}{dt} = -\frac{d[\text{B}]}{dt} = \frac{kT}{h} K^{\ddagger}[\text{A}][\text{B}] \qquad \textbf{56}$$

This expression becomes of value when $K^{\ddagger}$ is given a thermodynamic interpretation.

We proceed by treating the formation of the activated complex $(\text{AB})^{\ddagger}$ and the equilibrium constant $K^{\ddagger}$ in terms of the *free energy of activation,* according to Eq. **16** of Chap. 8, $K^{\ddagger}$ can be interpreted as

$$(\Delta G^\circ)^{\ddagger} = -RT \ln K^{\ddagger}$$

or $$K^{\ddagger} = e^{-(\Delta G^\circ)^{\ddagger}/RT} \qquad \textbf{57}$$

For the reaction at a given temperature the free energy of activation can be interpreted in terms of an *entropy of activation* and an *enthalpy of activation,* according to Eq. **2** of Chap. 7, as

$$(\Delta G^\circ)^{\ddagger} = (\Delta H^\circ)^{\ddagger} - T(\Delta S^\circ)^{\ddagger}$$

Substitution of this relation in Eq. **57** yields

$$K^{\ddagger} = e^{+(\Delta S^\circ)^{\ddagger}/R} e^{-(\Delta H^\circ)^{\ddagger}/RT} \qquad \textbf{58}$$

The transition-state theory interpretation of the rate constant is now obtained by recasting the constant terms of Eq. **56** to

$$k_2 = \left(\frac{kT}{h} e^{+(\Delta S^\circ)^{\ddagger}/R}\right) e^{-(\Delta H^\circ)^{\ddagger}/RT} \qquad \textbf{59}$$

With the recognition that the temperature variation of k_2 depends primarily on the exponential term of Eq. **59,** this equation agrees in form with the empirical Arrhenius expression.

The exponential temperature-dependent term now involves an enthalpy rather than an energy of activation. For liquid systems the value to be assigned to $(\Delta H^\circ)^{\ddagger}$ can be shown to be less than E_a by the usually small quantity RT. For gaseous systems the difference is nRT, where n is the molecularity of the reaction. Thus again, as in the simple Arrhenius approach, we are led to the idea of an activation energy.

Application of the idea that an activation energy stands in the way of the elementary reaction steps in an enzyme-catalyzed reaction leads to a diagram of the type shown in Fig. 11-4. Here it is assumed that the greatest energy-demanding step is the conversion of the enzyme-substrate complex to an enzyme-product species. The activation energies for other

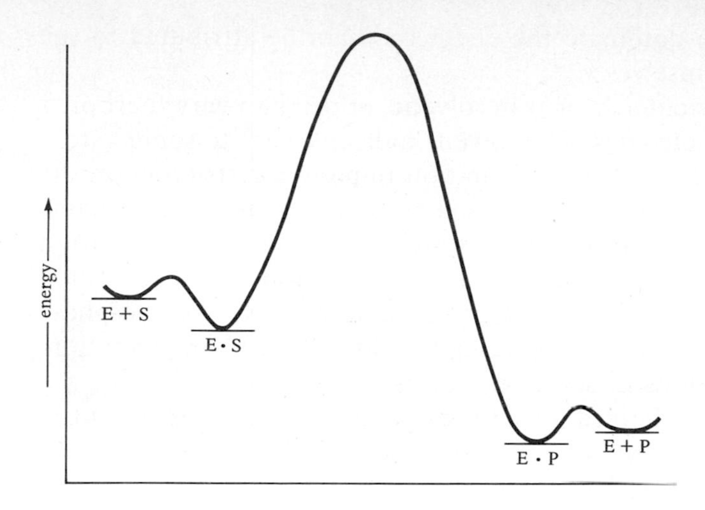

FIGURE 11-4
An energy-reaction coordinate profile for an enzyme-catalyzed reaction that proceeds through the formation of an enzyme-substrate complex. (E is enzyme, S is substrate, P is product.)

steps in the overall reaction are shown to be small and could, in fact, be zero. There is here, even in the energy profile of Fig. 11-4, a basis for a variety of rates that depends on both the enzyme and the substrate.

The tendency of the reaction to proceed depends not only on this energy term, which can be thought of as affecting the entropy of the thermal surroundings, but also on the entropy change due to the molecules involved in the reaction itself. This term enters, as Eq. **59** shows, in the preexponential term. If we avoid treating reactions in which gases are involved, so that the difference between $(\Delta H^\circ)^\ddagger$ and an energy term can be ignored, we can relate the empirical Arrhenius factor A by

$$A = \frac{kT}{h} e^{(\Delta S^\circ)^\ddagger/R} \qquad \textbf{60}$$

Since $(\Delta S^\circ)^\ddagger$ is the entropy change in going from the reagents to the activated complex, and since little can easily be said about the properties of the activated complex, the transition-state theory tends to avoid any definite quantitative predictions. In spite of the ill-defined nature of the transition state, a number of conclusions can be drawn concerning $(\Delta S^\circ)^\ddagger$, and they can be compared with the kinetic results.

The formation of an activated complex from reactant molecules is accompanied by the conversion of translational-like and rotational-like degrees of freedom of the reactants to vibrational degrees of freedom of the transition-state species. The more widely spaced energy levels for this latter type of molecular motion imply a smaller entropy and thus a negative value for $(\Delta S)^\ddagger$. In addition, however, solvent molecules are involved, and the effect on them can dominate the effect that can be attributed to the reacting molecules themselves.

All solutes (particularly ions) in solvents other than very inert ones interact with solvent molecules. The solvent molecules are to some extent oriented about the solute, and this orientation imposes a restriction on the motion of some of the solvent molecules. This solvation is an appreciable factor in determining the entropy of the system. Changes in this solvation entropy must therefore be considered in the formation of the activated complex. Therefore the uniformly negative value of $(\Delta S)^\ddagger$ for gas-phase dimerization-type reactions, corresponding to a loss of freedom of motion, does not hold for reactions of solvated species in solution.

When oppositely charged ions, for example, react to form a neutral molecule, the extent of solvation is greatly reduced. Then the activated complex can be expected to be formed with a decrease in solvation and a corresponding positive entropy of activation. An example of such a situation is provided by the displacement of water molecules in the reaction

$$Cr(OH_2)_6^{3+} + CNS^- \rightarrow Cr(OH_2)_5CNS^{2+} + H_2O$$

which has a $(\Delta S)^\ddagger$ value of $+30$ cal/mol deg.

On the other hand, the formation of an activated complex that carries charges when the reagents do not will lead to a large negative value for $(\Delta S)^\ddagger$ corresponding to the additional loss of freedom of motion by the solvating molecules. The displacement of bromide in the reaction step

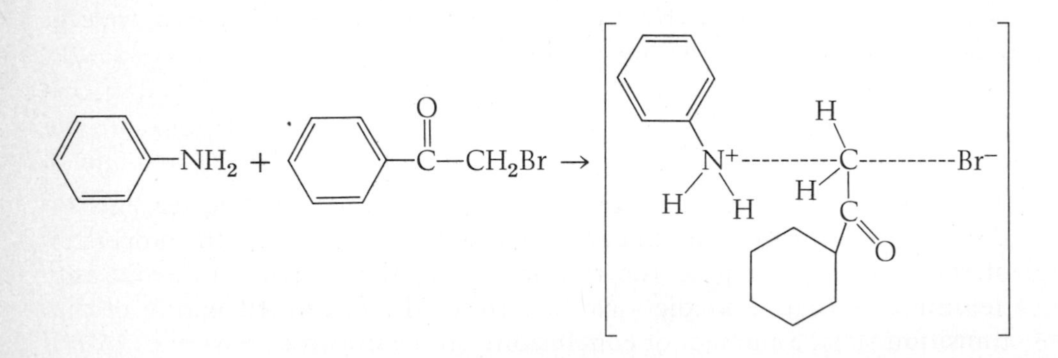

is an example of such a situation. The entropy of activation for this reaction has been reported as -50 cal/mol deg, a very large negative value.

Although not all entropies of activation-of-solution reactions can be rationalized so easily, this type of argument shows how the transition-state theory gives a valuable framework within which observed rate constants can be discussed.

PROBLEMS

1 The estimates of the rates of encounters and the duration of an encounter, in Sec. 11-1, were based on the assumption of relatively small reacting molecules with diffusion coefficient and radius values of 10^{-5} cm²/sec and 4 Å, respectively. What values would have been obtained for the rate constants of Eqs. **5** and **6** and for the cage lifetime of Eq. **7** if representative macromolecule values of the diffusion coefficient and the molecular radius of 5×10^{-7} cm²/sec and 20 Å had been used?

2 A representative species in solution reacts according to the diffusion-controlled rate law of Eq. **5.** An alternative reaction possibility is that in which one of the reacting molecules dissociates or falls apart in a way that corresponds to a first-order rate law. For this second reaction to proceed as rapidly as the first, how many vibrations of the bond that breaks can occur, on the average, before the bond does break? Assume reagent concentrations are 1 *M* and the bond-vibration frequency is 3×10^{13} Hz, that is, $\bar{\nu} = 1000$ cm^{-1}.

3 Deduce a rate law for a reaction $A + 2B \rightarrow C$ using the stationary-state method and the mechanism

$$A + B \rightleftarrows M$$

$$M + B \rightarrow C$$

What relative values of the rate constants for the elementary reactions would lead to rate laws that are simple products of reagent concentrations?

4 What rate laws might be expected for a reaction $A + B \rightarrow C$ if the following mechanism is expected?

$$2A \rightleftarrows M$$

$$M + B \rightarrow A + C$$

5 Verify that the rate law of Eq. **18** follows from the given mechanism for the reaction of $Co(CN)_5OH_2^{2-}$ with I^-.

6 The treatment of the mechanisms of enzyme-catalyzed reactions in Sec. 11-3 introduces the turnover number as the maximum number of product molecules that an enzyme molecule can produce per minute. The largest reported value is that of 36,000,000 for carbonic anhydrase. An upper limit to turnover numbers is determined by the time it takes for a product molecule and the enzyme to get away from each other after a successful conversion step so that the enzyme can go about its business again. Assuming a representative macromolecule value for D of 5×10^{-7} cm²/sec and molecular radii of 20 Å, and taking a separation to be complete when the molecules have parted by an extra distance of 40 Å, how long, on the average, will it take for an enzyme and its product molecule to separate from each other? On this basis, what is the maximum number of product molecules than an enzyme molecule can produce per minute?

7 The formation of urea from a solution containing ammonium and cyanate ions follows the rate law

Rate = $k[NH_4^+][OCN^-]$

The most reasonable reaction mechanism is based on an initial reversible acid-base reaction

$NH_4^+ + OCN^- \rightleftarrows NH_3 + OCNH$

This is then followed by the slow formation of product according to

$NH_3 + OCNH \rightarrow OC(NH_2)_2$

Use the stationary-state method to develop a rate law on the basis of this mechanism. On what basis does this lead to the observed rate law of Example 10-1? How is the observed rate constant to be interpreted?

12 Rates of Transport Processes

The rates with which solute molecules travel or are transported through a solvent or a membrane are important in many biological processes. Experimental studies of macromolecules in solution also depend on such transport.

Although transport can be included in a general study of kinetics, it usually implies processes that do not entail any chemical transformations. In most experimental studies of transport in macromolecular solutions, this distinction between physical and chemical can be made. In biological systems, as in the passage of materials through membranes, chemical changes often accompany the physical transport, at least in a transitory way. The studies of this chapter will provide background for these more complex processes without actually dealing with them.

12-1 DIFFUSION

In chemical kinetics we dealt with the change of chemical species in a reaction mixture with time. In diffusion studies, we deal with the change in the location of solute with time. The conceptually simplest (but not experimentally the most convenient) illustration of the nature of this time dependence is given by the changes exhibited by an initially thin sheet of solute (Fig. 12-1). As time progresses, the concentration of the solute would correspond to the curves that are shown. The rate at which these curves develop, i.e., the rate of spreading out of the solute, depends, for a given solvent and temperature, on the nature of the solute.

The driving force for diffusion is the rate of change of concentration along the diffusion direction, i.e., the concentration gradient $\partial c/\partial x$, where c is the concentration and x the one dimension we deal with here. The curved delta ∂ indicates partial derivatives, as in Sec. 7-3. Since diffusion produces solutions with concentrations that depend both on position and time, we must keep clear which variable we are treating in derivatives and integrals. The symbol ∂ instead of d indicates that here only x is being treated as a variable. That the second of the two variables is held constant can be assumed or can be shown by writing $(\partial c/\partial x)_t$.

Experimental studies of diffusion show that a proportionality exists between the rate of flow across any cross section with area A and the concentration gradient $\partial c/\partial x$ at that cross section. Thus we write

$$\text{Rate of diffusion in x direction} \propto A\frac{\partial c}{\partial x}$$

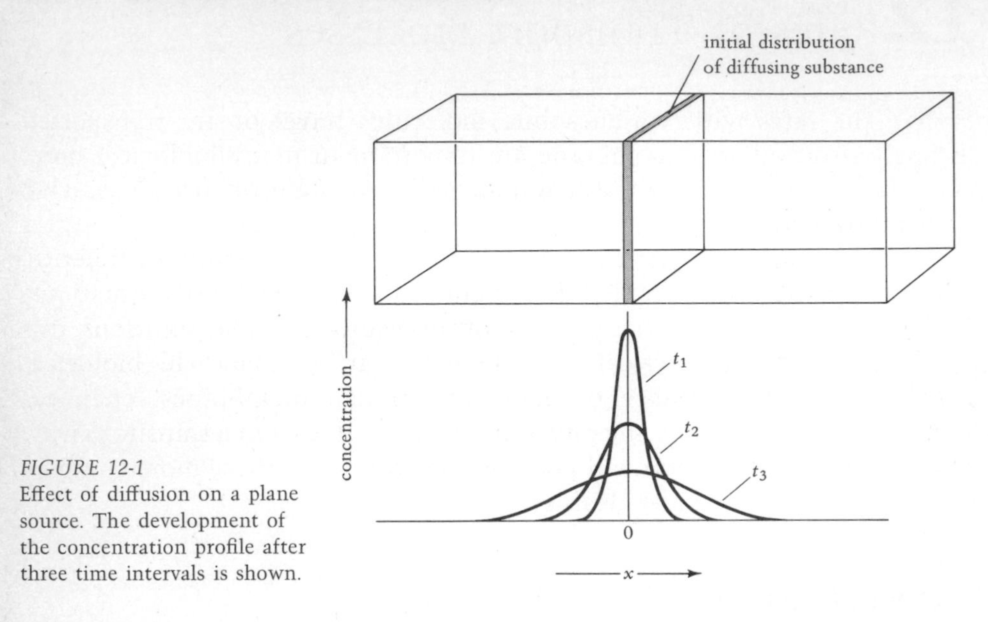

FIGURE 12-1
Effect of diffusion on a plane source. The development of the concentration profile after three time intervals is shown.

The diffusion direction is that in which the concentration decreases, i.e., in which $\partial c/\partial x$ is negative. This proportionality can be written as an equality by the introduction of the proportionality constant D as

$$\text{Rate of diffusion in x direction} = -DA\frac{\partial c}{\partial x} \qquad \mathbf{1}$$

The negative sign allows D, the *diffusion coefficient*, to take on positive values. Equation **1** is based on observations of diffusion processes and is known as *Fick's first law of diffusion*. The proportionality constant D depends on the nature of both the diffusing solute and the solvent. If the rate of diffusion is expressed in terms of the same quantity measure that is used in c, such as grams, molecules, or moles, suitable units for D are seen to be square centimeters per second.

If diffusion through a *unit* cross-sectional area is considered, Eq. **1** becomes simply

$$\text{Rate of diffusion} = -D\frac{\partial c}{\partial x} \qquad \mathbf{2}$$

It is often more convenient to have a description of the effect of diffusion on the concentration in a volume element, as in Fig. 12-2. For this we need to calculate the net rate with which the diffusing substance accumulates in the volume element $A\,dx$. This can be done by calculating the rate with which the substance enters the volume element at x and the

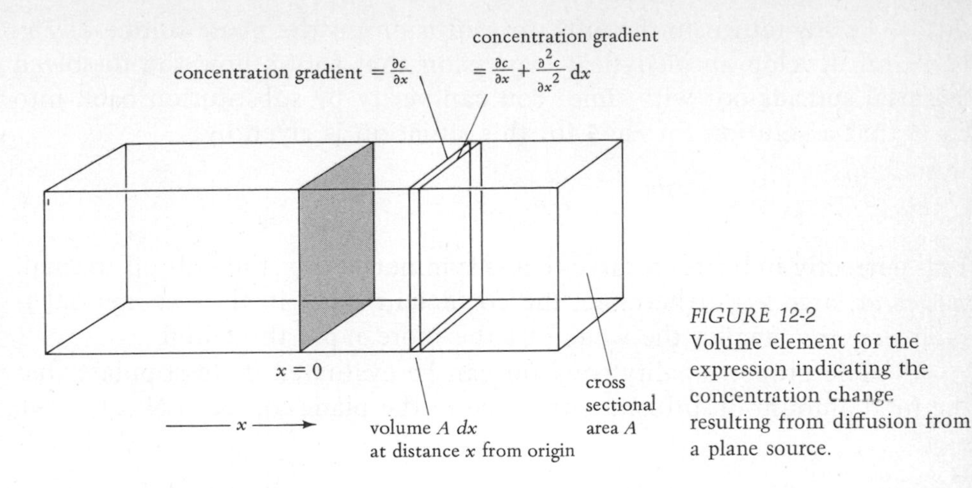

FIGURE 12-2
Volume element for the expression indicating the concentration change resulting from diffusion from a plane source.

rate with which it leaves at x + dx. The concentration gradient $\partial c/\partial x$, in general, will vary along x, and the rate with which it varies will be

$$\frac{\partial}{\partial x}\left(\frac{\partial c}{\partial x}\right) = \frac{\partial^2 c}{\partial x^2}$$

Now Rate of entry at x $= -DA\,\frac{\partial c}{\partial x}$ **3**

Rate of leaving at x + dx $= -DA\left(\frac{\partial c}{\partial x} + \frac{\partial^2 c}{\partial x^2}\, dx\right)$ **4**

Subtraction of Eq. **4** from Eq. **3** gives

$$\text{Rate of accumulation} = -DA\,\frac{\partial c}{\partial x} - \left[-DA\left(\frac{\partial c}{\partial x} + \frac{\partial^2 c}{\partial x^2}\, dx\right)\right]$$

$$= DA\,\frac{\partial^2 c}{\partial x^2}\, dx$$

Division of the rate of accumulation, i.e., the rate with which the amount changes in the volume element, by the volume $A\, dx$ of the element gives the rate of change of concentration $\partial c/\partial t$ as

$$\frac{\partial c}{\partial t} = D\,\frac{\partial^2 c}{\partial x^2} \qquad \mathbf{5}$$

This relation is known as *Fick's second law.* Equation **5** is the basis for the deduction of most of the consequences of diffusion.

Let us return to the effect of diffusion on the plane source of Fig. 12-1 and develop an analytical expression that shows how the dissolved material spreads out with time. You can verify by substitution back into Eq. **5** that a solution for Eq. **5** for this situation is given by

$$c = \frac{\text{const}}{t^{1/2}} e^{-x^2/4Dt} \qquad \mathbf{6}$$

This correctly indicates a curve that is symmetric in x and falls off to small values at large x. Furthermore, the coefficient of x^2 in the exponential is such that the smaller the value of t the more rapid the falloff.

The proportionality constant can be evaluated if we stipulate that the total amount of diffusing substance in the plane source is N. Then we have

$$N = \int_{-\infty}^{+\infty} c\, dx = \frac{\text{const}}{t^{1/2}} \int_{-\infty}^{+\infty} e^{-x^2/4Dt}\, dx \qquad \mathbf{7}$$

Definite integrals of this type are tabulated and give

$\int_{-\infty}^{+\infty} e^{-ax^2}\, dx = \sqrt{\pi/a}$. In this way we obtain

$$N = 2 \text{ const } (\pi D)^{1/2}$$

and $$c = \frac{N}{2(\pi D t)^{1/2}} e^{-x^2/4Dt} \qquad \mathbf{8}$$

Given the initial amount N of the substance set at $x = 0$ and the value of the diffusion coefficient, one can calculate the amount in any volume element, i.e., the concentration, at any position at a later time. Curves with the shape of those of Fig. 12-1 are obtained.

A simple and revealing measure of the tendency of particles to move as a result of this diffusion process is given by the average squared net distance the particles travel in a one-dimensional diffusion process. (The average net distance is not helpful. The particles are just as likely to move to the right as to the left, so that the average net distance is always zero. In the average of the net *squared* distance, movements to the right and left add up rather than cancel out.)

The amount of diffusing substance that after time t has reached an element dx at x will be $c dx$. The squared net distance traveled by the molecules of the diffusing substance is x^2. The average squared net distance is calculated by adding up, by means of an integration, the x^2 values contributed by the molecules of the sample and dividing by the total number of molecules. Thus

$$\overline{x^2} = \frac{\int_{-\infty}^{+\infty} x^2 c\, dx}{N}$$

$$= \frac{1}{2(\pi D t)^{1/2}} \int_{-\infty}^{+\infty} x^2 e^{-x^2/4Dt}\, dx \qquad \mathbf{9}$$

Again the integral is a tabulated one

$$\int_{-\infty}^{+\infty} x^2 e^{-ax^2}\, dx = \frac{1}{2a}\sqrt{\frac{\pi}{a}}$$

Substitution of this integral value in Eq. **9** gives

$$\overline{x^2} = 2Dt$$

and $$\sqrt{\overline{x^2}} = \sqrt{2Dt} \qquad \mathbf{10}$$

This simple result can be used to estimate the extent of movement that can be accomplished by a diffusional motion. For example, the value of D for sucrose in water at 25°C is reported to be 4.0×10^{-6} cm^2/sec. Thus

$$\overline{x^2} = 8.0 \times 10^{-6} t$$

and $$\sqrt{\overline{x^2}} = 2.83 \times 10^{-3} \sqrt{t}$$

for x in centimeters and t in seconds.

For example, in 1 h, or 3600 sec, the value of $\sqrt{\overline{x^2}}$ is 0.17 cm; this is a measure of the average net distance that sucrose molecules travel in a 1-h one-dimensional diffusion experiment.

Experimental studies that lead to values of the diffusion coefficient are not made by starting with the planar source of Fig. 12-1. Instead a solution is brought in contact with the solvent, as in Fig. 12-3. Furthermore, the progress of the solute into the solvent is not measured, but optical methods are used to reveal the *gradient*, i.e., the rate of change, of the concentration in the neighborhood of the initial boundary.

The development of this concentration gradient can be deduced from an expression for the concentration as a function of time. This concentration-versus-time relation is obtained by recognizing that the solute of each dx element of the solution will diffuse in the same manner as in the plane-source treatment, i.e., will follow Eq. **8.** To find the concentration at some point P as in Fig. 12-4, it is only necessary to integrate over all such planar elements. If y locates the point P and x locates a representative element of the diffusing solution, as in Fig. 12-4, the integration is over

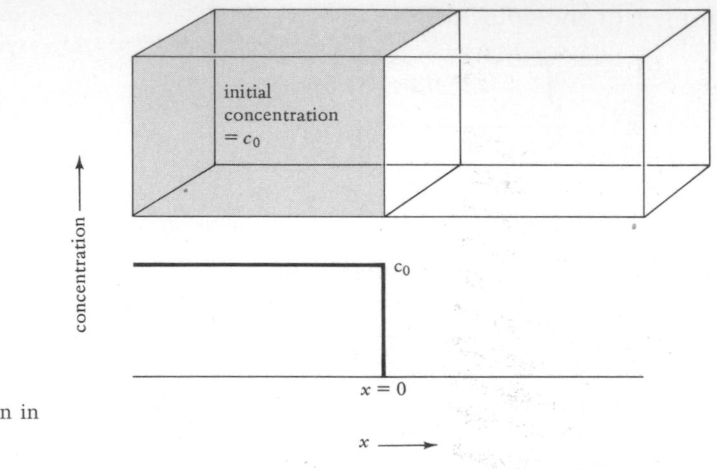

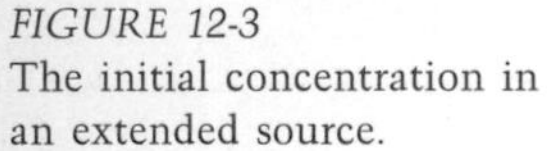

FIGURE 12-3
The initial concentration in an extended source.

all values of x from y to ∞. This development, with the assumption of an initial solute concentration of c_0 and an amount of diffusing substance in the volume element dx of $c_0\, dx$, leads to

$$c = \frac{c_0}{2(\pi Dt)^{1/2}} \int_{x=y}^{x=\infty} e^{-x^2/4Dt} \qquad \mathbf{11}$$

The integral can be put in a form so that tabulated values, known as the error-function integrals, can be used to give c at any values of y and t if the value of D is known.

Often more useful for treating experimental results are values of the magnitude of $\partial c/\partial y$. The derivative equivalent to the integral of Eq. **11** leads, if the sign of the gradient is ignored, to

$$\frac{\partial c}{\partial y} = \frac{c_0}{2(\pi Dt)^{1/2}} e^{-y^2/4Dt} \qquad \mathbf{12}$$

This gradient expression is the basis for most determinations of the diffusion coefficient D. The gradient curves that develop according to this expression are illustrated in Fig. 12-5.

In the commonest optical system one measures the gradient of the refractive index. This can be taken to be proportional to the concentration gradient. One way to proceed is to recognize that the maximum in the gradient curve, which occurs at $y = 0$, has $\partial c/\partial y = c_0/2(\pi Dt^{1/2})$ and thus

$$\frac{dn}{dy} = (\text{const}) \frac{c_0}{2(\pi Dt)^{1/2}} \qquad \mathbf{13}$$

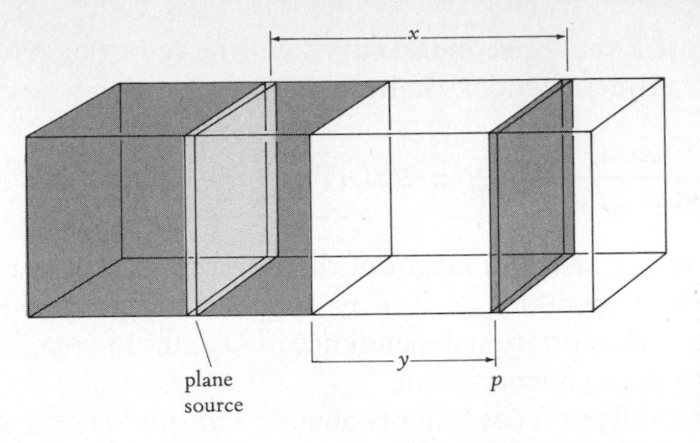

FIGURE 12-4
Diffusion from an extended source treated in terms of diffusion from plane sources.

where n is the refractive index. Furthermore, the area under the gradient curve is

$$\int_{-\infty}^{+\infty} \frac{c_0}{2(\pi Dt)^{1/2}} e^{-y^2/4Dt}\, dy = \frac{c_0}{2(\pi Dt)^{1/2}} (4\pi Dt)^{1/2} = c_0 \qquad \textbf{14}$$

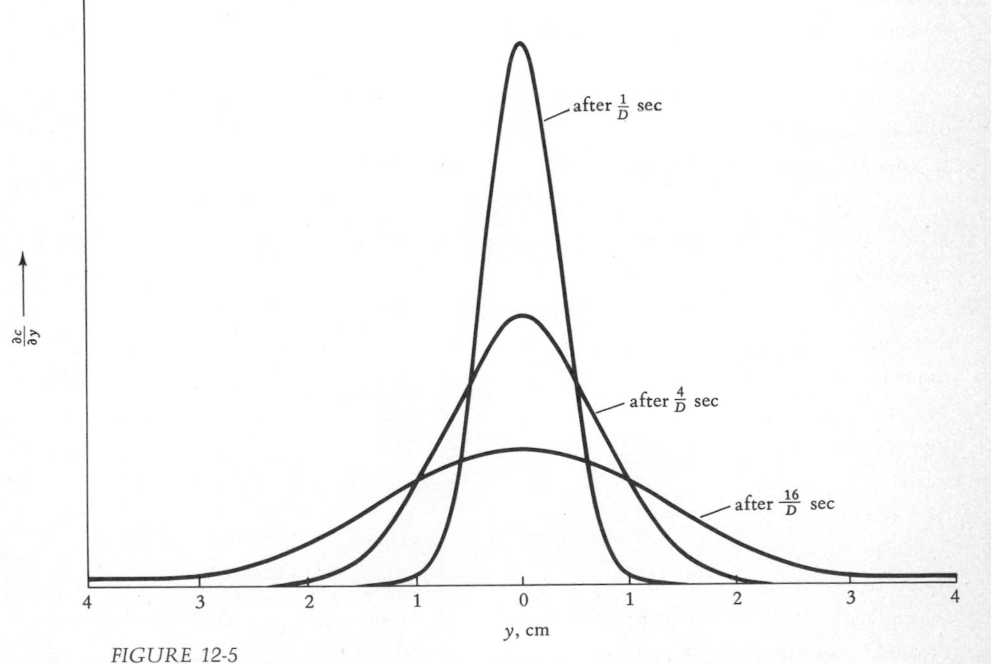

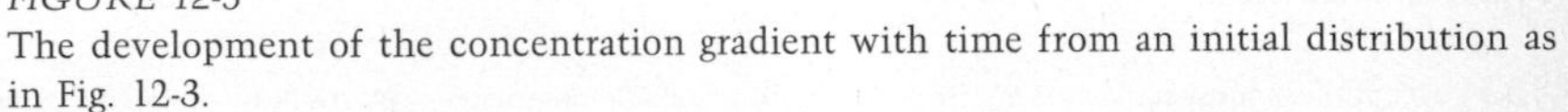

FIGURE 12-5
The development of the concentration gradient with time from an initial distribution as in Fig. 12-3.

Thus, the area under the refractive-index curve will be $(\text{const})c_0$. Finally, we can write, for the refractive-index-gradient curve

$$\frac{\text{Area}}{\text{Height}} = \frac{(\text{const})c_0}{(\text{const})c_0/2(\pi Dt)^{1/2}} = 2(\pi Dt)^{1/2} \qquad \mathbf{15}$$

Thus, measurements of the area and height of the refractive-index-gradient curve at various times yield values of D. Any concentration dependence can be attributed to the concentration dependence of D and can be removed by an extrapolation to zero concentration.

Some results for diffusion coefficients obtained in this way are given in Table 12-1.

TABLE 12-1
Physical Data, for 20°C and Water Solutions of Some Proteins†

protein	**molecular weight**	**diffusion coefficient** $D \times 10^7$, cm²/sec	**sedimentation coefficient** $s \times 10^{13}$, sec	**frictional ratio** f/fo	**specific volume** v, g/ml
Cytochrome *c* (bovine heart)	13,370	11.4	1.17	1.19	0.71
Myoglobin (horse heart)	16,900	11.3	2.04	1.11	0.74
Chymotrypsinogen (bovine pancreas)	23,240	9.5	2.54	1.19	0.73
β-Lactoglobulin (goat milk)	37,100	7.48	2.85	1.26	0.751
Serum albumin (human)	68,500	6.1	4.6	1.29	0.734
Hemoglobin (human)	64,500	6.9	4.46	1.16	0.75
Catalase (horse liver)	247,500	4.1	11.3	1.25	0.73
Urease (jack bean)	482,700	3.46	18.6	1.19	0.73
Fibrinogen (human)	339,700	1.98	7.63	2.34	0.71
Myosin (cod)	524,800	1.10	6.43	3.63	0.73
Tobacco mosaic virus	40,590,000	0.46	198	2.03	0.75

† All data except those for specific volume from A. L. Lehninger, "Biochemistry," Worth Publishers, Inc., New York, 1970.

12-2 A MOLECULAR INTERPRETATION OF DIFFUSION

To see how these experimental diffusion coefficients can be related to properties of the system, and particularly of the solute macromolecules, we take a molecular view of the diffusion process.

Consider diffusion across a distance interval dx over which the concentration changes from c to $c - dc$. The force that drives the molecules to the more dilute region can be related to the difference in the molar free energy of the solute at concentration c and at concentration $c - dc$. If Henry's law is assumed, the discussion of Sec. 7-4 leads to the free-energy difference *per molecule* of

$$G_{c-dc} - G_c = \frac{RT}{\mathfrak{N}} \ln \frac{P_{c-dc}}{P_c} = \frac{RT}{\mathfrak{N}} \ln \frac{c - dc}{c} \qquad \mathbf{16}$$

or

$$dG = \frac{RT}{\mathfrak{N}} \ln \left(1 - \frac{dc}{c}\right)$$

$$\cong -\frac{RT}{\mathfrak{N}} \frac{dc}{c} \qquad \mathbf{17}$$

where the relation $\ln(1 - y) = -y$ for small y has been used.

This free-energy difference corresponds to the work done in the transfer of one macromolecule across the distance dx and can therefore be written as a force times the distance dx. Thus dG = driving force $\times\, dx$, or

$$\text{Driving force} = \frac{dG}{dx} = -\frac{RT}{\mathfrak{N}} \frac{1}{c} \frac{dc}{dx} \qquad \mathbf{18}$$

A frictional force sets and balances this diffusion force when some constant velocity is reached. The frictional force exerted by a viscous solvent fluid of viscosity η has been derived for a macroscopic sphere of radius r by Stokes as

$$\text{Frictional force} = 6\pi r\eta \frac{dx}{dt} \qquad \mathbf{19}$$

It appears suitable to apply this expression to the motion of reasonably spherical macromolecules. The diffusion velocity increases, therefore, until the force of Eq. **19** just balances that of Eq. **18.** Then

$$6\pi r\eta \frac{dx}{dt} = -\frac{RT}{\mathfrak{N}} \frac{1}{c} \frac{dc}{dx}$$

or

$$c \frac{dx}{dt} = -\frac{RT}{6\pi r \mathfrak{N} \eta} \frac{dc}{dx} \qquad \mathbf{20}$$

If we agree that c implies a mass-per-unit-volume measure of concentration, it can be seen that the product $c\,dx/dt$ can be interpreted as the rate with which the diffusing substance moves through a unit cross section at x. This follows, as Fig. 12-6 suggests, from the fact that dx/dt, the average diffusion velocity in the x direction, is the distance the diffusing molecules will travel per unit time. Thus all molecules within a distance dx/dt of the cross section at x will cross in this unit time. These molecules are in a volume equal to dx/dt times the unit cross-sectional area. The mass of these molecules is the product of this volume and the concentration expressed as mass per unit volume. Thus $c\,dx/dt$ is the amount per unit time, i.e., the rate, with which the solute passes through the cross section. According to Eqs. **2** and **20,** we can now write

$$-D\frac{\partial c}{\partial x} = -\frac{RT}{6\pi r \mathfrak{N} \eta}\frac{\partial c}{\partial x}$$

This leads to the identification

$$D = \frac{RT}{6\pi r \mathfrak{N} \eta} \qquad \textbf{21}$$

Measurements of D and η could therefore lead to a value of the radius r for the macromolecule. Such a procedure is a little unsatisfactory. Molecules do not necessarily obey Stokes' law, even if they are spherical. Furthermore, macromolecules will generally be solvated and in moving through the solution will, to some extent, carry along this solvation layer. Equation **21** is important, however, in that it provides a way of determining the *effective* value of the group of terms $6\pi r \mathfrak{N} \eta$ for a solute characterized by molecules with radius r and a solvent characterized by viscosity η.

One can also investigate the extent to which the diffusing molecules act like Stokes' law spheres. The frictional effect of such spheres can be expressed as the drag force per unit velocity. If this now is represented by f_0, we have, according to Eq. **19,**

$$f_0 = 6\pi r \eta \qquad \textbf{22}$$

A rearrangement of Eq. **21** shows that we have attempted to equate this frictional force to an effective frictional force by

$$6\pi r \eta = \frac{RT}{D\mathfrak{N}} \qquad \textbf{23}$$

Now let f be the effective frictional force as given by

$$f = \frac{RT}{D\mathfrak{N}} \qquad \textbf{24}$$

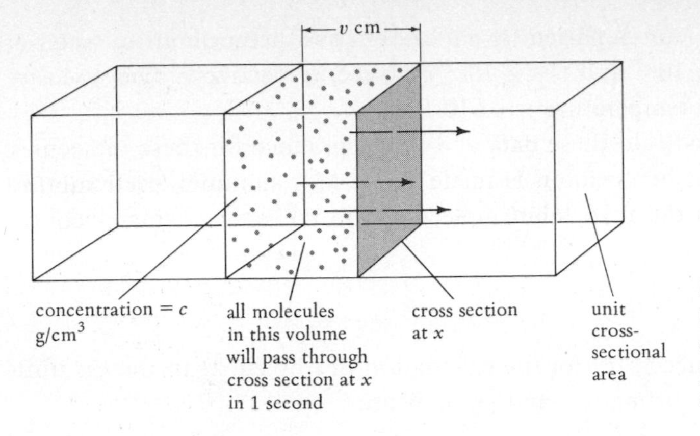

FIGURE 12-6
Relation of velocity dx/dt and concentration to amount passing through a unit cross-sectional area.

The extent to which the diffusing molecules perform like spheres that obey Stokes' law can be seen by the closeness of the so called *frictional ratio* f/f_0 to unity. We expect globular proteins to give values of this ratio near unity and rodlike molecules to give values considerably greater than unity.

An example is provided by hemoglobin which, with a molecular weight of 64,500 g/mol and a specific volume of 0.75 ml/g, is calculated on the basis of a spherical shape to have a radius of 26.8 Å. With the water viscosity of 0.0100 P at 20°C

$$f_0 = 6\pi r\eta = 5.05 \times 10^{-8}$$

Further, with the value of D of 6.9×10^{-7} from Table 12-1,

$$f = \frac{RT}{D\mathfrak{N}} = 5.86 \times 10^{-8}$$

The frictional ratio for hemoglobin is thus

$$\frac{f}{f_0} = 1.16$$

The closeness to unity again confirms the general spherical shape of such globular protein molecules. Some representative values of the frictional ratio are included in Table 12-1.

Finally, it is important to recognize that Eq. **21** is useful even if it is not satisfied when molecular radii are inserted for r. It is used as a means of expressing effective radii for diffusing molecules or effective values of the $6\pi r\eta$ term that Stokes' law has introduced.

Example 12-1

Both myoglobin and hemoglobin are globular proteins that seem suitable for the spherical-particle–Stokes' law development that led to Eq. **21.** The diffusion coefficients reported for myoglobin and hemoglobin in water at 20°C are 11.3×10^{-7} and 6.9×10^{-7} cm²/sec, respectively. The viscosity of water at this temperature is 0.010 P.

What radii do these data and Eq. **21** produce for these molecules? It is known that hemoglobin is made up of four subunits, each subunit being similar to the myoglobin molecule. Are the radii estimates consistent with this?

Solution

The quantities necessary for the calculation of r by Eq. **21** in the cgs units of ergs, dynes, centimeters, and seconds are

$$R = 8.314 \times 10^7 \text{ erg/deg mol}$$

$$T = 293 \text{ K}$$

$$\mathfrak{N} = 6.02 \times 10^{23} \text{ molecules/mol}$$

$$\eta = 0.0100 \text{ g/cm sec}$$

$$D = \begin{cases} 11.3 \times 10^{-7} \text{ cm}^2/\text{sec} & \text{for myoglobin} \\ 6.9 \times 10^{-7} \text{ cm}^2/\text{sec} & \text{for hemoglobin} \end{cases}$$

Substitution in Eq. **21,** with the recognition of the unit identity, 1 dyn = 1 g cm/sec² (which can be seen from the $f = ma$ relation) yields, for myoglobin

$$r = \frac{RT}{6\pi\mathfrak{N}\eta D}$$

$$= 19.0 \times 10^{-8} \text{ cm} = 19.0 \text{ Å}$$

Similarly, for hemoglobin, one obtains

$$r = 31.1 \times 10^{-8} \text{ cm} = 31.1 \text{ Å}$$

The radius of hemoglobin, as expected, is found to be greater than that of myoglobin, and both are reasonably consistent with the unit-cell volume values found by the method of Sec. 2-4.

The volumes of the molecules can be calculated as

$$V_{\text{myoglobin}} = \tfrac{4}{3}\pi(19.0)^3 = 28{,}700 \text{ Å}^3$$

$$V_{\text{hemoglobin}} = \tfrac{4}{3}\pi(31.1)^3 = 126{,}000 \text{ Å}^3$$

According to this calculation, the ratio of the molecule volumes is 4.4, and, particularly with the recognition that there would be some free volume within the hemoglobin molecule, this value is consistent with the idea that the hemoglobin molecule consists of four myoglobinlike units.

12-3 THE ULTRACENTRIFUGE AND THE SEDIMENTATION-VELOCITY METHOD

In Sec. 12-2 the tendency of a solute to diffuse across a concentration gradient was treated. Macromolecules in solution can be made to alter their distribution in space by subjecting them to forces other than concentration gradients. If the macromolecule differs in density from the solvent, the simplest demonstration of this consists of allowing a solution to stand so that the force of gravity acts. A greater and more easily observed effect can be produced by using an ultracentrifuge, in which a sample of the macromolecule solution rotates at a very high speed, in the neighborhood of 10,000 to 80,000 rev/min. The ultracentrifuge, some features of which are shown in Fig. 12-7, is a very important tool for macromolecule research.

Two essentially different types of experiments can be performed to determine the behavior of solutions of macromolecules upon ultracentrifugation. In one, the rate of movement of the solute during the centrifugation is observed. In the other, the sample is centrifuged until an equilibrium distribution is obtained.

The first method starts with a well-defined boundary, or layer, of solution near the center of rotation and follows the movement of this layer toward the outside of the cell as a function of time. Such a method is termed a *sedimentation-velocity* experiment.

A particle of mass m at a distance x from the center of rotation experiences a force given by

$$f_{\text{centrif}} = m'x\omega^2 \qquad \mathbf{25}$$

where ω is the angular velocity in radians per second and m' is the effective mass of the solute particle, i.e., the actual mass corrected for the buoyancy effect of the solvent.

To express this buoyancy effect we first recognize that v, the specific volume of the solute, is the volume of 1 g of solute. The volume of m g of solute is mv, and the mass of this volume of solvent is $mv\rho$, where ρ is the solvent density. Thus the effective mass m' of the solute is $m - mv\rho = m(1 - v\rho)$. We now can rewrite Eq. **25** as

$$f_{\text{centrif}} = m(1 - v\rho)x\omega^2 \qquad \mathbf{26}$$

This force is balanced for some constant-drift velocity dx/dt by a frictional force given by Stokes' law, Eq. **19**, as

$$f_{\text{fric}} = 6\pi r\eta \frac{dx}{dt} \qquad \mathbf{27}$$

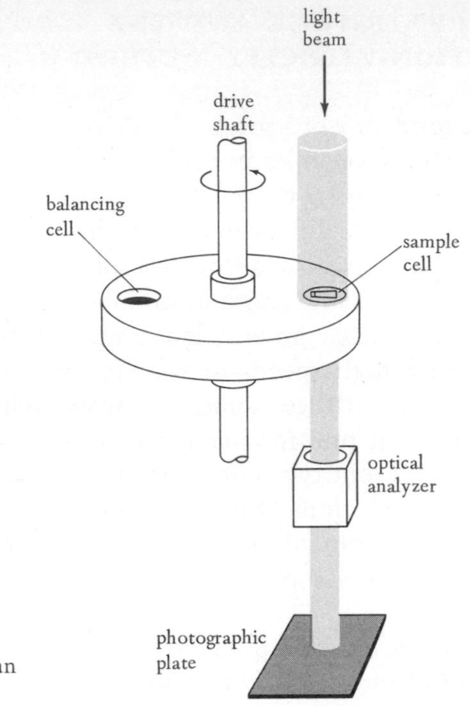

FIGURE 12-7
The basic features of an ultracentrifuge.

Equating these two force expressions will lead us to the constant-drift velocity. A rearrangement of the equality

$$m(1 - v\rho)x\omega^2 = 6\pi\eta r \frac{dx}{dt} \tag{28}$$

that collects the dynamic variables gives

$$\frac{dx/dt}{x\omega^2} = \frac{m(1 - v\rho)}{6\pi r\eta} \tag{29}$$

The collection of dynamic terms on the left side of Eq. **29** is reported to describe the results of sedimentation-velocity experiments. This collection $(dx/dt)/x\omega^2$ can be looked on as the velocity with which the solute moves per unit centrifugal force field. The *sedimentation coefficient* s is introduced as

$$s = \frac{dx/dt}{x\omega^2} \tag{30}$$

The experimental results can therefore be tabulated as values of s. The value of s for many macromolecules comes out to be of the order of 10^{-13} sec.

A convenient unit having the value 10^{-13} sec has therefore been introduced, called a *svedberg*, in honor of T. Svedberg, who did much of the early work with the ultracentrifuge.

According to Eq. **29,**

$$s = \frac{dx/dt}{x\omega^2} = \frac{m(1 - v\rho)}{6\pi r\eta} \tag{31}$$

Rearrangement and multiplication by Avogadro's number gives

$$M = \mathfrak{N}m = \frac{6\pi r\eta\mathfrak{N}s}{1 - v\rho} \tag{32}$$

Now the troublesome terms involving η and r can be replaced by their effective values, such as appear in the measurable quantity D of Eq. **21,** to give the desired result

$$M = \frac{RTs}{D(1 - v\rho)} \tag{33}$$

Thus, measurements of the sedimentation and diffusion coefficients and of the densities of the solvent and solute allow the deduction of the molecular weight of the macromolecules. The entry of the diffusion coefficient might lead you to expect that the results are somewhat affected by the molecular shape. The necessary data for such calculations for a few macromolecular materials are included in Table 12-1.

A particular advantage of the sedimentation-velocity technique is that a macromolecule solution containing two or more types of macromolecules is separated according to the molecular weights of the components. Figure 12-8 shows the type of sedimentation diagram obtained for a system containing a number of macromolecular species.

Better resolution can be obtained by allowing the sedimentation to occur in a *density-gradient* solution, prepared, for example, by filling the centrifuge tube layer by layer with solutions of decreasing sucrose concentration. As the macromolecular substance or mixture of substances is centrifuged, it will then move through a solvent with gradually increasing density. The result is more stable macromolecular zones and a better "spectrum" of the components. The technique is thus a modification of the sedimentation-velocity method.

Example 12-2

About how long would it take the boundary of a myoglobin sample to move a distance of 1 mm in a centrifuge operating at 60,000 rev/min if the initial distance of the sample boundary from the center of rotation is 6 cm? The sedimentation constant s reported for myoglobin is 2.04 svedbergs at 20°C.

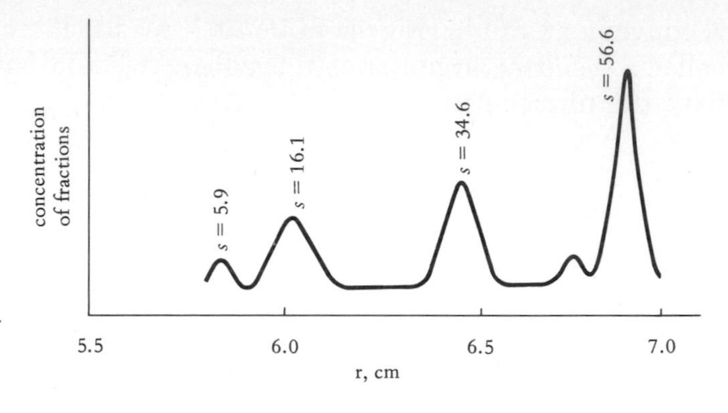

FIGURE 12-8
Separation of *Limulus* hemocyanin into fractions with different sedimentation constants, given in svedbergs. Centrifugal force is 120,000 times gravity, and time after reaching full speed is 35 min. [*From T. Svedberg, Proc. R. Soc. Lond.*, **B127**:*1 (1939).*]

Solution

The velocity of movement dx/dt can be obtained from Eq. **30**; that is, $s = (dx/dt)/x\omega^2$. First, we recognize that ω is 2π times the number of revolutions per second. Thus

$$\omega = \frac{2\pi(60{,}000)}{60} = 6.28 \times 10^3 \text{ rad/sec}$$

Then, with $s = 2.04$ svedbergs $= 2.04 \times 10^{-13}$ sec, we obtain,

$$\frac{dx}{dt} = x\omega^2 s$$

$$= 6(6.28 \times 10^3)^2(2.04 \times 10^{-13})$$

$$= 4.8 \times 10^{-5} \text{ cm/sec} = 1.7 \text{ mm/h}$$

Thus, it will take approximately 0.6 h, or 35 min, for the myoglobin boundary to move an additional millimeter away from the center of rotation.

Example 12-3

Sedimentation-velocity studies of human hemoglobin in water at 20°C yield a sedimentation coefficient of 4.46 svedbergs. The specific volume v of hemoglobin is estimated to be 0.75 ml/g, and the density ρ of the solvent is 1 g/ml. The diffusion coefficient for hemoglobin under the same conditions is 6.9×10^{-7} cm^2/sec. Use these data to estimate the molecular weight of hemoglobin.

Solution

Equation **33** provides the necessary relation. We have

$$M = \frac{RTs}{D(1 - v\rho)}$$

$$= \frac{(8.314 \times 10^{7})(293)(4.46 \times 10^{-13})}{(6.9 \times 10^{-7})(1 - 0.75 \times 1.00)}$$

$$= 63{,}000 \text{ g/mol}$$

12-4 THE SEDIMENTATION EQUILIBRIUM METHOD

A second general centrifuge technique called *sedimentation equilibrium,* allows the centrifugation process to proceed until an equilibrium distribution of the solute throughout the cell is obtained. Thermodynamics has introduced free energy as a convenient quantity for the study of equilibrium, and it can be used here to deal with the equilibrium-concentration gradient that develops. In particular, the centrifugal and diffusional contributions to the free energies G_{x1} and G_{x2} at the radial positions x_1 and x_2 are calculated. At equilibrium, the values of G_{x1} and G_{x2} must be equal.

The free-energy difference, as a result of the centrifugal effect, for particles at x_1 and at x_2 is obtained by finding the work required to move the particles from x_1 to x_2. The free energy is more negative at larger values of x and, on an individual molecule basis, is expressed by

$$\Delta G_{\text{centrif}} = -\int_{x_1}^{x_2} m' x\omega^2\, dx$$

$$= -\frac{m'\omega^2}{2}(x_2^2 - x_1^2) \qquad \textbf{34}$$

With m' replaced by $m(1 - v\rho)$, this becomes

$$\Delta G_{\text{centrif}} = -\frac{m(1 - v\rho)\omega^2(x_2^2 - x_1^2)}{2} \qquad \textbf{35}$$

This free-energy factor expresses the tendency of all the particles to congregate at large values of x, where the centrifugal free energy is low.

Balance is brought about by the diffusion tendency. Following Eq. **16,** this is, per molecule,

$$\Delta G_{\text{diffus}} = \frac{RT}{\mathfrak{N}} \ln \frac{c_2}{c_1} \qquad \textbf{36}$$

where c_2 and c_1 are concentrations at x_2 and x_1.

At equilibrium, the decrease in ΔG given by Eq. **35** just balances the increase given by Eq. **36,** and for the process of moving solute from x_1 to x_2, one has the equilibrium condition

$$\Delta G_{\text{centrif}} + \Delta G_{\text{diffus}} = 0$$

On rearrangement after substitution of Eqs. **35** and **36,** this yields

$$M = \mathfrak{N}m = \frac{2RT \ln (c_2/c_1)}{(1 - v\rho)\omega^2(x_1^2 - x_2^2)} \qquad \mathbf{37}$$

Thus, if measurements of the relative concentrations are made at two positions after equilibrium has been obtained, one can use Eq. **37** to calculate a value for the mass of the individual particles or the mass of an Avogadro's number of particles, i.e., the molecular weight. This is the most accurate of the sedimentation methods for the determination of the molecular weight. Notice, for example, that Stokes' law was not used in the deduction of Eq. **37** and that the diffusion coefficient is absent from this equation. The principal drawback to the method is the time (as much as several days) it can take for equilibrium to be established.

Example 12-4

Calculate the molecular weight of a macromolecular substance from the following sedimentation-equilibrium data.†

Rotor speed	15,000 rev/min
Temperature	12.4°C
$1 - v\rho$	0.277
Concentration (arbitrary units):	
At $x = 6.827$	3.52
At $x = 7.093$	13.52

Solution

For insertion in Eq. **37** we need

$$\omega = \frac{15{,}000}{60} \times 2\pi = 1.57 \times 10^3 \text{ rad/sec}$$

$$T = 285.6 \text{ K}$$

and $R = 8.314 \times 10^7$ ergs/deg mol

Substitution in Eq. **37** then yields

$$M = \frac{2(8.314 \times 10^7)(285.6)(2.303) \log (13.52/3.52)}{0.277(1.57 \times 10^3)^2(7.093^2 - 6.827^2)}$$

$$= 25{,}300 \text{ g}$$

A more recent variation on the sedimentation-equilibrium technique makes use of a density gradient. Most simply a solution is prepared that contains the macromolecular species and a heavy salt, e.g., cesium

†From a more extensive listing given in C. H. Chervenka, "A Manual of Methods," Beckman Instruments, Inc.

chloride. Centrifugation until equilibrium is reached produces a strong CsCl gradient and thus a density gradient throughout the tube. If the CsCl concentration has been appropriately chosen, the macromolecular substance will form a rather narrow band, as illustrated in Fig. 12-9. At this point the specific volume v of the macromolecular substance is equal to that of the solvent. Thus the term $1 - v\rho$ of Eq. **26** is unity, and there is no tendency for the macromolecular substance to move along the tube. Nearer the center of rotation, toward the top of the tube, the density ρ is less, and sedimentation will drive the macromolecules towards the band. Beyond the band, toward the bottom of the tube, the density ρ, will be such that $v\rho$ will be greater than unity. The term $1 - v\rho$ is then negative, and macromolecules will be buoyed up toward the band.

12-5 ELECTROPHORESIS

In the studies of the preceding sections, macromolecule movements in solution have been due to diffusion and to the difference in density of the macromolecule and the solvent. Many of the macromolecules of interest, e.g., proteins, exist in solutions as charged species, as the studies of Chap. 4 showed. It follows that movement of such particles can be produced by the application of an electric field. Such a process, known as *electrophoresis,* is the most important example of the more general electric-field motion effects described as *electrokinetic.*

Studies of the migration of charged solute particles under the influ-

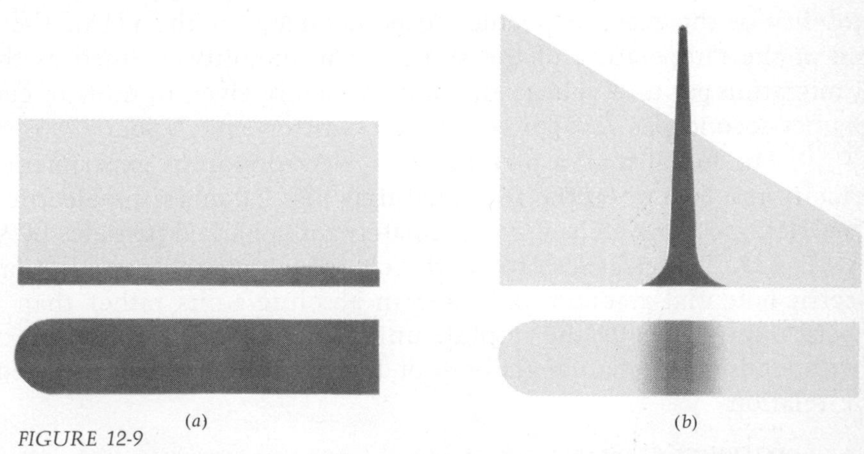

FIGURE 12-9
Concentration distribution of DNA and CsCl along the axis of the centrifuge tube at (*a*) the beginning of the CsCl equilibrium density-gradient centrifugation and (*b*) at the end of the run, i.e., after attainment of equilibrium. (*From W. Szybalski, Fractions, 1968, no. 1, Beckman Instruments, Inc.*)

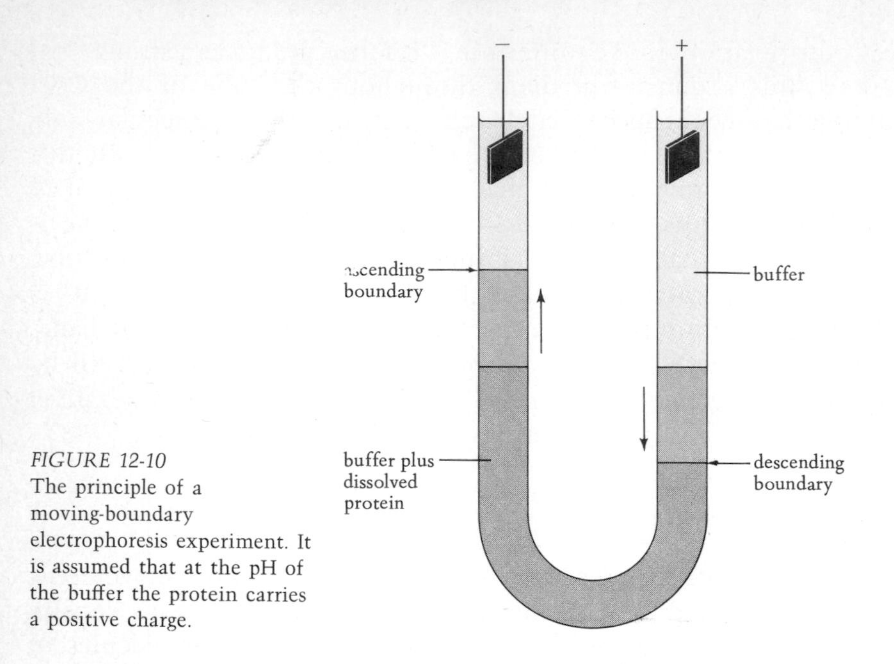

FIGURE 12-10
The principle of a moving-boundary electrophoresis experiment. It is assumed that at the pH of the buffer the protein carries a positive charge.

ence of an applied electric potential can be carried out in a *moving-boundary* electrophoresis apparatus shown schematically in Fig. 12-10. Depending on the charge carried by the particles, there will be a general movement of these species toward one electrode or the other. As a result, one of the original boundaries will move up, and the other will move down. Measurement of the movement of the boundary with time allows the *mobility* of the macromolecules to be deduced, at the pH of the buffer and at the temperature of the system. The mobility, defined as the rate of migration per unit voltage gradient, is usually given in units of centimeters per second per volt per centimeter, written cm^2/V sec.

The mobility of a particle in an electrophoresis experiment is the velocity reached when the frictional drag just balances the electric force. The frictional drag is given approximately for spherical particles by Stokes' law, Eq. **19.** The electric force is the charge eZ on the particle times the electric-potential gradient expressed in absolute units rather than in the practical units of volts, the absolute unit being $\frac{1}{300}$ V. The mobility u, which corresponds to a potential gradient of 1 V/cm can then be estimated from the relation

$$6\pi r\eta u = eZ(\tfrac{1}{300})$$

or

$$u = \frac{eZ/300}{6\pi r\eta} \tag{38}$$

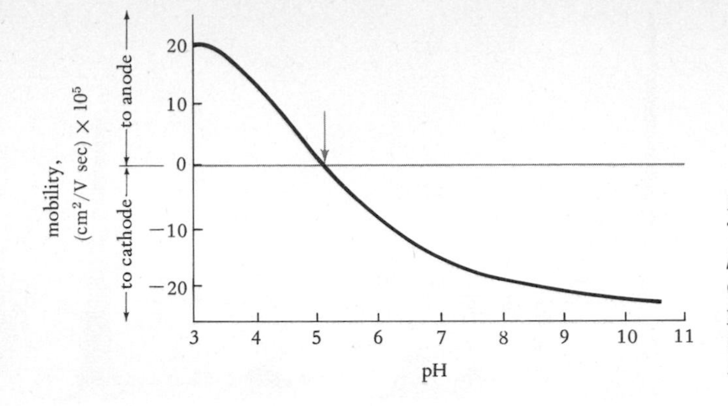

FIGURE 12-11
The mobility of β-lactoglobulin as a function of pH, showing the isoelectric point to be at a pH of 5.1. [*From K. O. Pedersen, Biochem. J.*, **30:**1961 (1936).]

The order of magnitude of protein mobilities can be estimated by introducing the representative and numerical values

$$e = 4.80 \times 10^{-10} \text{ esu}$$
$$Z = 5$$
$$r = 20 \text{ Å} = 20 \times 10^{-8} \text{ cm}$$
$$\eta = 0.010 \text{ g/cm sec}$$

Then

$$u = 20 \times 10^{-5} \text{ cm}^2/\text{V cm}$$

On this basis, for a voltage gradient of 10 V/cm, the boundaries would be expected to move about 7 cm in the course of 1 h.

The observed magnitude of protein mobilities and their dependence on pH is illustrated by the mobility-pH curve for β-Lactoglobulin given in Fig. 12-11. The magnitude of the mobility at various pH values can be compared with typical mobilities of simple ions of the order of 70×10^{-5} cm²/V sec. Remember, however, that simple ions are in fact extensively hydrated and therefore not to be pictured as the small spheres shown in Fig. 2-23. Protein molecules, particularly in the multicharged forms that exist at high or low pH, are, by comparison, not as resistant to movement as might be expected.

An important feature illustrated in Fig. 12-11 is the existance of a pH for which the mobility is zero. This pH is known as the *isoelectric point,* sometimes referred to as pH_I, or even as pI. A closely related quantity is the isoionic point defined (Sec. 4-5) as the pH at which the macromolecule carries no net charge as a result of proton gains or losses in acid-base reactions. The isoelectric and the isoionic points can differ because of the net charge that might result from the ion atmosphere forming about the

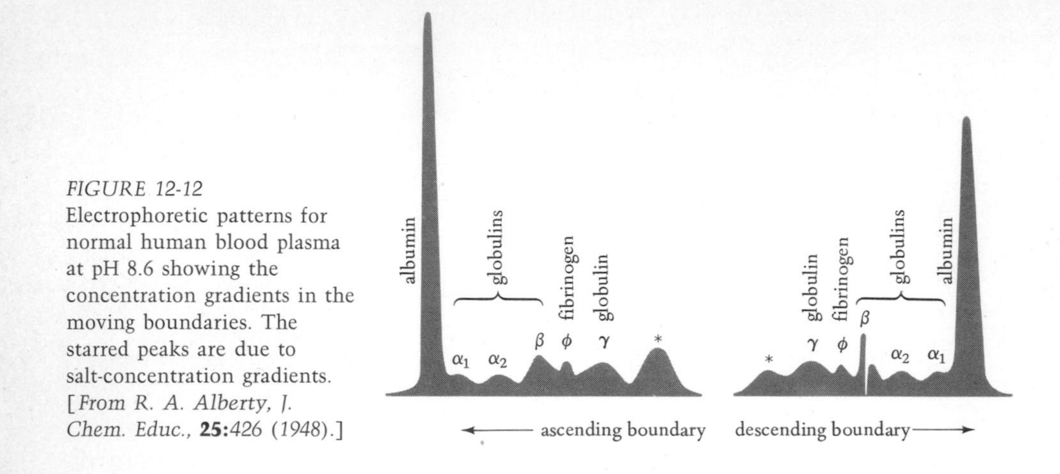

FIGURE 12-12
Electrophoretic patterns for normal human blood plasma at pH 8.6 showing the concentration gradients in the moving boundaries. The starred peaks are due to salt-concentration gradients. [*From R. A. Alberty, J. Chem. Educ.*, **25:**426 (1948).]

protein molecule when it is in a buffer solution. In practice the pH of the isoelectric and that calculated for the isoionic point are usually little different.

Different proteins have different isoelectric points and at a given pH show different mobilities. Electrophoresis is therefore a valuable tool for separating biological fractions into pure components. It supplements the ultracentrifuge, which separates according to molecular weight. Electrophoresis experiments can show that even if a sample is homogeneous with respect to molecular weight, it may contain different components having different electrical properties. Figure 12-12 shows the concentration gradients that develop in the rising and falling solution-solvent boundaries when a system of several species is subjected to moving-boundary electrophoresis.

Although the moving-boundary apparatus of Fig. 12-10 provides accurate mobility data, its use requires large amounts of material and considerable care to form and maintain sharp boundaries. Other techniques depending on solid or gelatinous supporting media are now preferred for the many analysis and separation uses to which electrophoresis is put. Such methods, known as *zone electrophoresis,* have generally supplanted the moving-boundary method. Simplest, and often used as a medium, is filter paper moistened with buffer solution and arranged as in Fig. 12-13. Also used are a variety of gels such as starch, agar, or polyacrylamide.

A final variation on electrophoretic techniques is known as *isoelectric focusing*. In this method a pH gradient is established throughout the electrophoresis cell such that the solution at the negative-electrode end of the cell is relatively basic and at the positive end is relatively acidic. As a result, any species will migrate toward the electrode of opposite charge, but in so doing it will be subject to changing pHs that will make it tend to lose its charge. The process continues until the species is concentrated

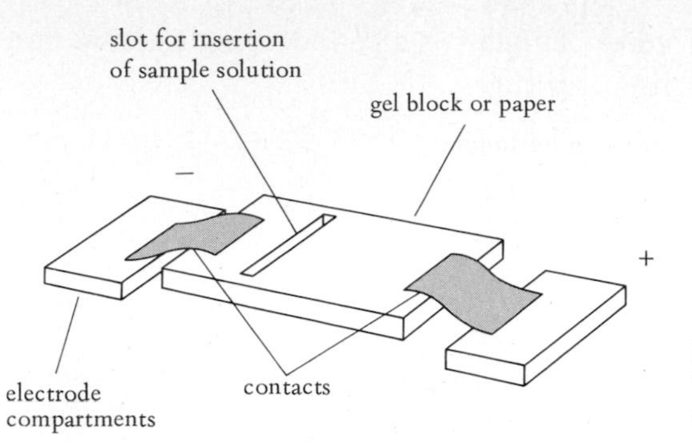

FIGURE 12-13
The basic features of zone electrophoresis.

at the position where the pH is that of its isoelectric point. If there are several species, each will concentrate at the position corresponding to its isoelectric point. If a suitable pH gradient is established, species with very small charge differences or isoelectric-point differences can be separated.

PROBLEMS

1 *a* For one of the substances for which the diffusion coefficient is given in Table 12-1, calculate and plot the concentration gradient as a function of y, as in Fig. 12-5, for two different times.

b In view of the form of Eq. **12,** compare the influences of different values of D and t on the development of the concentration-gradient curve.

2 The diffusion coefficient of oxygen dissolved in water is 200×10^{-7} cm²/sec at 25°C. How long, on the average, does it take an oxygen molecule to go a distance of 1 mm in a particular direction?

3 The diffusion coefficient of sucrose in water at 20°C is 40×10^{-7} cm²/sec. Estimate its sedimentation coefficient and compare with the value of the macromolecules of Table 12-1.

4 For spherical particles of a given density, what dependence on molecular mass would be expected for the diffusion and the sedimentation coefficients?

5 From the data of Table 12-1, verify the frictional ratio for one example other than myoglobin or hemoglobin. Comment on the implications of the result.

6 The force of gravity can be expressed as 980 dyn/g. How many times greater is the force of a centrifuge operating at 75,000 rev/min with the sample 6 cm from the center of rotation, i.e., how many g's does the centrifuge produce?

7 Calculate the frictional ratio for bovine serum albumin given the following data, all applying to 25°C:

Molecular weight	66,500 g/mol
Specific volume	0.734 ml/g
Diffusion coefficient	6.97×10^{-7} cm²/sec

8 The following data were obtained† in a boundary sedimentation-velocity study with a 52,000 rev/min centrifuge.

time, min	radius to boundary
0	12.380
16.3	12.696
32.3	12.993
48.3	13.297
64.3	13.611
80.3	13.923

In view of Eq. **30,** which can also be written

$$s = \frac{1}{\omega^2}\frac{d \ln x}{dt} \qquad \text{or} \qquad s = \frac{2.303}{\omega^2}\frac{d \log x}{dt}$$

use a graphical treatement to obtain a value for the sedimentation constant.

9 The following are additional data obtained in the sedimentation-equilibrium study on which the example of Sec. 12-4 was based:

x, cm	c, arbitrary units
6.827	3.52
6.917	5.52
6.979	7.52
7.026	9.52
7.026	11.52
7.093	13.52

Use a graphical procedure to show that these data correspond to the concentration-radius relation expected on the basis of Eq. **37.**

10 Use data from the preceding problem, other than those used in Example 12-4, to deduce a value for the molecular weight (or use the slope of the best straight line in the graphical treatment developed to answer Prob. 9).

11 What would happen in a isoelectric-focusing study if the electrodes were connected improperly, i.e., the electrode at the basic end of the cell were made positive and that at the acidic end were made negative?

REFERENCES

VAN HOLDE, K. E.: "Physical Biochemistry," pp. 79–132, Prentice-Hall, Inc., Englewood Cliffs, N.J., 1971. Treatments of diffusion, sedimentation, and electrophoresis.

†"A Manual of Methods," Beckman Instruments, Inc.

Ultracentrifuge Studies

COSTES, J. H.: Ultracentrifugal Analysis, chap. 10 in S. J. Leach (ed.), "Physical Principles and Techniques of Protein Chemistry," Academic Press Inc., New York, 1970.

VAN HOLDE, K. E.: Sedimentation Equilibrium, *Fractions,* 1967, no. 1, Spinco Division of Beckman Instruments, Inc., Palo Alto, Calif.

SZYBALSKI, WACLAW, Equilibrium Sedimentation in Density Gradients, *Fractions,* 1968, no. 1, Spinco Division of Beckman Instruments, Inc., Palo Alto, Calif.

Appendixes

I Thermodynamic Properties of Selected Inorganic Compounds

The state symbols are *g* for gas, *l* for liquid, *c* for crystal, *aq* for aqueous solution and *eq buf* for an equilibrium mixture of species in an aqueous solution buffered to pH 7. All data apply to a pressure of 1 atm. Data for gases and aqueous solutions have been corrected for nonideality and are the values the indicated species would have if they behaved ideally at a pressure of 1 atm or a concentration of 1 *M*.

Selected Values of Thermodynamic Properties in the Standard States at 25°C†

			kcal/mol		cal/deg mol	
name	**formula**	**state**	ΔH°_h	ΔG°_f	S°	C_P°
Hydrogen	H_2	g	0.0	0.0	31.21	6.89
		aq	−1.0	4.2	13.8	
Hydrogen ion	H^+	aq	0.0	0.0	0.0	0.0
		eq buf	0.0	−9.55		
Hydroxyl ion	OH^-	aq	−54.96	−37.59	−2.59	−35.5
Water	H_2O	l	−68.32	−56.69	16.72	18.00
		g	−57.80	−54.64	45.11	8.02
Group I:						
Lithium	Li	c	0.0	0.0	6.70	5.65
Lithium ion	Li^+	aq	−66.55	−70.22	3.4	
Sodium	Na	c	0.0	0.0	12.2	6.79
Sodium ion	Na^+	aq	−57.28	−62.59	14.4	
Sodium chloride	NaCl	c	−98.23	−91.78	17.3	11.88
Potassium	K	c	0.0	0.0	15.2	6.97
Potassium ion	K^+	aq	−60.04	−67.46	24.5	
Potassium chloride	KCl	c	−104.17	−97.59	19.76	12.31
Group II:						
Magnesium	Mg	c	0.0	0.0	7.77	5.71
Magnesium ion	Mg^{2+}	aq	−110.41	−108.99	−28.2	
Calcium	Ca	c	0.0	0.0	9.95	6.28
Calcium ion	Ca^{2+}	aq	−129.77	−132.18	−13.2	
Calcium carbonate	$CaCO_3$	c	−288.45	−269.78	22.2	19.57
Group III:						
Boron	B	c	0.0	0.0	1.56	2.86
Aluminum	Al	c	0.0	0.0	6.77	5.82
Aluminum ion	Al^{3+}	aq	−125.4	−115.0	−74.9	
Group IV:						
Carbon	C	diamond	0.45	0.68	0.58	1.45
		graphite	0.0	0.0	1.36	2.07
Carbon monoxide	CO	g	−26.42	−32.81	47.30	6.96
Carbon dioxide	CO_2	g	−94.05	−94.26	51.06	8.87
		aq	−98.90	−92.26	28.1	
		eq buf	−97.49	−93.37		
Carbonic acid	H_2CO_3	aq	−167.22	−148.94	44.8	
		eq buf	−165.81	−150.06	71.3	

† Most of the data are from Frederick D. Rossini, Donald D. Wagman, William H. Evans, Samuel Levine, and Irving Jaffe, Selected Values of Chemical Thermodynamic Properties, *Natl. Bur. Stand. Circ.* 500, 1952. Additional values are from the compilation given by R. C. Wilhoit in H. D. Brown (ed.), "Biochemical Microcalorimetry," Academic Press, Inc., New York, N.Y., 1969.

Selected Values of Thermodynamic Properties in the Standard States at 25°C (*Continued*)

name	formula	state	kcal/mol		cal/deg mol	
			ΔH_h°	ΔG_f°	S°	C_P°
Bicarbonate ion	HCO_3^-	aq	−165.39	−140.26	21.8	
Carbonate ion	CO_3^{2-}	aq	−161.84	−126.17	−13.6	
Group V:						
Nitrogen	N_2	g	0.0	0.0	45.78	6.96
		aq	−2.52	4.32	22.8	41
Ammonia	NH_3	g	−11.02	−3.96	46.03	8.52
		aq	−19.19	−6.35	26.6	
		eq buf	−31.67	−9.58	−4.9	
Ammonium ion	NH_4^+	aq	−31.67	−18.97	27.1	19.1
Ammonium hydroxide	NH_4OH	aq	−87.50	−63.04	43.3	−16.4
Nitric oxide	NO	g	21.60	20.72	50.38	7.14
Nitrogen dioxide	NO_2	g	8.09	12.39	57.47	9.06
Nitrous oxide	N_2O	g	19.49	24.76	52.58	9.25
Dinitrogen tetroxide	N_2O_4	g	2.31	23.49	72.73	18.90
Dinitrogen pentoxide	N_2O_5	c	−10.0	32	27.1	
Nitrous acid	HNO	aq	−28.4	−12.82		
Nitric acid	HNO_3	l	−41.40	−19.10	37.19	26.26
		aq	−49.37	−26.43	35.0	
Phosphoric acid	H_3PO_4	c	−305.7	−267.5	26.42	25.35
		aq	−307.9	−273.1	37.8	
		eq buf	−309.96	−280.44	57.3	
Dihydrogen phosphate ion	$H_2PO_4^-$	aq	−310.38	−270.73	21.6	
Hydrogen phosphate ion	HPO_4^{2-}	aq	−309.38	−260.91	−8.0	
Phosphate ion	PO_4^{3-}	aq	−305.3	−244.02	−53.8	
Pyrosphosphoric acid	$H_4P_2O_7$	aq	−542.2	−486.8	67.7	
		eq buf	−544.2	−502.3	113.0	
Trihydrogen pyrophosphate ion	$H_3P_2O_7^-$	aq	−544.1	−484.7	54	
Dihydrogen pyrophosphate ion	$H_2P_2O_7^{2-}$	aq	−544.6	−481.6	42	
Hydrogen pyrophosphate ion	$HP_2O_7^{3-}$	aq	−544.0	−472.4	13	

Selected Values of Thermodynamic Properties in the Standard States at 25°C (*Continued*)

name	formula	state	kcal/mol		cal/deg mol	
			ΔH_h°	ΔG_f°	S°	C_P°
Pyrophosphate ion	$P_2O_7^{4-}$	aq	−545	−459.6	−27	
Magnesium hydrogen phosphate	$MgHPO_4$	aq	−416.7	−373.65	14.5	
Group VI:						
Oxygen	O_2	g	0.0	0.0	49.00	7.02
		aq	−2.89	3.9	26.5	40
Water	H_2O	l	−68.32	−56.69	16.72	18.00
		g	−57.80	−54.64	45.11	8.02
Hydroxide	OH^-	aq	−54.96	−37.59	−2.59	−35.5
Sulfur	S	rhombic	0.0	0.0	7.62	5.40
		monoclinic	0.07	0.02	7.78	5.65
Sulfur dioxide	SO_2	g	−70.76	−71.79	59.40	9.51
Sulfur trioxide	SO_3	g	−94.45	−88.52	61.24	12.10
Hydrogen sulfide	H_2S	g	−4.81	−7.89	49.15	8.12
		aq	−9.4	−6.54	29.2	
Sulfurous acid	H_2SO_3	aq	−145.5	−128.59	56	
Sulfuric acid	H_2SO_4	aq	−216.90	−177.34	4.1	4.0
Hydrogen sulfate	HSO_4^-	aq	−211.70	−179.94	30.32	
Sulfate	SO_4^{2-}	aq	−216.90	−177.34	4.1	4.0
Group VII:						
Fluorine	F_2	g	0.0	0.0	48.6	7.52
Fluoride ion	F^-	aq	−78.66	−66.08	−2.3	−29.5
Chlorine	Cl_2	g	0.0	0.0	53.29	8.11
Chloride ion	Cl^-	aq	−40.02	−31.35	13.2	−30.0
Hydrogen chloride	HCl	g	−22.06	−22.77	44.62	6.96
		aq	−40.02	−31.35	13.2	−30.0
Bromine	Br_2	g	7.34	0.75	58.64	8.60
		l	0.0	0.0	36.4	
Bromide ion	Br^-	aq	−28.90	−24.57	19.29	−30.7
Iodine	I_2	g	14.88	4.63	62.28	8.81
		c	0.0	0.0	27.9	13.14
		aq	5.0	3.93		
Hydrogen iodide	HI	g	6.2	0.31	49.31	6.97
Iodide ion	I^-	aq	−13.37	−12.35	26.14	−31.0
Triodide ion	I_3^-	aq	−12.4	−12.31	41.5	

II Thermodynamic Properties of Some Organic and Biochemical Compounds†

The state symbols are *g* for gas, *l* for liquid, *c* for crystal, *aq* for aqueous solution and *eq buf* for an equilibrium mixture of species in an aqueous solution buffered to pH 7. All data apply to a pressure of 1 atm. Data for gases and aqueous solutions have been corrected for nonideality and are the values the indicated species would have if they behaved ideally at a pressure of 1 atm or a concentration of 1 *M*.

† Most of the data are from the appendix to R. C. Wilhoit, Thermodynamic Properties of Biochemical Substances chap. 2 in H. D. Brown (ed.), "Biochemical Microcalorimetry," Academic Press, Inc., New York, 1969.

Selected Values of Thermodynamic Properties in the Standard States at 25°C

			kcal/mol		cal/deg mol	
name	**formula**	**state**	ΔH_f°	ΔG_f°	S°	C_P°
Hydrocarbons:						
Methane	CH_4	g	−17.89	−12.14	44.50	8.54
Acetylene	C_2H_2	g	54.19	50.0	48.00	10.50
Ethylene	C_2H_4	g	12.50	16.28	52.45	10.41
Ethane	C_2H_6	g	−20.24	−7.86	54.85	12.585
Benzene	C_6H_6	g	19.82	30.99	64.34	19.52
		l	11.72	41.30	29.76	
Alcohols:						
Methanol	CH_3OH	g	−48.10	−38.70	56.8	
		l	−57.04	−39.75	30.3	19.5
		aq	−58.77	−41.88	31.63	
Ethanol	C_2H_5OH	l	−66.20	−41.63	38.49	26.76
		aq	−68.60	−43.25	35.8	
2-Propanol	C_3H_7OH	l	−75.97	−43.09	43.16	36.86
		aq	−79.07	−44.27	36.7	
Glycerol	$C_3H_8O_3$	l	−160.3	−114.7	48.9	51.8
		aq	−161.7	−118.9	58.8	57
Aldehydes and ketones:						
Acetaldehyde	C_2H_4O	g	−39.68	−31.78	63.15	13.4
		aq	−50.75	−33.24	30.9	
Acetone	C_3H_6O	l	−58.99	−36.70	47.5	30.3
		aq	−52.99	−38.17	39.0	

Selected Values of Thermodynamic Properties in the Standard States at 25°C (*Continued*)

			kcal/mol		cal/deg mol	
name	**formula**	**state**	ΔH_f°	ΔG_f°	S°	C_P°
Carboxylic acids:						
Formic acid	$HCOOH$	g	−86.67	−80.24	60.0	
		l	−97.8	−82.7	30.82	23.67
		aq	−98.0	−85.1	39.1	
Formate ion	$HCOO^-$	aq	−98.0	−80.0	21.9	
Acetic acid	CH_3COOH	l	−115.73	−93.08	38.2	29.5
		aq	−115.98	−96.58	49.1	37.0
		eq buf	−116.06	−99.79	59.6	
Acetate ion	CH_3COO^-	aq	−116.06	−90.08	27.0	
Butyric acid	$C_4H_3O_2$	l	−127.9	−90.6	54.1	42.9
		aq	−127.3	−95.5	72.6	
		eq buf	−128.0	−98.6	76.7	
Butyrate ion	$C_4H_2O_2^-$	aq	−128.0	−88.92	48.2	
Myristic acid	$C_{14}H_{28}O_2$	c	−200.1			
Palmitic acid	$C_{16}H_{32}O_2$	c	−212.9	−75.3	108.8	110.1
		aq		−68.8		
		eq buf		−71.5		
Palmitate ion	$C_{16}H_{31}O_2^-$	aq		−62		
Succinic acid	$C_4H_6O_4$	c	−224.86	−178.62	42.0	36.8
		aq	−218.02	−178.45	64.4	55
		eq buf	−217.19	−184.75	88.3	
Hydrogen succinate ion	$C_6H_5O_4^-$	aq	−217.23	−172.71	47.8	23.0
Succinate ion	$C_6H_4O_4^{2-}$	aq	−217.18	−165.02	22.2	−29.3
Fumaric acid	$C_4H_4O_4$	c	−193.75	−156.13	39.7	34.0
		aq	−185.2	−154.35	62.4	
		eq buf	−185.8	−163.57	91.3	
Hydrogen fumarate ion	$C_4H_3O_4^-$	aq	−185.1	−150.13	48.6	
Fumarate ion	$C_4H_2O_4^{2-}$	aq	−185.8	−143.85	25.2	
L(+)-Lactic acid	$C_3H_6O_3$	c	−165.88	−125.06	34.3	30.5
		aq	−164.01	−128.77	53.0	
		eq buf	−164.11	−133.21		

Selected Values of Thermodynamic Properties in the Standard States at 25°C (*Continued*)

name	formula	state	kcal/mol ΔH_f°	kcal/mol ΔG_f°	cal/deg mol S°	cal/deg mol C_P°
L(+)-Lactate ion	$C_3H_5O_3^-$	aq	−164.11	−123.50	35.0	
DL-Lactic acid	$C_3H_6O_3$	l	−161	−124	45.9	50.5
L-Malic acid	$C_4H_6O_5$	c	−263.7			
		aq		−213.1		
		eq buf		−221.1		
L-Hydrogen malate ion	$C_4H_5O_5^-$	aq		−208.4		
L-Malate ion	$C_4H_4O_5^{2-}$	aq		−201.4		
DL-Malic acid	$C_4H_6O_5$	c	−264.2			
Citric acid	$C_6H_8O_7$	c	−369.0			
		aq	−364.7	−297.37	78.7	76.6
		eq buf	−362.28	−307.99	122.5	
Dihydrogen citrate ion	$C_6H_7O_7^-$	aq	−363.5	−293.10	68.4	44.9
Hydrogen citrate ion	$C_6H_6O_7^{2-}$	aq	−362.92	−286.61	48.6	0.2
Citrate ion	$C_6H_5O_7^{3-}$	aq	−362.12	−277.89	22.0	−60.9
Pyruvic acid	$C_3H_4O_3$	l	−140.0			
		aq	−145.2	−116.3	43	
		eq buf	−142.5	−122.7		
Pyruvate ion	$C_3H_3O_3^-$	aq	−142.5	−112.9	41	
α-Ketoglutaric acid	$C_5H_6O_5$	c	−245.3			
α-Ketoglutarate ion	$C_5H_4O^{2-}$	aq		−189.63		
Esters:						
Ethyl acetate	$C_4H_8O_2$	l	−115.2	−80.7	62.8	25.2
		aq		−84.8		
Sugars:						
α-D-Glucose	$C_6H_{12}O_6$	c	−304.60	−217.63	50.7	52.31
		aq	−301.88	−218.58	63.1	
β-D-Glucose	$C_6H_{12}O_6$	c	−303.07	−217.23	54.5	
		aq	−302.16	−218.88	63.1	
α,β-D-Glucose	$C_6H_{12}O_6$	aq	−302.05	−219.16	64.4	73
α-Lactose	$C_{12}H_{22}O_{11}$	c	−531.0			
		aq	−533.55	−374.02	94.2	
β-Lactose	$C_{12}H_{22}O_{11}$	c	−534.6	−374.5	92.3	98.1
		aq	−533.82	−374.32	94.3	
α,β-Lactose	$C_{12}H_{22}O_{11}$	aq	−533.72	−374.60	95.5	

Selected Values of Thermodynamic Properties in the Standard States at 25°C (*Continued*)

name	formula	state	kcal/mol		cal/deg mol	
			ΔH_f°	ΔG_f°	S°	C_P°
α-Maltose	$C_{12}H_{22}O_{11}$	aq	−534.96	−376.10	96.4	
β-Maltose	$C_{12}H_{22}O_{11}$	aq	−534.83	−375.76	95.7	
α,β-Maltose	$C_{12}H_{22}O_{11}$	aq	−534.91	−376.36	97.5	
Sucrose	$C_{12}H_{22}O_{11}$	c	−531.0	−369.1	86.1	101.7
		aq	−529.60	−370.8	96.5	151.3
Amino acids:						
Glycine	$C_2H_5O_2N$	c	−128.4	−90.27	24.74	23.71
		eq buf	−125.0	−90.80	37.9	
Glycine ion	$C_2H_6O_2N^+$	aq	−126.0	−94.00	45.3	41.0
Glycine dipolar ion	$C_2H_5O_2N^{+-}$	aq	−125.0	−90.80	37.9	8.8
Glycinate ion	$C_2H_4O_2N^-$	aq	−114.4	−77.46	28.8	−13.1
DL-Alanine	$C_3H_7O_2N$	c	−134.7	−88.9	31.6	29.1
L-Alanine	$C_3H_7O_2N$	c	−134.5	−88.48	30.88	29.22
		eq buf	−132.6	−88.71	38.0	
L-Alanine ion	$C_3H_8O_2N^+$	aq	−133.35	−91.91	46.0	69
L-Alanine dipolar ion	$C_3H_7O_2N^{+-}$	aq	−132.6	−88.71	38.0	33.7
L-Alaninate ion	$C_3H_6O_2N^-$	aq	−121.8	−75.25	29.1	17.1
DL-Valine	$C_5H_{11}O_2N$	c	−147.7	−86.0	43.3	
L-Valine	$C_5H_{11}O_2N$	c	−147.7	−85.8	42.72	40.35
		eq buf	−147.7	−85.72	42.3	
L-Valine ion	$C_5H_{12}O_2N^+$	aq	−146.33	−88.84	57.4	−130.8
L-Valine dipolar ion	$C_5H_{11}O_2N^{+-}$	aq	−146.27	−85.72	42.3	93
L-Valinate ion	$C_5H_{10}O_2N^-$	aq	−135.62	−73.47	41.8	79.8
DL-Leucine	$C_6H_{13}O_2N$	c	−155.3	−85.7	49.5	46.7
L-Leucine	$C_6H_{13}O_2N$	c	−154.6	−85.2	50.10	49.8
		eq buf	−153.77	−84.19	49.4	
L-Leucine ion	$C_6H_{14}O_2N^+$	aq	−154.16	−87.37	58.9	−157.8
L-Leucine dipolar ion	$C_6H_{13}O_2N^{+-}$	aq	−153.77	−84.19	49.6	121
L-Leucinate ion	$C_6H_{12}O_2N^-$	aq	−143.55	−70.89	39.3	107.0
L-Cysteine	$C_3H_7O_2NS$	c	−127.3	−81.9	40.6	41.4
		eq buf		−81.01		
L-Cysteine ion	$C_3H_8O_2NS^+$	aq	−83.51			
L-Cysteine dipolar ion	$C_3H_7O_2NS^{+-}$	aq		−80.98		

Selected Values of Thermodynamic Properties in the Standard States at 25°C (*Continued*)

name	formula	state	kcal/mol		cal/deg mol	
			ΔH_f°	ΔG_f°	S°	C_P°
L-Hydrogen cysteinate ion	$C_3H_6O_2NS^-$	aq		−69.55		
L-Cysteinate ion	$C_3H_5O_2NS^{2-}$	aq		−54.87		
L-Cystine	$C_6H_{12}O_4N_2S_2$	c	−249.6	−163.9	67.06	62.60
		eq buf		−159.7		
L-Cystine ion	$C_6H_{14}O_4N_2S_2{}^{2+}$	aq		−163.6		
L-Cystine dipolar ion	$C_6H_{13}O_4N_2S_2{}^{2+,-}$	aq		−162.1		
	$C_6H_{12}O_4N_2S_2{}^{2+,2-}$	aq		−159.3		
L-Cystinate ion	$C_6H_{10}O_4N_2S_2{}^{2-}$	aq	−181.9	−134.4		
L-Methionine	$C_5H_{11}O_2NS$	c	−178.0	−121.5	55.32	69.36
L-Methionine ion	$C_5H_{12}O_2NS^+$	aq				
L-Methionine dipolar ion	$C_5H_{11}O_2NS^{+-}$	aq	−177.9			
DL-Aspartic acid	$C_4H_7O_4N$	c	−233.5	−174.3	36.9	
L-Aspartic acid	$C_4H_7O_4N$	c	−232.44	−174.32	40.66	37.11
		eq buf	−225.48	−176.03	69.7	
L-Aspartic acid ion	$C_4H_8O_4N^+$	aq	−228.29	−175.40	54.8	
L-Aspartic acid dipolar ion	$C_4H_7O_4N^{+-}$	aq	−226.44	−171.62	51.7	
	$C_4H_6O_4N^{+,2-}$	aq	−225.48	−166.32	37.2	
L-Aspartate ion	$C_4H_5O_4N^{-2}$	aq	−216.5	−152.65	21.5	
L-Asparagine	$C_4H_8O_3N_2$	c	−188.9	−126.9	41.7	38.4
L-Asparagine dipolar ion	$C_4H_8O_3N_2{}^-$	aq	−183.1	−125.7	57.1	
L-Asparagine monohydrate	$C_4H_{10}O_4N_2$	c	−259.5	−183.4	50.10	49.7
L-Glutamic acid	$C_5H_9O_4N$	c	−241.2	−174.7	44.98	41.88
		eq buf	−234.7	−172.53	59.5	
L-Glutamic acid ion	$C_5H_{10}O_4N^+$	aq	−234.6	−175.58	70.2	
L-Glutamic acid dipolar ion	$C_5H_9O_4N^{+-}$	aq	−234.7	−172.53	59.5	
	$C_5H_8O_4N^{+,2-}$	aq	−234.2	−166.7	41.6	
L-Glutamate ion	$C_5H_7O_4N^{2-}$	aq	−224.6	−153.8	30.5	
L-Glutamine	$C_5H_{10}O_3N_2$	c	−197.4	−127.2	46.63	43.93
		eq buf	−192.4	−126.2	60	
L-Glutamine dipolar ion	$C_5H_{10}O_3N_2{}^{+-}$	aq	−192.4	−126.2	60	
L-Arginine	$C_6H_{14}O_2N_4$	c	−148.6	−157.0	59.9	55.8

Selected Values of Thermodynamic Properties in the Standard States at 25°C (*Continued*)

			kcal/mol		cal/deg mol	
name	**formula**	**state**	ΔH_f°	ΔG_f°	S°	C_P°
L-Arginine dipolar ion	$C_6H_{14}O_2N_4{}^{+-}$	aq	−147.1			
Glycylglycine	$C_4H_8O_3N_2$	c	−178.3	−117.47	45.4	39.1
		eq buf	−175.49	−117.85		55.5
Glycylglycine ion	$C_4H_9O_3N_2{}^+$	aq	−175.84	−122.10	68.5	69
Glycylglycine dipolar ion	$C_4H_8O_3N_2{}^{+-}$	aq	−175.49	−117.61	55.3	38
Glycylglycinate ion	$C_4H_7O_3N_2{}^-$	aq	−164.89	−106.54	53.1	
Nitrogen-containing compounds:						
Urea	CH_4ON_2	c	−79.58	−47.07	25.00	22.26
		aq	−75.92	−48.47	42.0	19.7
Creatinine	$C_4H_7ON_3$	c	−56.8	−6.8	40.0	33.2
		aq		−5.53		
		eq buf		−5.53		
Creatinine ion	$C_4H_8ON_3{}^+$	aq		0.99		
Creatine	$C_4H_9O_2N_3$	c	−128.2	−63.1	45.3	41.1
		aq		−61.95		
		eq buf		−61.95		
Creatine ion	$C_4H_{10}O_2N_3{}^+$	aq		−65.58		
	$C_4H_8O_2N_3{}^-$	aq		−42.5		
Phosphate compounds:						
Glucose 1-phosphoric acid	$C_6H_{13}O_9P$	aq		−427.7		
		eq buf		−437.1		
Glucose 1-hydrogenphosphate ion	$C_3H_{12}O_9P^-$	aq		−426.2		
Glucose 1-phosphate ion	$C_3H_{12}O_9P^{-2}$	aq		−417.33		
Glucose 6-phosphoric acid	$C_6H_{13}O_9P$	aq		−429.6		
		eq buf		−438.9		
Glucose 6-hydrogenphosphate ion	$C_6H_{12}O_9P^-$	aq		−427.6		
Glucose 6-phosphate ion	$C_6H_{11}O_9P^{-2}$	aq		−419.1		

Answers to Selected Problems

CHAPTER 1

1-1 9.7 cm^3

1-2 1.14 liter

1-6 0.0409 mole/liter; 2.46×10^{19} molecules/cc

1-7 28.3

1-9 (*a*) $x_{H_2} = 0.64$, $x_{N_2} = 0.36$
(*b*) $P_{total} = 1.07$ atm; $P_{H_2} = 0.68$ atm $P_{N_2} = 0.39$ atm

1-11 26.6

1-12

	energy, erg/molecule	velocity, cm/sec	momentum, g cm/sec
He	6.17×10^{-14}	1.36×10^5	9.04×10^{-19}
Hg	6.17×10^{-14}	0.192×10^5	63.9×10^{-19}

1-13 0.35 cm/sec

1-14 (*a*) 1.000, 1.000
(*b*) 1.069, 2.326

1-15 115.5 atm

1-17 4.8 Å = 4.8×10^{-8} cm

1-18 6700 Å, 4.5×10^{14} Hz

1-19 $\lambda = 3.2 \times 10^{-3}$ cm, $\nu = 0.93 \times 10^{13}$ Hz

1-20 Visible: ~6000 K; infrared: ~300 K

1-22 (*a*) 12.9×10^{-39} g cm^2
(*b*) 0, 0.86×10^{-16}, 2.58×10^{-16} erg
(*c*) 1.31×10^{10} Hz and 2.6×10^{10} Hz; 2.3 cm and 1.15 cm
(*d*) microwave

1-23 (*a*) 3.5×10^{-6} sec
(*b*) 2.9×10^5 per sec
(*c*) 4.7×10^{12} rad/sec, 0.75×10^{12} rev/sec
(*d*) 2.6×10^6

CHAPTER 2

2-2 5.87 Å

2-3 1.87 Å

2-4 35°56′ and 76°10′

2-5 632, 121, 211, 412

2-7 (*a*) $2°43\frac{1}{2}'$
(*b*) d is a multiple of 16.2 Å
2-8 15.9 Å
2-9 14,600 g

CHAPTER 3

3-2 1, 1, 4, 9, 24
3-4 (*a*) 0.12 Å (*b*) 0.276 Å (*c*) 2.72×10^{-28} Å
3-5 $\epsilon(n = 1) = 1.50 \times 10^{-11}$ erg, $\epsilon(n = 2) = 6.01 \times 10^{-11}$ erg Absorption would occur at a wave length of 440 Å, in the ultraviolet.
3-6 $v = nh/2ml$; velocity jumps $= 2.2 \times 10^{-33}$ cm/sec $= 4.9 \times 10^{-35}$ mi/hr
3-7 (*a*) 4.18×10^{9}
(*b*) 9.86×10^{-24} erg
(*c*) $\frac{1}{2}kT$ is larger by a factor of 2.09×10^{8}
3-9 (*a*) 1.14×10^{-23} g (*b*) 19.0×10^{5} dyne/cm
3-10 (*a*) ν(O—H) $= 11.11 \times 10^{13}$ Hz; ν(O—D) $= 8.17 \times 10^{13}$ Hz
(*b*) μ(O—H) $= 1.575 \times 10^{-24}$ g; μ(O—D) $= 2.95 \times 10^{-24}$ g
(*c*) k(O—H) $= 7.67 \times 10^{5}$ dyne/cm; k(O—D) $= 7.77 \times 10^{5}$ dyne/cm
3-12 14,000 G
3-13 (*a*) 0.0211 G
(*b*) 90 Hz
(*c*) 0.035 G, 150 Hz
3-14 For $\nu = 100$ MHz, $\Delta\epsilon = 6.6 \times 10^{-19}$ erg, and $\Delta\epsilon/kT = 1.6 \times 10^{-5}$
3-16 (*a*) 33,000 ml/g

CHAPTER 4

4-1 [HA] = 0.298, $[H^+] = 2.1 \times 10^{-3}$
4-2 6.3×10^{-5}
4-3 $[H^+] = 3.5 \times 10^{-3}$, $[OH^-] = 2.8 \times 10^{-12}$
4-4 $CO_3^{=}$, $K_b = 2.0 \times 10^{-4}$
4-5 (*a*) 2
(*b*) 1.6
(*c*) 0
(*d*) 11.3
4-6 (*a*) $[H^+] = 10^{-2}$, $[OH^-] = 10^{-12}$
(*b*) $[H^+] = 10^{-9}$, $[OH^-] = 10^{-5}$
(*c*) $[H^+] = 1.74 \times 10^{-4}$, $[OH^-] = 5.75 \times 10^{-11}$
(*d*) $[H^+] = 7.9 \times 10^{-13}$, $[OH^-] = 1.3 \times 10^{-2}$
4-7 0.026
4-8 0.013
4-9 11.9

4-10 9.4

4-11 No acid added, pH = 11.9; 0.05 mole acid added. pH = 10.8; at equivalent point, pH = 6.0

4-18 3.9 Å

4-19 202

4-20 1700

CHAPTER 5

5-1 −275 kcal

5-2 For hexane, sucrose, and stearic acid, fuel values in kcal/g are 11.5, 4.0, and 9.6 respectively.

5-3 For car, 0.99 kcal/lb mi; for man, 0.5 kcal/lb mi.

5-4 For oxidation of sucrose, 1350 kcal/mol sucrose, 112 kcal/mol oxygen. For oxidation of stearic acid, 2712 kcal/mol stearic acid, 104 kcal/mol oxygen. For protein, 460 kcal/"mole" protein, 98 kcal/mol oxygen (for oxidation to urea).

5-6 (*a*) 916
(*b*) 923 cal
(*c*) 794 cal

5-7 930 cal

5-8 3.69 kcal absorbed

5-9 $\Delta H_1 = +0.79$ kcal/mol, $\Delta H_2 = +0.05$ kcal/mol

5-11 −112.26 kcal/mol

5-13 (*a*) −57.2 kcal
(*b*) 9.77 kcal

CHAPTER 6

6-3 ΔS(thermal surr.) = +29.2 cal/deg; ΔS(universe) = 0; ΔS(system) = −29.2 cal/deg.

6-4 (*a*) ΔS(thermal surr.) = −74.3 cal/deg; ΔS(universe) = 0; ΔS(system) = +74.3 cal/deg
(*b*) ΔS(thermal surr.) = 0; ΔS(system) = +74.3 cal/deg, ΔS(universe) = +74.3 cal/deg

6-5 Entropies of vaporization, in cal/mol deg: methane, 19.8; ethyl ether, 20.2; carbon tetrachloride, 20.5; benzene, 20.8; mercury, 22.5; hydrogen, 10.5; acetic acid, 14.9; water, 26.0.

6-6 ΔS(fusion) = 5.2 cal/mol deg; ΔS(vap) = 26.0 cal/mol deg

6-8 32.05 cal/mol deg

6-9 ΔS(univ) = −21.6 cal/deg

CHAPTER 7

7-2 $\Delta G = 5.4$ kcal

7-4 2.05 kcal

7-5 -56.68 kcal/mol
7-7 -925 cal/mol
7-8 0.034 mol/liter = 0.15 g per 100 g water
7-9 $G_{real} - G_{ideal} = 33$ cal/mol
7-10 1.6×10^{-10}
7-11 1.0×10^{-14}
7-12 (*a*) 0.0176 atm
(*b*) -6.35 kcal/mol
(*c*) -6.35 kcal/mol
7-13 $\Delta G_f(CaCO_3$ in seawater$) = -269.14$ kcal/mol
7-14 0.68
7-15 0.133
7-16 0.43
7-17 (*a*) 0.92 (*b*) 0.75

CHAPTER 8

8-1 23.9 mm Hg
8-2 1.78×10^{-10}
8-3 1.4×10^{28}
8-4 (*a*) $CO(g) + \frac{1}{2}O_2(g) \rightleftarrows CO_2(g)$
(*b*) -61.45 kcal, $K = 1.1 \times 10^{45}$
(*c*) 6×10^{-48} atm
8-5 For eq. buf., ratio of glucose 6-phosphate to glucose 1-phosphate is 21.
8-6 10^{-7}
8-7 (*a*) 4×10^6
(*b*) -13 kcal
(*c*) 13 kcal
(*d*) -2 kcal; 2 kcal of work
8-8 11×10^6
8-9 $(-9 - 0.02)$ kcal
8-10 1.4×10^{-4}
8-12 5.4×10^{-5}, glutamic acid ion
8-13 99°C
8-14 -25.5 kcal
8-15 (*b*) 121
(*c*) 166
8-16 At 25°C, $K = 0.41$; at 37°C, $K = 0.61$.
8-17 2.4×10^{-14}

CHAPTER 9

9-2 (*a*) 31.26 mm; $\Delta P_{calc} = 0.56$ mm, $\Delta P_{obs} = 0.62$ mm; error in $\Delta P = 10\%$
(*b*) $P_{calc} = 1.11$, $\Delta P_{obs} = 1.32$; error in $\Delta P = 16\%$
9-3 Mole fraction = 0.0168; 0.95 mol
9-4 $\pi/C = 68$(extrap); $M = 340$
9-5 0.30 mol/1,000 g water
9-6 7.8 atm
9-7 27 atm
9-8 (*a*) $\bar{M}_n = 150{,}000$, $\bar{M}_m = 166{,}000$
(*b*) $\bar{M}_n = 133{,}000$, $\bar{M}_m = 150{,}000$
9-9 $\bar{M}_n = 53{,}000$, $\bar{M}_m = 78{,}750$

CHAPTER 10

10-1 At $t = 24$, $k = 0.0341$ hr^{-1}
10-2 0.00247 min^{-1}
10-3 0.71
10-4 $k = 0.0485$ day^{-1}; $t_{1/2} = 14.3$ days
10-5 10^{-6}
10-6 6,900 yr
10-7 6.5 liter mol^{-1} min^{-1}
10-8 (*a*) First
(*b*) 0.023 min^{-1}
10-9 Rate (mol/liter sec) = $60[I^-][OCl^-]$
10-10 Rate (mol/liter sec) = $60[I^-][OCl^-]/[OH^-]$
10-11 $K_m = 0.28$ mmol/liter; $k_2[E_T] = 7.1$
10-14 (*a*) 4.2
(*b*) 1.4
10-15 (*a*) 0.055 sec
(*b*) $E_a = 5{,}200$ cal; $A = 7 \times 10^{10}$ liter/mole sec

CHAPTER 11

11-1 Rate constant for Eq. **5** = 0.25×10^{-11}; rate constant for Eq. **6** = 1.5×10^{9}. Duration of encounter = 130×10^{-10} sec
11-2 5,000
11-6 Time for separation = 5×10^{-8} sec; maximum number of product molecules $\cong 1{,}200 \times 10^{6}$/min

CHAPTER 12

12-2 80 sec

12-3 0.4 svedberg
12-6 0.2×10^6
12-7 1.16
12-8 8.24 svedberg
12-10 25,300

Index

LIST OF THE ATOMIC MASSES OF THE ELEMENTS

element	symbol	atomic number	atomic mass*
Actinium	Ac	89	(227)
Aluminum	Al	13	26.98
Americium	Am	95	(243)
Antimony	Sb	51	121.75
Argon	Ar	18	39.95
Arsenic	As	33	74.92
Astatine	At	85	(210)
Barium	Ba	56	137.34
Berkelium	Bk	97	(249)
Beryllium	Be	4	9.012
Bismuth	Bi	83	208.98
Boron	B	5	10.81
Bromine	Br	35	79.91
Cadmium	Cd	48	112.40
Calcium	Ca	20	40.08
Californium	Cf	98	(251)
Carbon	C	6	12.011
Cerium	Ce	58	140.12
Cesium	Cs	55	132.91
Chlorine	Cl	17	35.45
Chromium	Cr	24	52.00
Cobalt	Co	27	58.93
Copper	Cu	29	63.54
Curium	Cm	96	(247)
Dysprosium	Dy	66	162.50
Einsteinium	Es	99	(254)
Erbium	Er	68	167.26
Europium	Eu	63	151.96
Fermium	Fm	100	(253)
Fluorine	F	9	19.00
Francium	Fr	87	(223)
Gadolinium	Gd	64	157.25
Gallium	Ga	31	69.72
Germanium	Ge	32	72.59
Gold	Au	79	196.97
Hafnium	Hf	72	178.49
Hahnium	Ha	105	(260)
Helium	He	2	4.003
Holmium	Ho	67	164.93
Hydrogen	H	1	1.0080
Indium	In	49	114.82
Iodine	I	53	126.90
Iridium	Ir	77	192.2
Iron	Fe	26	55.85

*Based on mass of ^{12}C at 12.000. Values in parentheses represent the most stable known isotopes.